THE COMETS OF GOD

Also by Jeffrey Goodman:

The Genesis Mystery

American Genesis

We Are The Earthquake Generation

THE COMETS OF GOD

NEW SCIENTIFIC EVIDENCE FOR GOD

Recent archeological, geological and astronomical
discoveries that shine new light on the Bible
and its prophecies

JEFFREY GOODMAN, Ph.D.

ARCHEOLOGICAL RESEARCH BOOKS, L.L.C.

TUCSON

Published by ARCHEOLOGICAL RESEARCH BOOKS, L.C.C.
1800 South Persia Place, Tucson, ARIZONA 85748-8144

Copyright © 2011 by Jeffrey Goodman

ALL RIGHTS RESERVED. No part of this book may be reproduced, stored in a retrieval system or transmitted in any form or by any means - electronic, mechanical, photocopy, recording, scanning, or other - except for brief quotations in newspapers, magazines, reviews or articles, without the prior written permission of the publisher.

Unless otherwise noted, Scripture quotations are from *The Holy Bible*, King James Version, (KJV).

International Standard Book Number: 978-0-9844891-2-1

Library of Congress Control Number: 2010904478

Books by Archeological Research Books are available at special discounts for bulk purchases, fund raising, sales promotion or educational use. Special editions can also be created.

WEB SITE: www.thecometsofgod.com

EMAIL: jg@thecometsofgod.com

Printed in the United States of America

Book Design by Jeffrey Goodman

ACKNOWLEDGMENTS

Many thanks to my "most excellent" editor Dr. Roseanna Mitchell who had both the classical and scientific background to work with such diverse material and the kindness to stick with me from beginning to end; my friend Brandon Brygider, a geophysicist and attorney who alerted me to relevant research and provided important insights, my wife and secretary Anna who provided a second level of editing (that helped to convert my jargon into English), and my daughter Hannah, who provided computer software troubleshooting and the cover design. Also many thanks to John D'Andrea who helped with the illustrations, Dave and Joyce Hawley, Jacquelyne Brady, and the late Clifford Brygider whose encouragement and financial aid kept this project moving foward at key times. Graphics by Lynne East Graphic Design, Tucson.

In dedication to my wife Anna

CONTENTS

PREFACE ix

ONE   Are There Correct Answers and How Can We be Sure? 1

TWO   Stars That Fall From Heaven 21

THREE   The Lord of Hosts 65

FOUR   Scientific Explanations for Old Testament Catastrophes - Part 1: Noah's Flood and Sodom and Gomorrah 101

FIVE   Scientific Explanations for Old Testament Catastrophes - Part 2: The Exodus, Joshua's Great Victory, Debra and Barak's Victory and the Blast that Killed 185,000 Assyrians 153

SIX   The Beginning of the End: The First Four of the Seven Trumpets and Seven Vials of the Book of Revelation 193

SEVEN   The Three Woes – The Fifth, Sixth and Seventh Trumpets and Vials 231

EIGHT   The Tale of the Tower – The Truth about What Really Happened at the Tower of Babylon 287

NINE	Armageddon, The Seven Seals, The Day of the Lord, and A Rock Cut Without Hands 325
TEN	The Sign of the Son of Man in Heaven 393
ELEVEN	What the Bible Knew First 407

APPENDIX A Sumerian/Babylonian Cometary Gods 459

APPENDIX B Russia Is Not Magog 467

NOTES 487

CITATIONS 509

BIBLIOGRAPHY 545

INDEX 567

ILLUSTRATIONS 587

About the Author 596

PREFACE

> While everyone is entitled to their
> own opinions,
> everyone is not entitled to their own set of facts.

Scientists say that science is about understanding the laws of nature. Few scientists would say that knowing God or understanding the Bible contributes to understanding the laws of nature. Nevertheless, the Bible contains testable scientific information about the laws of nature and God's use of the power of nature to express his wrath. When properly translated and understood, the Bible contains a great deal of information about nature, and natural events that profoundly affect life on Earth. If what the Bible has to say about nature and the reasons for past miraculous catastrophic events involving nature are proven correct wouldn't this provide evidence for God and give man a peek into the mind of God? What if the Bible is in part, a message to mankind that reveals God's purpose and plans by communicating scientific information about important aspects of nature?

This book presents scientific evidence for the existence of the God of the Bible based on heretofore unrecognized scientific information recorded in the Bible, long before man discovered this scientific information on his own. The scientific information examined here predominately involves the subject of comets, one of the least understood cosmic aspects of nature. Admittedly, few people have recognized when the Bible is talking about comets, the giant iceballs of space. To be able to recognize the scientific information about comets contained in Scripture, we must know something about astronomy and geology, as well as the original meaning of key words in the Bible when it talks about the objects of heaven. For example, the correct understanding of key words used in the Bible's account of Noah's Flood transforms this story from being highly questionable in terms of science to becoming a brilliant scientific account of a cometary disaster.

With advances in scientific research, the Biblical passages about past catastrophes and catastrophes yet to come can now be properly studied. A number of scientific instruments have been used to discover information pertinent to the Biblical text. These instruments include the telescope and the microscope, and cameras, seismographs, and gravity meters mounted on space satellites. Scientific information has also come from soil (core) samples recovered from suddenly abandoned archeological sites, peat bogs, thick tsunami deposits, and sediments on the bottom of the ocean. All these samples have a chemical signature or thumbprint that speaks of cosmic impact. Most importantly, new light has been shed upon the Biblical text by the recent discovery of two large impact craters. These craters point to two separate disastrous cosmic impacts that took place during Old Testament times. Early versions of this manuscript anticipated the discovery of these craters. One crater was found at the bottom of the Indian Ocean below 12,500 feet of water, and the other crater was found in southern Iraq in a marshy area that was drained by Saddam Hussein. These craters testify to two cosmic impacts that took place during Biblical times where populations in the greater Near Eastern region and even world-wide populations experienced impact-driven disasters including earthquakes and tsunamis of unprecedented sizes. In addition to the catastrophic events caused by these two impacts being accounted for in the Bible, they should also be described in writings of other (literate) ancient Near Eastern cultures, and indeed they are.

Hidden within the Bible is a textbook of information on what is now the most important subject to mankind in all of astronomy and even in all of science – comets and cometary impacts. Why comets? Aside from the Sun, scientists now realize that the greatest force that has affected life on the Earth is not terrestrial but extraterrestrial, and comes from comets, and their cousins, asteroids and meteorites. Indeed, many scientists now acknowledge that the greatest <u>threat</u> to the survival of civilization as we know it comes from comet and asteroid impacts. We find that the catastrophic events recounted in the Bible show that God works through the laws of nature (which he says he established) by using comets as his instruments to implement some of his plans. Readers will be surprised to learn that the Bible refers to comets as God's "mighty ones," "weapons of wrath," and "ministers of flaming fire." Some of God's plans include God revealing himself to mankind via comets, and God bringing correction and judgment via comets.

In the Bible, God even asks mankind to test him about the things that have taken place in the past and about things prophesied to take place in the future.[*] With man's current knowledge of comets

---

[*] *Isaiah 42:8-9 NIV* says, "I am the Lord; that is my name . . . See,

and the Biblical accounts of cometary activity and descriptions of comets, we have a body of testable information involving an aspect of nature that leads to scientific evidence for God as manifesting influence in heaven and upon the Earth. Not only do recent scientific findings provide independent corroborative evidence for some of the catastrophic events of the Old Testament, but in doing so they provide scientific insights for the Bible's detail-specific prophecies of catastrophic events yet to come. Just as the Bible contains a textbook of scientific information about comets, and correctly identifies comets as the causes for most of the Old Testament Biblical catastrophes, the Bible's prophecies for the future also involve comets and impacts. The Bible's scientific insight into natural phenomena involving the different types of comet impact speaks of a three and a half year period when the Earth will undergo seven separate rounds of comet bombardment. We can now clearly recognize that the major cause for the catastrophic events prophesied to occur during the end times will involve comet impacts.

Examining scripture in light of modern advancements in archeology, history, linguistics, geology, and astronomy increases our understanding, and appreciation of the accurate and amazing scientific information in the Bible. We now have scientific proof for the existence of the God of the Bible and scientific insight about the events prophesied for our future.

---

the former things ('once predicted' – *Tanakh*) have taken place, and new things I declare: before they spring into being, I announce them to you." *Isaiah 45:11-12* says, "Ask me of things to come . . . and concerning the work of my hands command ye me. I have made the Earth, and created man upon it: I, even my hands, have stretched out the heavens, and all their **host** ('*stars*' in *Septuagint* and '*starry hosts*' in NIV – in particular **comets**) have I commanded."

ONE
———————————————————

Are There Correct Answers and How Can We Be Sure?

After the 1991 Gulf War and again after the 2001 terrorist attack of September 11th, many people became interested in what the Bible had to say about the future. When large scale disasters occur, many people wonder if the event is connected to what the last book of the Bible, the *"Book of Revelation,"* says will precede the end of the world. Fears about such things as earthquakes, global economic crisis, wars, nuclear attack, and plagues suddenly became more tangible.

A *Time/CNN* opinion poll found "that more than one-third of Americans say they are paying more attention now to how the news might relate to the end of the world, and have talked about what the Bible has to say on the subject. Fully 59% say they believe the events in *Revelation* are going to come true...." [1] Interestingly, this figure is significantly up from the numbers in a *Newsweek* opinion poll run several years earlier that "found that 40% of American adults do believe that the world will one day end, as *Revelation* describes in the

Battle of Armageddon." [2]

In recent years the increased interest in how the world will end can in part be attributed to the best selling books of the *Left Behind* series. The *Left Behind* books presented an imaginative fictional account of end times Bible prophecy. The phenomenal success of the *Left Behind* books drew the public's attention to the possibility of a planetary Doomsday. Interestingly, the *Left Behind* series was not the first to capture a large audience's attention regarding the end times. The *Left Behind* series of books addressed the same end times apocalyptic Bible prophecies as the 1970's best selling non-fiction book, *The Late Great Planet Earth*. In the 1970's Hal Lindsey's *The Late Great Planet Earth* broke sales records with over 15 million copies sold (as of today, over 30 million copies have been sold). Now over twenty years later, the fictionalized *Left Behind* series by Tim LaHaye and Jerry Jenkins, with over 40 million copies sold, reflected a renewed and increased interest in what the Bible says about how the world will end.

"Eschatology," the branch of theology that is concerned with the last things, has traditionally been the domain of evangelical Christians; however, only about half of the *Left Behind* readers were evangelical. New interest in what the Bible says will happen during a three and a half year period of great tribulation now comes from a broad spectrum of people. Christians from many different denominations as well as non-Christians want to know what the Bible says the future will bring. Even atheists are interested, if only to argue against the topic and against theological tales.

Today, as more people than ever are concerned about what the Bible has to say regarding the end times, it is important to precisely understand what is being conveyed in the Bible. Concern over the spectra of end times events has even entered the political arena and influenced decisions. Former president Ronald Reagan, former Secretary of Defense Caspar Weinberger, former presidents George H. Bush, and George W. Bush have all made direct reference to Biblical end times scenarios. For many Americans it is a foregone conclusion that the Bible says the world will end someday. However, understanding or knowing the exact nature and sequence of the disastrous events that are to precede this end as foretold in the Bible is another question. Without knowing the origin and true nature of these events, how can we recognize if a disastrous event is actually fulfilling Bible prophecy? How can people honestly say, "This is the finger of God" as was said after one of the plagues in *Exodus 8:19*?

Since the first century, numerous interpretations have been put forth to explain the apocalyptic events prophesied for the future, and the order in which they are to occur. However, these interpretations have had to continually be adjusted in order to reflect new information and modern scientific approaches. As difficult as it is to believe, at one time the Bible's Scriptures were interpreted to support the theories that the Earth was flat, that the Sun revolved around the Earth, and that the cosmos was a perfect creation with little danger to fear from the heavens or sky. Invention of the telescope in the 17th century encouraged growth in man's scientific knowledge about Earth and the heavens, and new insights into the Bible's text were gained as well. During the 19th century, because of excavation and the discovery of ancient writings, archeological knowledge about the ancient Biblical lands increased and new insights into the Biblical text evolved as well. Additionally, insights into *Revelation*'s end times text have come with the fulfillment of certain prophecies. For example, as history has unfolded, empires (such as the Babylonian and Roman Empires) have risen and fallen, the nation of Israel was reborn in 1948, and a united Europe (the European Community) has come into being. These geopolitical events are now generally recognized as fulfilling Biblical prophecy. However, different and competing interpretations for a number of *Revelation*'s end time events are still being presented. While some of these interpretations are plausible, none seem to account for all of the details given about the many specific and perplexing apocalyptic events; nor do they fit with all of the other perplexing events called for in *Revelation*. Also, many of these interpretations are in conflict with Biblical principles such as multiple witnesses, sowing and reaping, and there being nothing new under the Sun (which requires precedent for future events - *Ecclesiastes 1:9* and *3:15*).

These competing interpretations have been the subject of heated debate, particularly those involving catastrophe on a global scale and worldwide warfare. Since the 1950's, the advent of the atomic bomb has made nuclear weapons and nuclear war the centerpiece of most Bible prophecy interpretations. However, what is described in *Revelation* requires far greater energy than the detonating of all the nuclear arsenals of the world at once. Nevertheless, impressed and intrigued by the awesome power released by thermonuclear blasts in the 1950's, many argued that here, at last, were sufficient means to accomplish the massive destruction called for in apocalyptic Scripture. This became the reigning theory in spite of the fact that nuclear explosions cannot produce the amount of shaking and ground movement nor the horrific global waves of destruction described in *Revelation*. In addition, it is important to recognize that there is no Biblical precedent for man's weapons (being used or) serving as God's instruments. To the contrary, the destruction of the Egyptian army at the Reed (Red) Sea, Joshua's great victory over the Amorites in the Valley of Gibeon, and the blast that killed 185,000 invading Assyrians outside the gates of Jerusalem during the reign of Hezekiah all speak of the God of the Bible using nature or seeming supernatural means to accomplish His ends. Nevertheless, the nuclear theory has persisted. A 1984 Yankelovich poll found that 39% of those responding said Biblical prophecies of the Earth's destruction by fire referred to a nuclear war. [3]

Today's best selling prophecy books present readers with some very trying interpretations and explanations for some of the cryptic verses in the *Book of Revelation*. These prophecy books present no logical explanations for determining when verses are literal or symbolic. Rather than trying to determine the true meaning of the cryptic verses and bringing them into alignment with reality, these interpretations add to the confusion by calling for unprecedented and unworldly nightmarish scenes.* For example, the imagery of the Fifth Trumpet (*Revelation 9:1–11*) has been interpreted to describe never seen before demon possessed, locust like scorpions that fly out of a bottomless pit to sting and torment men with unimaginable suffering for five months. The imagery of the Sixth Trumpet (*Revelation 9:14-21*) has been interpreted to describe never seen before creatures, in this case an army of 200 million evil spirits astride horse like creatures riding across the sky to kill one-third of the world's population with the fire and smoke and brimstone that spews out of their mouths. A popular alternate interpretation for the Sixth Trumpet says that the imagery represents helicopters and cruise missiles, and a Chinese army of 200 million men riding astride mobilized ballistic missile launchers that kill one-third of the world's

---

*   *I Corinthians 14:33* says, "God is not the author of confusion."

population with fire and smoke and brimstone. The question here is why should one believe in an improbable 200 million man army when there are so many other details given for the Sixth Trumpet that simply do not fit with this ballistic missile interpretation?

Along with the questionable interpretation of *Revelation*'s events and what causes them, nearly all prophecy books of today have incorrectly assumed one of the "bad guys" of the end times would be Russia. Most prophecy books still claim that based on its origins, Russia is the evil empire of *Ezekiel 38-39* that will invade the Holy Land during the end times. This idea can be traced back to a small group of 18th and 19th century theologians, who based their assertions about Russia's origins on historical references that had been purposefully altered (see Chapter 9 and Appendix B for details). Even now, in spite of what professional archeologists and historians have to say, and in spite of the fact that Russia is not one of the nations of *Ezekiel 38-39*, this theory persists. Many years ago one had to be a scholar to correctly know about Russian origins. However, over the last ten years, the irrefutable truth about their origins has been embraced and celebrated by the Russians, featured in a number of television documentaries, published in *National Geographic* magazine, and covered in high school world history text books.

Actually, archeologists and historians know the exact identity of the nations of *Ezekiel 38-39* based on the dealings these ancient nations had with the Assyrian Court. In fact, the records of the ancient Assyrian Court talk about these ancient nations the same way that the Bible does. [4] Correctly identifying the nations of *Ezekiel 38-39* is essential to understanding who will be at the Battle of Armageddon, an important end times battle and what is to take place at this Battle.

Years ago Dewey Beegle, a professor at Wesley Theological Seminary, wrote that many Christians were uneasy about the prophetic systems being taught to them, and were searching for better alternatives. [5] Even now, despite the popularity and abundance of prophecy books, many Christians are still searching for better answers about what the Bible has to say regarding the different events that are to take place during the end times. Interpretations of end time prophecy that are mainly based on current events; call for the appearance of armies of demon (evil spirit) possessed creatures; misidentify the nations of *Ezekiel 38-39*; ignore and contradict Biblical principles; or make shaky connections between nuclear weapons and portions of apocalyptic Scripture, have left some Christians and theologians unconvinced. Among those who are skeptical about apocalyptic disaster brought by nuclear warfare are those who see prophetic fulfillment as solely God's responsibility and say, "God does not need man's modern inventions" to work his will. [6]

An article in *Time* magazine titled "Is the Bible Fact or Fiction?" noted that as knowledge continues to increase, Jewish and Christian believers around the world have become more attuned than ever before to the significance of scientific evidence "in establishing the reality of the events underlying their faith." [7] As people are looking to strengthen their faith and understand the Bible's prophecies for the end times, there is a strong desire for better explanations and scientifically based answers as to what end times Bible prophecies are really saying. Yet, most current Bible prophecy books do not reflect the latest archeological, geological, and astronomical knowledge and discoveries that relate to Bible prophecy. Dr. Mark A. Noll, Professor of Christian Thought at Wheaton College in his book *The Scandal of the Evangelical Mind* (the Evangelical Book of the Year for 1995 and "*Christianity Today's*" Book of the Year for 1995) in part explained this scarcity of scientific information. He wrote, "Goaded on by the questionable use of science in the larger culture, fundamentalists and their evangelical successors dropped the nineteenth century conviction that the best theology should understand and incorporate the best science. . . they gave up the nineteenth century belief that it was important to make a positive adjustment to an era's best science."[8]

While 19th century theologian, Charles Hodge, a professor at Princeton Seminary cautioned scientists about advancing cosmological theories that contradicted the central teachings of the Bible; he also defended "the proposition that the Bible must be interpreted by science." [9] Hodge wrote that:

> Nature is as truly a revelation of God as the Bible (*Psalm 19:1*); and we only interpret the Word of God by the Work of God when we interpret the Bible by science.... When the Bible speaks of the foundations, or of the pillars of the Earth, or of the solid heavens, or of the motion of the Sun, do not you and every other sane man, interpret this language by the facts of science? For five thousand years the Church understood the Bible to teach that the Earth stood still in space, and that the Sun and stars revolved around it. Science has demonstrated that this is not true. Shall we go on to interpret the Bible so as to make it teach the falsehood that the Sun moves around the Earth, or shall we interpret it by science, and make

the two harmonize? [10]

In a 1941 Symposium entitled, "Science, Philosophy and Religion" the great physicist Albert Einstein said, "Science without religion is lame, and religion without science is blind." Accordingly, we must not be afraid to look at the Bible through the modern eyes of science, where science is about understanding nature and the reasons for things. We must go beyond beliefs based on 18th and 19th century ideas and interpretations of *Revelation* in order to see what the Scriptures reveal for us now. It is important to remember that the Bible itself says that certain words and prophecies were sealed or kept secret. The Bible says certain words and prophecies were closed up until the end times, when knowledge will have increased enough to properly understand what was being said (*Daniel 12:4, 9, Jeremiah 23:19-20 NAS,* and *Jeremiah 30:23-24 NAS*). If the Bible is what it claims to be, then new discoveries from science and history should constantly be confirming the Bible and making end times prophecy more understandable and more believable. For faith based believers who take the Bible to be without error, it is not a case of science contradicting Scripture nor of science threatening the prophetic system they may subscribe. Instead it is a case of science helping to harmonize specific scriptural texts with science and reality. * New

---

* Note that the discussion of the role of science here is apart from the ongoing controversy of a "young Earth" versus an "old Earth." The question being addressed here is how the disasters of the end times are to happen. The contenders represent phenomena for which we have had the benefit of direct observation; phenomena such as the nuclear destruction of Hiroshima and Nagasaki, the Good Friday Alaska Earthquake, the eruption of Mount Saint Helens, the tidal wave that flooded Krakatoa, the multiple impacts of Comet Shoemaker-Levy 9 on the planet Jupiter, and the Indonesian tsunami of December, 2004. The great Jewish Biblical scholar

interpretations of Scriptures should not be rejected solely because they may require change or revision to beliefs about events in the *Book of Revelation*. One's personal beliefs must not be held so tightly that the actual truth found in the Bible is strangled. Therefore, the question here is not whether we should see what science tells us about end time events. Rather the question must be why science is not being consulted when interpreting the Bible and understanding events in the *Book of Revelation*? Ironically, in this book a number of instances are cited where the Bible gives amazingly accurate scientific information long before modern science came to discover or verify this important information.

So, is there scientific evidence that will help explain the apocalyptic Scriptures and in turn help establish credibility for the events prophesied to occur? Are there any common denominators or patterns in play? If science decodes some of *Revelation's* cryptic passages to reveal detailed scientific information, what does this say about the inspiration behind the Bible's end times prophecies?

People want credible explanations about what the Bible says will take place during the end times. Believers want scientific explanations that they can share with unbelieving friends and not worry about being viewed as irrational. Over the years, various fields of science have made significant advancements, and instead of these findings proving the Bible fictional, science has moved to substantiate the reality of the Bible's stories. In order to look at the book of *Revelation's* cryptic and horrifying passages and find solid scientific information and incredible scientific insights, we must

---

Maimonides (1135-1204, Spain) in *The Guide for the Perplexed* said, "Conflicts between science and religion result from misinterpretations of the Bible."

ARE THERE CORRECT ANSWERS 11

enhance our study of the full context of these passages. Indeed, disaster scenarios like those found in *Revelation* have already drawn the interest and concern of scientists. Since the types of disaster found in the *Book of Revelation* are now being discussed in popular science books, and portrayed in Hollywood movies, one could say that the writing is now on the wall about how the world will end.*

The prophecies of *Revelation* involve a number of very frightening images and scenarios. Here we encounter things far scarier than the "lions, tigers, and bears" that worried poor Dorothy in the Land of Oz. For example, *Revelation* tells of a great dragon being cast onto the Earth; a beast with seven heads and ten horns rising up out of the sea; and a beast with two horns rising up out of the Earth. Also described are bizarre locust like creatures with teeth and the power of scorpions coming upon the Earth, and an army of 200 million horsemen out of whose mouths come fire and smoke, and brimstone, riding across the heavens and killing a third part of all mankind. *Revelation* also tells of painful sores falling upon men; a great earthquake such as never seen by humanity before; stars falling from heaven; a bottomless pit; the air rolling back as a scroll when it is rolled together; lightnings; thunders; and the Sun turning black as sackcloth. Finally, *Revelation* also tells of the Moon becoming as blood; the armies of the world gathering together to battle at a place called Armageddon; giant 100 pound hailstones mixed with fire and blood being cast upon the Earth; the cities of the nations falling; mountains disappearing; and every island fleeing away.

These are extraordinary events! Who wouldn't want to know

---

*   The Bible's *Book of Daniel* (5:5-31) tells how writing mysteriously appeared on one of the palace walls in Babylon. The writing prophesied the fall of the Babylonian Empire, which then came to pass.

the causes for all of these frightful incidents? What types of social and physical events would produce such strange imagery? To find explanations and answers with some degree of credibility, we must begin deciphering or decoding the strange imagery used in *Revelation*. If we are uncertain about the imagery, then how can we be certain of the events that are prophesied to occur during the end times? Could part of the decoding process of this imagery come by determining *exactly* how the words in these descriptions were used when the Book of *Revelation* was written? In these verses, which words are to be taken literally, and which words are to be taken symbolically? What insight will we gain by understanding how the contemporaries of the people of the Bible wrote about catastrophic physical events and dreadful political events? Have we also considered that the Biblical authors, who are said to have been inspired by God, were at times communicating more accurate scientific information than they or their initial audiences realized? Most important in this process is determining the ***original*** meaning of certain key words used in the Bible.

In this book, the latest scientific information will provide answers to questions raised by the *Book of Revelation*. For the first time science, ancient literature, physical evidence, linguistics, and clues hidden within the Bible itself are all brought together to present unrivalled and compelling explanations for the prophecies of the *Book of Revelation*. By using forensic methods, the cryptic symbols of *Revelation* are revealed, showing how they "fit" together to present a clear and detailed picture of events prophesied for the end times. Recent information from the fields of geology, archeology, and astronomy is used to understand Scripture and reveal amazing scientific explanations for the prophesied events. Which is more intriguing - that so many of the verses in *Revelation* actually convey scientific information, or that the Bible contained this wealth of reliable (scientific) information 1,500 years before the so-called modern scientific community discovered it?

Most Christians today would readily acknowledge that Jesus was spoken of throughout the Old Testament (*Luke 24:27*). However, few seem to recognize the numerous Scriptures throughout the Old Testament that speak of the end times. Fewer still have looked closely at the great catastrophes of the Old Testament to see what they can tell us about the catastrophes of the end times. In addition, newly discovered physical evidence on Earth, including a two mile wide impact crater associated with enormous dust deposits and an eighteen mile wide impact crater beneath the sea associated with widespread tsunami deposits, corroborates some of these Old Testament events, and for the first time ties these events to the historical record.

The most recently discovered impact site is called the Burckle Crater. It lies on the sea floor 12,500 feet below the surface of the Indian Ocean, and it is comparable in size to the entire island of Manhattan! On November 13, 2006 a feature story appeared in the *New York Times* titled "Ancient Crash, Epic Wave – Did Catastrophe fall from above in 2807 BC?" This article announced the discovery of an eighteen mile wide impact crater. [11] The article told how the impact that caused the crater also generated a series of mountain high tsunamis that went out in all directions from the impact site. Physical evidence of these tsunamis comes from chevron shaped deposits of marine sediment and impact debris that were carried inland by the powerful waves of these mega tsunamis. Some of these chevrons are over 600 feet high. *(See Illustration D)*

Dr. Ted Bryant, a geomorphologist and Associate Dean of

science at the University of Wollongong in Australia said, "We're not talking about any tsunami you have ever seen . . . Aceh (the deadly Indonesian tsunami of 2004) was a dimple. No tsunami in the modern world could have made these features. End-of-the world movies do not capture the size of these waves."[12] Dr. Bryant notes that "There are chevrons around the Indian Ocean that all point back to this one crater site."[13] Chevron deposits in Madagascar, Western Australia, and India all point to the Burckle Crater.

Four of these chevron shaped deposits lie in southern Madagascar, and study of these deposits showed that they contain unsorted marine sands, shells and micro fossils mixed with materials characteristic of a cosmic impact, such as impact glass, impact spherules, and shattered rock fragments. These deposits also contained high levels of iron, nickel and chrome, metals associated with cosmic impacts. Tiny bits of these metals were even found fused to the types of marine micro fossils found in the area, such as foraminifera.[14] Core samples taken near the impact crater have a marine and cosmic composition similar to samples taken from the distant chevrons.[15]

Burckle Crater and the chevron shaped tsunami deposits were found by Dr. Dallas Abbott, a research scientist at Columbia's Lamont-Doherty Observatory in New York and her scientific associates. These scientists call themselves the "Holocene Impact Working Group." Members of this international research group are recognized experts in geology, geophysics, geomorphology, tsunamis, tree rings, soil science, and archeology. Up until the last decade, astronomers generally believed that cosmic impacts were very rare and only occurred once every 500,000 to one million years, but scientists in the Holocene Impact Working Group "calculate that catastrophic impacts could happen every 1,000 years." [16]

Based on the ages of the sediments found in three separate deep sea cores taken near the crater, Dr. Abbott and her associates estimate that the Burckle Crater impact occurred around 2,500 to 3,000 BC. Dr. Marie Agnès Courty, a geoscientist at the European Center for Prehistoric Research in Tautavel, France has been studying soil deposits from around the world for evidence of dust from cosmic impacts and she believes that there was a major impact around this same time period – 4,800 years ago. [17] In addition to Burckle Crater, the chevrons and the dust deposits, there are a number of other lines of physical evidence that call for a large scale catastrophic event of some sort around 2500 to 3000 BC. More importantly, a date of 2500 to 3000 BC is around the time the Bible gives for Noah's Flood.

Dr. Bruce Masse, an archeologist at the Los Alamos National Laboratory in New Mexico is a member of the Holocene Impact Working Group and he thinks that the Burckle Crater impact may tie to one of the many ancient stories, legends, and myths about a flood from around the world. However, critics do not see the relevance of a flood story that does not give definitive information about the flood being caused by a cosmic impact. While a flood story may tell of devastating rains, hurricane force winds, darkness, and people dying, it still needs to clearly tie the flood to a description of an impact that would have caused towering walls of water to wash over the land. Dr. Masse recognizes that "extraordinary proof" is needed to connect a particular flood story to the Burckle Impact and tsunami.

Among the flood stories of the world, only the Biblical Flood story gives "extraordinary proof" that can connect it to the Burckle Impact and tsunami. Beyond the fact that the date of the Biblical

Flood and the date of the Burckle Impact match, when properly translated, the Bible in fact talks about a cosmic impact and a tsunami. This is because the Bible *specifically* says that the Flood was caused by comet bombardment and then tells of a number of events that can only be caused by towering walls of water from an impact driven tsunami washing across the land. *(See Illustration F)*

When properly translated, Scripture tells of comets being "loosed" and "broken up," and describes how the waters of towering tsunami waves quickly covered all the high hills and mountains, and then how the waters left behind by the tsunami left the land flooded. This of course, brings to mind the devastation brought by the 2004 Indonesian tsunami, where the thirty foot waves coming in left behind a flood with less than ten feet of standing water.

Amazingly, a passage in the Babylonian Talmud dating to the third century AD says that when the Holy One wanted to bring a flood upon the world he used two stars or comets to bring the Flood. This in effect, tells how the Flood was the result of an impact driven tsunami and supports the translation presented in Chapter 4. This idea is not new; a number of astronomers and geologists have long called for a comet impact being the cause of the Flood. For example, Dr. Benny Peiser of Johns Moores University in Liverpool, England, an expert on cosmic impact, while discussing the Biblical Flood said, "Before we did not know of any phenomena that could trigger such a massive flood disaster . . . Now recent research on ocean impacts shows that a large body hitting one of the world's oceans could trigger the kind of massive tsunami capable of wiping out coastal cities." [18] Astronomers Victor Clube of Oxford, and Bill Napier of the Royal Observatory in Edinburgh, "see a clear astronomical association" between the break up of a "giant comet" and "the Biblical Flood at the start of the third (millennium BC)." [19] These comments by scientists about the Flood being caused by cosmic impact were all made before the discovery of the Burckle Crater.

A second new impact site is called the "Amarah Crater," and it is over twice the size of famed Meteor Crater in Arizona. In an article now available on the web entitled, "Meteor Clue to End of Middle East Civilization" from the *Sunday Telegraph* (November 4, 2001), British Science Correspondent Robert Matthews wrote:

> Studies of satellite images of southern Iraq have revealed a two mile wide circular depression which scientists say bears all the hallmarks of an impact crater. If confirmed, it would point to the Middle East being struck by a meteor with the violence equivalent to hundreds of nuclear bombs. [20] *(See Illustration B)*

The Amarah Crater is a "smoking gun" that establishes that a cosmic impact of enormous size took place in the Holy Land region during Biblical times. The explosion was more than 1,000 times as powerful as the atomic blast of about fifteen kilotons of TNT that destroyed Hiroshima in 1945. It was one of the greatest catastrophic events ever witnessed by mankind in recorded history. The Amarah Crater was disguised as a big lake in the midst of a vast swampy area in Southern Iraq. The crater was found by Dr. Sharad Master, a geologist with the Impact Cratering Research Group at the University of Witwatersrand, South Africa. Dr. Master reported on his findings in the November 2001 issue of the journal, "*Meteorites and Planetary Science.*" He wrote that the "strikingly circular shape" of this lake was in marked contrast with the "highly irregular shapes" of the other lakes in the marsh region, which were also much smaller.[21] Only after Saddam Hussein had the lake drained in his campaign against the Marsh Arabs, did satellite photos reveal the lake's true identity. His description reveals that the circular depression was enclosed by an elevated rim and an annulus, a ring like feature produced by material ejected by an impact, which are the textbook features of an impact crater.[22] Dr. Mike Baillie, a professor of paleoecology at Queens University in Belfast, says that just a single comet impact large enough to create the Iraqi Crater "would have caused a mini nuclear winter (colder temperatures and less rainfall) with failed harvests and famine, bringing down any agriculture based populations."[23]

While these two new impact craters, the Burckle and the Amarah, are currently causing great excitement in the scientific community, they are generally unknown to those who have tried to prove the reality of the miraculous events of the Bible and to those who have worked to interpret end times Bible prophecy.

As a result of new evidence and information, a clearer and more scientifically sound interpretation of what really happened during some of these Old Testament events is possible. Then the connection between these past events and the end times events described in the *Book of Revelation* becomes clearer. For example, when Jesus spoke of the end times, he said "the stars will fall from the sky, and the heavenly bodies will be shaken" (*Matthew 24:29 NIV*), and he likened the day of his return to the days of Noah and the Flood and to the day of Lot and the destruction of Sodom and Gomorrah (*Luke 17:26-30*). When we know the ancient meaning of the word "star," the common denominator among the three events Jesus spoke of becomes evident: cosmic impact! As we see the prophecies of *Revelation* connected to a number of other prophecies found in both the Old and New Testament, *THE COMETS OF GOD* systematically builds a multi-dimensional case for interpretations of end times events that is difficult to dismiss.

In *THE COMETS OF GOD* we see how Scriptures throughout the Bible work together to reveal what things are to happen during the end times, and why these things are to happen. This provides a more comprehensive picture of what the Bible says is God's plan for mankind, God's activity in history, and God's plans for the end times. [24] The disastrous end time events appear to be part of a number of long standing Biblical patterns that relate to the God of the Bible's plans for man and the Earth. The more we understand these patterns and plans, the better we recognize the scientific causes of disastrous events prophesied for the end times.

Once the causes of the disastrous events of *Revelation* and the nations of the end times are correctly identified, a distinctly different picture of the end time events emerges than provided by popular prophecy books. While a number of prophecy books have some correct information about the mysterious events of *Revelation*, they do not provide scientific connections or Scriptural support for their interpretations. Interpretations and explanations of *Revelation*'s disasters should involve connections to scientific realities, Old Testament catastrophes, other Old and New Testament prophecies, and be consistent with Biblical principles. This book's explanations of end times events will not ask its readers to have blind faith or to suspend their intellect. This book shows that what the scientific community now knows about mankind's history, Earth's geology and the astronomy of our solar system is now in alignment with end times Bible prophecy.

In one sense, comprehending the *Book of Revelation*, the

last book of the Bible is like dealing with a textbook. To be able to answer the test questions at the back of the book, one must know and understand what is in the front and middle of the book. In other words, the *Book of Revelation* represents the culmination of what has been taught throughout the Bible.

The new interpretations and explanations of Scripture presented in this book allows the prophecies of *Revelation* to be more understandable, more logical, and more believable. Many of the theologians and scientists who have gone over early drafts of this book have praised it for its originality and "goodness of fit" in interpreting important Biblical events and prophecies. One prominent Christian educator, Dr. Clifford Kelly, from *Focus on the Family* wrote, "After reading only one chapter of your work, I went back and reread chapters eight and nine of *Revelation*, texts I've studied for almost twenty years and found myself saying, 'Of course!' You really are onto something monumental here. . . ."[*]

---

[*] Personal communication from Dr. Clifford Kelly, Director of Academics for Dr. James Dobson's *"Focus on the Family Institute,"* letter dated May 1, 2001.

TWO

Stars That Fall From Heaven

And I beheld when he had opened the sixth seal, and, lo, there was a great earthquake; and the sun became black as sackcloth of hair, and the moon became as blood; And the stars of heaven fell unto the Earth, even as a fig tree casteth her untimely figs, when she is shaken of a mighty wind. And the heaven departed as a scroll when it is rolled together, and every mountain and island were moved out of their place.

Revelation 6:12-14

It is difficult to overstate the almost unimaginable energy that is released when a massive asteroid or comet hits the Earth ... Were one to land in Southern California, for example, all of Los Angeles along with several kilometers of the rock from the Earth's crust beneath it would be picked up and largely vaporized, lumps raining down on Hawaii and New York an hour or so later. Not that Honolulu or New York City would be left standing by then. Phenomenal seismic shocks following the impact would have already shaken them flat.

> *Rogue Asteroids and Doomsday Comets: The Search for the Million Megaton Menace That Threatens Life on Earth* by Duncan Steel, Ph.D. [1]

Throughout the ages many scholars have tried to determine the causes of different disasters referred to in the last book of the Bible. While there have been many ideas about the descriptions and many explanations for the events in the *Book of Revelation*, it is important to know that in the last decade scientists have provided valuable information for unraveling some of the mysterious passages written about the "effects" of these disastrous events. This scientific information has provided explanations few Bible scholars had considered until the last decade.

Yet, even now very little reexamination is being done by Bible scholars, and most end time authors still use a combination of explanations for these catastrophic events – nuclear warfare, demonic warfare, earthquakes, volcanic eruptions, and meteorite impacts. These explanations are often so contradictory and arbitrary that readers are unsure of how things really happen or why. For example, these explanations don't recognize that meteorites, that explode in the air or on the land or in the sea produce different types of catastrophic events. It has become apparent that we must go beyond traditional religious explanations and oversimplified generalizations if we sincerely want to grasp the reality and causes of the events in the *Book of Revelation*.

The information that different fields of scientific study can contribute must not be ignored or underestimated when trying to analyze and understand *Revelation's* mysterious passages. Beyond theology, specific knowledge of ancient cultures, history, geology, and astronomy all provide valuable information. Failure to look at *Revelation* from these different perspectives does not protect the authority of the Bible, rather the Bible is denied the opportunity to reveal its amazing specific information, and without this information we are left with a very distorted view.

Explanations involving nuclear catastrophe and demon possessed creatures have characterized most Bible prophecy books written since World War II. Such explanations overlook the scientific method, and the use of scriptural support. Yet, what has been an enigma over the centuries for so many people has recently become obvious to a certain select group. A number of scientists in the fields of astronomy and planetary science have readily recognized what is being described in the imagery and scenarios from the *Book of Revelation*. Consider what five scientists in these fields had to say:

1) Dr. Marcelo Gleiser, professor of physics, astronomy and natural philosophy at Dartmouth College, is unabashedly secular and believes the *Book of Revelation* is only "brilliant propaganda" to scare people into the Christian faith. Nevertheless, he recognized the *Book of Revelation's* scientific content in *The Prophet and the Astronomer: A Scientific Journey to the End of Time* where he

writes:
> The last book of the New Testament, the *Revelation of John*, tells of the destruction of Earth and the lower heavens by a succession of cosmic disasters, which include collisions with 'blazing stars,' causing the Sun to darken, the Moon to turn bloody red, and the stars to fall from the sky . . . Every one of the visions is punctuated by cosmic cataclysms, signifying not only the absolute power of God over nature but also his use of nature's power to express his wrath. [2]

2) Dr. John S. Lewis, a professor of planetary science at the University of Arizona and Co-Director of the NASA Space Engineering Research Center at the University of Arizona comments in his book *Rain of Iron and Ice: The Very Real Threat of Comet and Asteroid Bombardment*:
> . . . the description of future events in *Revelation* leans heavily upon the phenomenology of violent cosmic events . . . The central theme is clear and unambiguous: the events described in *Revelation* are of astronomical origin and describe real physical events, not mere portents or symbols. Did John (the author of the *Book of Revelation*) somehow know more about impact phenomena than any scientist before the present decade?[3]

3) Dr. Victor Clube, a Senior Research Fellow in Astrophysics at the University of Oxford and a past Director of the Royal Observatory in Edinburgh, and Dr. Bill Napier, an astronomer at the Royal

Observatory in Edinburgh, went out on a scientific limb of sorts in their book *The Cosmic Serpent: A Catastrophic View of Earth History* published in 1982. They report finding passages of the *Book of Revelation* such as the Seven Trumpets (*Revelation 8-11*) to be "impact inspired." These scientists noted that they were not championing *Revelation* but only ". . . remarking on the clarity of the astronomical associations." [4] They also wrote that:

> The apocalyptic literature of the Bible . . . is rich in allusions to what we can only see as astronomical events of the sort we have described (cometary impact). [5]

Indeed, in their book *The Cosmic Winter* written eight years later, they held to their theories and presented a special table listing the "possible comet impact elements" for the Seven Trumpets and the Seven Vials of the *Book of Revelation*. [6]

4) Astronomer Gerritt L. Verschuur, in his book *Impact: The Threat of Comets and Asteroids* noted that "The *Revelation of Saint John* speaks of the stars of heaven falling to Earth as a fig tree drops its fruit (*Revelation 6:13*) and similar tantalizing snippets" that relate to comet impact. Verschuur added his professional opinion that *Revelation 8* in the Bible "is clearly a description of an impact catastrophe." [7]

To these highly qualified scientists, who have no theological agendas to promote, the *Book of Revelation* clearly speaks of the Earth experiencing a series of catastrophic cosmic events. These scientists recognize cosmic impact in *Revelation* because the disasters are described in a context that provides a volume of accurate and detailed information about the different types of comet impact and their resulting effects.* It's amazing to find this information in a source as ancient as the Bible since, as Dr. Lewis noted, some of the Bible's information about impact phenomena was not known by any scientist until the last decade.

Since these scientists have recognized *Revelation's* many passages about "stars" that fall from heaven as references to cosmic impact, it is important to understand how the word "star" was used during ancient times. Misunderstandings of the original meaning of the word "star" have resulted in obvious modern errors by the misinterpretation of Bible prophecy and errors in the critique of Bible prophecy. For example, Gary DeMar in his book *End Times Fiction*, a critique of the *Left Behind Series*, wrote: "Revelation 6:13 says that 'the stars of the sky fell to the Earth.' How is this possible since the size of a star is many times larger than the Earth? A star hitting the Earth would vaporize it." [8] DeMar is unaware that to the ancient Hebrews and Greeks the word "star" was a non-specific term that was used to designate any of the luminous bodies seen in the heavens: comets, meteors, planets, and the sun like bodies we now call stars.† In ancient literature, when the word "star" was used,

---

* Note that scriptural context makes it clear that the "stars" *Revelation* says will fall are to be **literally** interpreted as being "stars," because they cause phenomena associated with literal stars (comets and asteroids) and cosmic impact. There is no contextual basis for the "stars" that fall in *Revelation* to be symbolically or figuratively interpreted as being intercontinental nuclear-tipped missiles (ICBM's), or as demonic warriors. A comet impact will do much greater damage upon the Earth than a nuclear missile.

† Another example of error resulting from not understanding the ancient meaning of the word "star" can be found in *The Most Revealing Book of the Bible*, by Vernard Eller who in 1974 wrote, "It is obvious that

the reader was given further information by the descriptions of its appearance and its actions. Consider the usage of the word "star" in the *Aeneid* by the great Roman poet Virgil of the first century BC:

> Scarce had the old man ceased from praying when a peal of thunder was heard on the left and a *star* gliding (fell) from the heavens amid the darkness, rushed through space followed by a train of light; we saw the *star* suspended for a moment above the roof, brighten our home with its fires, then tracing out a brilliant course, disappear in the forests of Ida; then a long trail of flame illuminated us, and the place around reeked with the smell of sulfur. Overcome by these startling ***portents***, my father arose, invoked the gods and worshipped the holy *star*.
>
> *Virgil, The Aeneid* [9]

Virgil's use of the word "star" clearly refers to a comet fragment or meteor that falling from the heavens streaks through the sky and hits the forest as a meteorite. Sometimes descriptive words were added to the word "star" to clarify the meaning. Heavenly comets

---

if even the tiniest of *stars* moved anywhere close to the Earth, the Earth would give way rather violently. Yet, in John's picture (*Revelation 8:10-11*) the Earth goes right on (and with people living on it after being hit by a "star"); and in subsequent scenes (for example: *Revelation 9:1*) he again has *stars* flying to Earth." (*The Most Revealing Book of the Bible*, Vernard Ellis, Eerdman Publisher, 1974, as quoted in *Prophecy and Prediction*, Dewey Beegle, 1978, p. 143.) In this quoted passage, Ellis doesn't understand that John in these verses is using the word "star" to designate "comets," not the gigantic luminous bodies that we today call "stars."

were referred to by the ancients as "hairy stars," meteors as "shooting stars," and planets as "wandering stars." Interestingly, the Greek word for "star" is the word *aster*; and from this word we get the word "disaster," which can literally mean "bad star." In 16th century Italian *disastro* referred to an "ill starred event." (So, a recitation of ancient and historic meanings and usages can be instructive in properly understanding *Revelation*.)

Comets are giant balls of ice and dust that represent the left over building blocks of the solar system. It is well known that comets and fragments from comets, also called cometary meteorites, have fallen to the Earth frequently in the past and still are today! In the ancient world comets were seen as both the "messengers" that warned of or preceded disasters as well as the actual agents of disaster.

A review of ancient literature reveals that any mention of stars that fall from heaven or gods attacking from heaven would be references to comets, or meteors striking the Earth. For example, the ancient Sumerian hymn, "*Hymnal Prayer of Enheduanna: The Adoration of Inanna in Ur,*" speaks of both Ishkur, the Bull of Heaven, and Inanna, the Queen of Heaven, saying that it is they "who rain flaming fire over the land." [10] The Babylonian "*Epic of Gilgamesh*" also speaks of Ishkur saying "With the **snort** of the Bull of Heaven, pits (small impact craters) were opened. Into them fell one hundred young men of Uruk (Erech)." [11] Another ancient Sumerian work called "*Lamentation Over the Destruction of Sumer and Ur*" tells about cometary gods who devastated the land. The text describes burning brought by "hailstones and flames" as a harrow (a comet fragment – the frozen water of comets contain combustible elements and compounds that can produce flames, especially highly flammable methane and sulfur) coming from above struck the city;

and heaven and Earth trembled, dust passed over the mountains, the sky darkened, large trees were uprooted, the forest growth ripped out, blood ran freely as mouths and heads were crushed, and "large stones one after another, fell with great thuds." [12] Is it surprising that *Revelation's* description of stars that fall from heaven to Earth, hailstones with fire, and darkened skies are really descriptions of impacts by comets and/or meteors?

Once we understand the basics about comets – their origin, composition, behavior, and what happens during and after different types of impact - it becomes obvious that the effects of cometary impacts are the cause of the catastrophic events described in *Revelation*, including **all** of the horrible events of the Seven Trumpets and the Seven Vials. Accordingly, the key to understanding the catastrophes described in *Revelation* lies in recognizing that they represent a series of comet impacts, just as scientists of today have recognized that multiple comet impacts can occur during a relatively short period of time. Even events calling for earthquakes describe earthquakes of a size that could only be caused by cosmic impact. It will surprise many to learn that the Fifth and Sixth Trumpets spoken of in the Bible also involve cometary impact, not nuclear warfare and not demonic warfare. Later, these two Trumpets will be looked at more closely, where we will not find demon possessed creatures wrecking havoc. Instead, we will discover a scientifically accurate picture of active comets. The Fifth and the Sixth Trumpets provide close ups of what active comets look like, how they work, and the pestilence they can bring.

The prophecies of *Revelation* clearly indicate a *series* of increasingly severe catastrophic episodes during a three and a half year period. Up until recently, it was generally believed that bombardments of the Earth by comets and asteroids only occurred as single incidents, and that these bombardments were separated by very long periods of time, ranging from ten thousand to one million years. However, scientists now know that bombardments of the Earth by comets and asteroids may be "concentrated into brief catastrophic periods in which multiple impacts occur." [13]

Based on recently discovered craters, scientists now believe that the dinosaurs became extinct when at least three cosmic impacts occurred during a relatively short period of time. [14] Nevertheless, the ultimate proof that multiple impacts can and do occur during a short period of time came as the world watched on television in 1994. Over the course of a week in July of that year, millions of TV viewers saw 21 separate large fragments (mini comets) from the break up of Comet Shoemaker-Levy 9 hit the (gaseous) surface of Jupiter. As each fragment hit Jupiter's upper atmosphere, it blew up in a giant fireball followed by a rising plume of debris, which later came crashing down to leave a black spot on Jupiter's banded face. [15] Soon there was a string or chain of black spots on Jupiter. Modern observation of these multiple impacts of Comet Shoemaker-Levy 9 opened astronomers' eyes to the probability that there have been other multiple impacts. Scientists quickly reassessed images of several of Jupiter's moons. Amazingly, the photographs revealed over a dozen cases involving similar chains of impact craters.[16] Here was more evidence that multiple impacts can occur during a short period of time. (Suspected topography resulting from multiple impacts include the Carolina Banks, Chesapeake Bay, and the Great Lakes basin).

Ironically, and dangerously closer to home, less than one month after the spectacle of Comet Shoemaker-Levy 9, Comet

Machholz 2 was discovered. This comet was found to consist of six recently separated fragments or components, all traveling in an orbit that brings them into the inner solar system and across the orbit of Earth every 5.23 years.[17] Where initially just one large comet made the near Earth fly by, there were now six smaller comets. Astronomers are concerned that these six comets (Comet Machholz 2 – A, B, C, D, E, and F), could eventually approach Earth and threaten to impact our planet.[18]

    The threat to the Earth as a potential victim of multiple impacts was made even clearer by observations of Comet Linear. Comet Linear was discovered in 1999 while it was making its first visit into the inner solar system (to go around the Sun). As it passed close to the Sun (0.74 AU), the added heat from the Sun began to melt the ice holding it together causing it to break up. Astronomers had a ringside seat as the comet unexpectedly began to disintegrate before their eyes. Images taken by the Hubble Space Telescope during August 2000 showed large fragments breaking off of the comet's nucleus. After a few weeks, Comet Linear was reduced to sixteen mini comets, each with a tail of its own, and countless pieces of debris, all of which were headed away from the sun and back out into space. Each mini-comet was larger than a football field.[19] The show put on by Comet Linear was a visual example of how frequently and suddenly the Earth could be subject to multiple impacts.

    After we learn about all of the horrible things that comets do to mankind and the Earth, it is obvious that the Seven Trumpets of *Revelation* describe the effects of seven separate rounds or episodes of comet impact, as the ultimate cause. During a round or episode of impact, more than one comet or comet fragment may hit the Earth. It is not unusual for comets and comet fragments to break up into multiple pieces as they come in to strike the Earth. The Seven Trumpets focus on how each of these seven impact episodes will affect the Earth, while the Seven Vials focus on how these *same* seven impact episodes will affect humanity. In essence, *Revelation* gives two witnesses for each round or episode of cometary impact. This is consistent with the Bible teaching that two or more witnesses are needed to establish a matter (*Deuteronomy 19:15, Genesis 41:32*), especially when "stars" cause such destruction..

Most popular prophecy books have neither recognized nor explained the parallels and relationship between the Seven Trumpets and the Seven Vials. Both the First Trumpet and the First Vial tell about an impact upon the land; the Second Trumpet and Second Vial tell of an impact upon the sea; the Third Trumpet and Third Vial tell of an impact upon a major river system. The Fourth Trumpet and Fourth Vial tell about an impact in the atmosphere; the Fifth Trumpet and Fifth Vial tell of a land impact that brings darkness and disease; the Sixth Trumpet and the Sixth Vial tell about an impact upon the great river Euphrates; and the Seventh Trumpet and the Seventh Vial tell about a massive impact that brings great hail and a mighty earthquake, and ultimately causes every island and mountain to disappear.

The authors of these prophecy books have not recognized that the Seven Vials are actually extensions of the same seven impacts described in the Seven Trumpets. Instead these authors have called for 14 separate events, the Seven Trumpets being followed by the Seven Vials. The obvious problem with this interpretation is that the Seven Trumpets tell of the world's population virtually being destroyed, and the planet being decimated. Now, if the Seven Trumpets destroy most of the world's population and decimate the planet, who and what would be left for the Seven Vials to destroy or conquer? The inability to correctly identify the causes of the events described in the Seven Trumpets and the Seven Vials has made it difficult for prophecy writers to recognize the obvious connections between these events. However, linguistics and the etymology of what certain words meant during ancient times allows a modern geoscientist to see beyond the mistaken translations and recognize the connections between what was being described and actual cosmic events.

The concept that the God of the Bible will use a series of cometary impacts to cause the disastrous events described in *Revelation* will require many people to reconsider long held ideas about what will happen during the end times. For the past fifty years many have believed that the massive disasters of *Revelation* would be the result of nuclear activity and that nuclear Armageddon was inevitable. By the time Hal Lindsey's bestsellers of the 1970's, *The Late Great Planet Earth* and *There's a New World Coming* were published, the nuclear war interpretation of *Revelation* was already well established. Lindsey's book took this interpretation a step further by tying together scientific developments in nuclear technology with world wide current events. "The prophecy books that proliferated in the 1970's and 1980's emulated Lindsey in making nuclear war a centerpiece of their scenarios."[20] Unfortunately, few prophecy writers considered that the massive destruction described in *Revelation* could actually be the result of the God of the Bible using the power of nature to express His wrath and not the result of nuclear war started by mankind.

There are also scientific reasons why the catastrophes of *Revelation* are not the result of nuclear war. While nuclear bombs exhibit great explosive power, they simply do not deliver the power to generate the shaking and ground movement necessary for the types of destructive events described in *Revelation*. Nuclear bombs do not move or melt mountains. Nuclear bombs certainly **cannot** make every island and mountain of the world suddenly disappear as called for in *Revelation 16:20*. On the other hand, except for radioactivity, a comet striking the Earth would produce the very same phenomena as a nuclear explosion, but on dramatically greater scale. For example, while a nuclear explosion could cause a moderate earthquake and the ground to shake locally, a six mile wide comet striking Earth at over 25,000 miles per hour would deliver the power to effect a deep penetrating blow. This blow would immediately cause an earthquake of unprecedented size. The shaking produced would be felt across the globe. It is interesting to note that the atomic bombs dropped on Hiroshima and Nagasaki (air bursts) barely caused the ground to shake. Significant shaking would have caused underground water pipes to bend or break, but this was not the case. [21] In comparison, even a small cosmic impact can produce seismic waves that would literally throw people around and cause the ground to roll up and down like the waves of the sea, as its massive energy is rapidly converted into rock crushing power and elastic rebound, like a soft ball hitting the bat in a home run. Most important, the impact of a comet that is several hundred miles or more in diameter *could* make every island and mountain of the world suddenly disappear as called for in *Revelation 16:20*. In Chapter 7 we see that planetary scientists have recently learned the impact of a huge comet could: 1) actually blast away a large part of the Earth's crust upon which the Earth's islands and mountains rest; and 2) actually penetrate the Earth's crust and release enough heat in the Earth's interior to cause the Earth's crust to melt, and every island and mountain to disappear.

Over time this sea of hot melted rock (magma) would cool and a new crust, in effect, a new earth would form. Scientists have come to call such mega impacts "planetary scale impacts." Based on the giant impact basins produced, scientists now recognize that "planetary scale impacts" have occurred on Mercury, Mars and the Moon.

Whether they be regional, intercontinental or planetary in scale, it should now be recognized by all that *only cosmic impacts can produce catastrophes of the size the Bible says will occur during the end times!*

The Richter scale determines the magnitude of an earthquake by the size of its seismic waves. The size of these seismic waves is reflected by the amount of shaking and movement produced by the earthquake. An increase of one whole number on this scale represents an earthquake that moves the Earth ten times more than earthquakes of the next lower number. An increase of one whole number on the scale also represents thirty times more energy being released. So a magnitude 8 earthquake represents ground movement that is ten times greater than that of a magnitude 7 earthquake, and 100 times greater than of a magnitude 6 earthquake.[*] While studies by geophysicists have made it clear that fault generated earthquakes have a physical limit of around "10" on the Richter Scale, cosmic impact generated earthquakes can produce shaking and ground movement that is several hundred thousand times greater, and could reach the equivalent of 12 or 14 on the Richter Scale! For example, an impact of great size, such as the six mile wide comet that

---

[*] While a whole number increase on the Richter Scale reflects a tenfold increase in the amplitude of ground motion at a given distance, this roughly translates to a 30-fold increase in the amount of energy released. A magnitude 8 earthquake represents 100 times the ground motion and 900 times the energy released by a magnitude 6 earthquake..

led to the extinction of the dinosaurs, could trigger an earthquake "100,000 times more powerful than the (7.9) earthquake that shook San Francisco in 1906 or the earthquake that devastated Indonesia in 2004." [22]

> For thus says the Lord of hosts, 'Once more in a little while, I am going to shake the heavens and the Earth, the sea also and the dry land.'
>
> *Haggai 2:6 (NAS)* and
> quoted in *Hebrews 12:26*

For a more detailed explanation about the maximum size of fault generated earthquakes, and why nuclear explosions cannot cause massive earthquakes or "set off" earthquakes along existing faults, see **NOTE ONE**.

In addition to an earthquake, a six mile wide comet hitting the land would produce a crater almost 120 miles in diameter and twenty four miles deep. A 300 mile wide ball of fire would arise from the explosion and vaporize everything it touched. There would also be a towering mushroom cloud or plume that would spread out, so that molten rock and incandescent dust ejected by the explosion would come to blanket the landscape for several hundred miles. The shock or blast wave produced by the explosion would at first cause the atmosphere to roll back. Then a thermal wave, a powerful super heated wind traveling at over 400 miles per hour would go out and set fire to everything in its path for a distance of 1,000 miles. [23] Here again we are talking about effects produced by a comet impact that are on a scale much greater than those produced by nuclear explosions.

> Their flesh shall consume away while they stand upon their feet, and their eyes shall consume away in their holes, and their tongue shall consume away in their mouths.
>
> Zechariah 14:12

> And the heavens departed as a scroll when it is rolled together....
>
> Revelation 6:14
(Sixth Seal)

Even if we put aside seismic effects, such as shaking and ground movement and just compare explosive power, there is an enormous difference between nuclear explosions and cometary impacts. The atomic bomb dropped on the city of Hiroshima had an explosive energy equivalent to the detonation of 15,000 tons of TNT (fifteen kilotons). In contrast, the 50 meter in diameter space rock that came in at over 25,000 miles per hours and created the 4,000 foot wide and 600 foot deep Meteor Crater in Arizona involved explosive energy equal to around **two thousand** Hiroshima size atomic bombs (20 to 40 megatons). According to Dartmouth astrophysicist, Marcelo Gleiser, a comet the size of Mt. Everest, the six mile wide comet that led to the extinction of the dinosaurs travelling at over 25,000 miles per hour, struck the planet with energy equivalent to an incredible "100 million megatons of TNT, ten thousand times more than all the thermonuclear bombs available at the height of the Cold War detonated together. [24] Nobel Prize winner, the late Physicist

Luis Alvarez and his son Walter Alvarez, geology professor at the University of California, Berkeley said this impact was "equivalent to the explosion of five billion (Hiroshima sized) atom bombs."[25] The bottom line is that there is no real comparison between the devastation that can be brought by nuclear events and the devastation that can be brought by cosmic events. It is impossible to compare the damage done by a thrown rock to that of a large mountain hurtling through space to smash into the Earth at thirty times the speed of a bullet. This almost unlimited power has alarmed astronomers and caused several of them to write books warning that impacts can bring a sudden and total end to civilization. Research astronomer Duncan Steel of the Anglo-Australian Observatory in his book *Rogue Asteroids and Doomsday Comets* writes:

> It is difficult to overstate the almost unimaginable energy that is released when a massive asteroid or comet hits the Earth. Merely stating that the explosive power is far greater than all the world's nuclear arsenals combined does not properly convey matters. The reader may think that such combined power might simply result in a larger area being flattened than that which a nuclear bomb devastates . . . In fact, the impact of a large asteroid or comet is quite different from that. Were one to land in Southern California, for example, all of Los Angeles along with several kilometers of the rock from the Earth's crust beneath it would be picked up and largely vaporized, lumps raining down on Hawaii and New York an hour or so later. Not that Honolulu or New York City

would be left standing by then. Phenomenal seismic shocks following the impact would have already shaken them flat.

> *Rogue Asteroids and Doomsday Comets: The Search for the Million Megaton Menace That Threatens Life on Earth* by Duncan Steel, Ph.D. [26]

Beyond the initial explosive effects, depending on where a comet impact takes place, there are also far reaching secondary types of physical devastation. A large impact that is on or very near the ground would not only produce a crater of substantial size, it would also put a tremendous amount of dust into the atmosphere. In fact, it would generate enough dust to blot out the sun for several months.

> . . . I saw a star (comet) fall from heaven unto the Earth . . . And he opened the bottomless (deep) pit; and there arose a smoke out of the pit (crater), as the smoke of a great furnace (mushroom cloud), and the sun and the air were darkened by reason of the smoke of the pit.
>
> *Revelation 9:1-2 (Fifth Trumpet)*

For large cometary impacts, along with the loss of sunlight, temperatures would drop and a "cosmic winter" would occur leading to global crop failure. While it is difficult to grasp how much dust a large comet impact would put into the environment, a team of ten of the world's leading astronomers and geoscientists said the impact that killed the dinosaurs was an "explosion equivalent to 250,000 Mount Saint Helen's eruptions," an explosion that would have released trillions of tons of debris and dust into the air.[27] How can one visually and mentally grasp such a tremendous amount of dust? Dr. David Morrison, space science chief at NASA's Ames Research Center, reflecting on images of cars and buildings being buried under a volcanic ash fall, said, "If you've seen pictures of what its been like in the Philippines with Mount Pinatubo erupting, just imagine that happening all over the Earth . . . The sky turns pitch black. Temperatures drop and kill crops. Most of the world would probably starve to death."[28] Even the dust raised by a small fifty to 100 meter impact would be several dozen times greater than that of a large nuclear bomb or a very large volcanic eruption, like Krakatoa or even the larger Tambora eruptions in recently recorded history.

People's natural fear instinct is triggered knowing about what could happen when an object from space hits the land. However, cosmic impacts that hit the ocean would be no less devastating. An impact that hits the ocean could instantly expose the seafloor miles below and trigger a giant tsunami (sea wave) at the surface down to the floor as well. The Indonesian Earthquake in 2004 set in motion a thirty foot high tsunami that resulted in over 200,000 deaths. An impact generated tsunami could travel at over 400 miles per hour and reach a height of 700 feet or more. Such a towering sea wave would travel hundreds of miles inland, destroying everything in its path and leaving deposits (tsunamites) far up mountain slopes. Scientists consulted by *Time-Life* say a six mile wide asteroid hitting the ocean "would give rise to a ring of seismic sea waves, or tsunamis, cresting nearly as high as the Rocky Mountains and three to four miles wide."

²⁹ If the asteroid were to strike in the middle of the Gulf of Mexico, "colossal water walls would travel outward" and simultaneously hit New Orleans, Tampa, Havana, and the Yucatan. These waves would "roll unchecked as far inland as Kansas City," surge across most of Mexico and Central America, covering Florida and the Caribbean Islands. The *Time-Life* experts explained that untold numbers of animals would drown and fish would suffocate from all the churning in the water. ³⁰

> . . . a great mountain burning with fire was cast into the sea; and the third part of the sea became blood; And the third part of the creatures which were in the sea, and had life, died; and the third part of the ships were destroyed.
>
> *Revelation 8:8-9 (First Trumpet)*

However, all cometary impacts do not take place on land or sea. The Tunguska impact in Siberia in 1908 is an example of a comet coming in from outer space that exploded after hitting the Earth's relatively dense atmosphere. An airburst at an altitude of forty miles could go unrecognized or be seen only as a flash of light or heard as a distant rumble of thunder, or felt in the rattling of windows. Nevertheless, an impact that takes place high up in the Earth's atmosphere would still be disastrous because it would literally blow a temporary hole in the ozone layer of the atmosphere. Since the ozone (O_3) layer of the atmosphere filters out most of the ultraviolet (UV) radiation from the sun, this would have dire effects on people caught moving around outside below the hole. Unfiltered ultraviolet radiation can cause severe sunburn in just seven minutes, and lead to cell damage, cataracts, and cancer.*

An airburst at a lower altitude would have especially serious effects. An airburst at an altitude of four to six miles would blacken the sky, and then literally burn or scorch men and set fires (everywhere) over a wide area when the intense heat of the fireball reached the ground. Literally speaking, the heavens would be ablaze and (flaming) fire would rain from the sky.† Since a small airburst does not produce significant cratering, it would be hard to recognize a cometary explosion as the cause of the disaster. Reports from surviving witnesses would be the main way of learning of a catastrophe caused by such an airburst. Eyewitness accounts of the Tunguska explosion are frightening.

> ...and power was given unto him to scorch men with fire. And men were scorched with great heat...
>
> *Revelation 16:8-9 (Fourth Vial)*

---

*   The danger from a temporary hole in the ozone is illustrated by events that occurred in Punta Arenas, a city of approximately 115,000 near the southern tip of Chile. In the fall of 1999 doctors reported an epidemic of sunburns. This occurred when the edge of the infamous Antarctic ozone hole passed over the city on successive Sundays. People were exposed to unfiltered radiation that was double the normal ultraviolet level and this caused the epidemic. Exposure to the Sun for just seven minutes was enough to burn skin. "Ozone Alert," Emma Young, New Scientist.com., October 7, 2000, and "Ozone Holes," Associated Press, October 20, 2000. Also see "Increase in Sunburn and Photosensitivity Disorders," Jaime Abarca, et.al, *Journal of the American Academy of Dermatology*, Part 1, February 2002.

†   If the air burst occurred over the ocean the surface waters of the ocean would boil. Life in the sea would be devastated by either the fireball or by the blast and thermal waves that followed.

STARS THAT FALL FROM HEAVEN

So far, we have seen that the disasters described in the book of *Revelation* are **not** and *could not* be caused by nuclear warfare. The disasters that authors of modern prophecy can't explain by nuclear warfare, they explain by demonic warfare. Nonetheless, like the erroneous idea of nuclear warfare as a cause of end times disasters, the disasters of *Revelation* are not caused by demonic warfare either. Popular prophecy books use demonic warfare to explain the Fifth and Sixth Trumpets of the *Book of Revelation*. However, once we understand the basics about the appearance, behavior, and composition of comets we gain insight as to just what exactly is being described in the controversial and perplexing passages of the Fifth and the Sixth Trumpets.

Defying scientific credibility, these haunting verses have been erroneously interpreted to involve an army of never before seen demon possessed, locust like creatures, and an army of 200 million evil spirits astride horse like creatures or an equally improbable army of 200 million men astride mobilized ballistic missile launchers who are led by four demonic angels. "End times" prophecy authors have presented these explanations even though there is no **precedent** in Scripture or history for these types of demonic creatures and bizarre events.* However, there are scientifically sound explanations for the <u>Fifth and Sixth</u> Trumpets that have clear cut precedents in events

---

* Scripture's call for precedent concerning events to occur in the future is expressed in *Ecclesiastes 1:9* ("The thing that hath been, it is that which shall be . . . there is no new thing under the sun.")and 3:15 (". . . that which is to be hath already been. . . ."). For example, the ten plagues of the Exodus out of Egypt provide precedent for the series of catastrophic events prophesied to occur during the end times (*Micah 7:15-16*). *Hebrews 12:26* tells how God "hath promised saying, Yet *once more* I shake not the Earth only, but also heaven." (Also Haman, the Agagite, who "sought to destroy all the Jews [*Esther 3:6 and 9:24*]" was a precedent for Hitler's efforts and what is prophesied to be the antichrist's efforts [*Revelation 12*]).

that occurred during Old Testament times. These explanations also fit with the pattern of all of the horrible scenarios of the Seven Trumpets and the Seven Vials as caused by seven cometary impact events.

Contrary to popular theories, the Bible's story of the Sixth Trumpet, from beginning to end, gives a detailed description of the physical appearance of active comets striking the Earth. The Scriptures of the Sixth Trumpet tell how active comets (not spiritual creatures and not Intercontinental Ballistic Missiles) when heated by the Sun issue fire, smoke, and brimstone out of their "mouths," that is, out of the vents in the crust that covers their icy bodies. The horse like imagery used in the Sixth Trumpet, where "the heads of the horses were as the heads of lions," is consistent with many ancient and classical descriptions of comets as "stars with a mane like a horse [or lion]," just as *kometes astares* means "hairy star" in Greek.[31] Amazingly, scientists did not discover that active comets had "mouths" or vents in their crust through which they outgas fire, smoke, and brimstone until a spacecraft flew close to Comet Halley in 1986. The resulting photographs of icy vents spewing jets of gas and dust took astronomers by surprise; "gas jets were unknown until 1986."[32] Yet, the Sixth Trumpet (*Revelation 9:19*) makes clear reference to these vents or mouths. These are remarkable observations prior to the scientific discoveries of the last 100 years. And, Scripture correctly recognizes the power of the gas and dust jetting out of a comet's mouth as written, "For their power is in their mouths..." *(See Illustration A)*

And the number of the army of the horsemen were ***two hundred thousand thousand (200 million)***: and I heard the number of them. And thus I saw the horses in the vision, and them that sat on them, having *breastplates of fire, and of jacinth, and brimstone*: and the heads of the horses were as the heads of lions, and out of their *mouths issued fire and smoke and brimstone*. By these three was the third part of men killed, by the fire, and by the smoke, and by the brimstone, which issued out of their mouths. For their **power is in their mouths, and in their tails**: for their tails were like unto serpents, and had heads, and with them they do **hurt**.

Revelation 9:16-19 (Sixth Trumpet)

Instead of describing frightening demonic creatures, the Sixth Trumpet brilliantly conveys, in surprisingly few words, (of competent scientific quality) vital information about comets we can only now begin to appreciate. Instead of questioning the credibility of the events described in *Revelation*, the correct interpretation helps to establish the scientific possibility of these events coming to pass.

In addition to accurately describing how active comets have mouths or vents through which they issue fire, smoke, and brimstone, the Sixth Trumpet correctly describes the overall appearance of active comets and tells of the death comets can bring. In the description, "tails like serpents" that "had heads," we have a picture of an active comet's head and tail. An active comet's tail is an extension of the material being jetted out of its mouth, which can act somewhat like the rudder of a boat in how it affects the track and speed of a comet's travel through space. Thus, *Revelation 9:16-19* is also scientifically consistent and technically correct when, referring to comets, it says, "their power is in their mouths and in their tails." When *Revelation 9:16-19* says that with their heads "they do hurt," it is a reference to the devastation that occurs when the head of an active comet impacts the Earth; where "hurt" indicates loss of life or limb.

Revelation 9:16-19 adds to the picture of incoming comets when it refers to riders having "breastplates of fire and of jacinth (deep blue) and brimstone." This is a perfect description of an active comet's coma, which a comet typically develops after it enters the inner solar system from deep space and is heated by the Sun. This "breastplate" or "coma" represents an enormous round to elliptical shaped *cloud* of gas and dust that forms around the comet's body. In other words, the body or nucleus of the comet is found within this "breastplate" or "coma" or "cloud." Spectral analysis of this "breastplate" shows that it indeed contains "brimstone" or sulfur gas, just as *Revelation 9:16-19* describes "breastplates of . . . brimstone." Also, the color of a comet's cloud is jacinth or deep blue, just as *Revelation 9:16-19* describes "breastplates of . . . jacinth.". What appears to be the body or nucleus of the comet can be seen within this cloud or breastplate. Also, when set against the blue background of the cloud, this nucleus has a "fiery" star like appearance just as *Revelation 9:16-19* describes "breastplates of . . . fire.[*] (See Illus. A)

Recognizing the scientifically accurate information about comets in the Sixth Trumpet points to an unfortunate mistranslation of one of the lines of the Sixth Trumpet from the original Greek. The mistranslation of this line has fueled the erroneous belief that the Sixth Trumpet deals with demonic warfare. Most English versions

---

[*] Astronomer's call this fiery appearing inner feature of the cloud a "false nucleus" or a "central condensation." This false nucleus contains the actual nucleus of the comet.

of the Bible incorrectly translate *Revelation 9:14* to read:

> Loose the four *angels* which are *bound in* the great river Euphrates.
>
> *Revelation 9:14 (Sixth Trumpet)*

The translation of this sentence is incorrect. It does not relate to the rest of the subject matter of the Sixth Trumpet which precisely described the heads, tails, comas, outgassing vents, and destructive power of comets. The correct translation of the sentence is easy enough to determine. In this sentence the Greek word (*aggelos*, #32 in *Strong's Concordance*) translated as "angels" can also be translated as "messengers"; and the Greek word (*epi*, #1909 in *Strong's Concordance*) translated as "in" can also be translated as "toward" or "for." So a retranslation of the sentence could read:

> Loose the four ***messengers*** which are ***bound for*** the great river Euphrates.*
>
> *Revelation 9:14 (Sixth Trumpet)*
> retranslation

In the ancient world comets were often referred to as messengers, so the sentence translated correctly as "Loose the four *messengers* or *comets* which are ***bound for*** the river Euphrates," relates to the rest of the subject matter of the Sixth Trumpet. In a logical sequence

---

* In the next chapter, Chapter 3 we will see that the question of who controls the comets of heaven, and who can "loose or bind" them is an important theme in the ancient Near East and in the Bible.

the Sixth Trumpet begins by telling about four comets *bound for* the Euphrates River, then describes what these comets look like, how they outgas, and how they will bring death and destruction. These four comets headed towards the Euphrates River are the reason why the Sixth Vial says that the waters of the Euphrates will dry up.

This clearer translation is also consistent with the next verse where *Revelation 9:15* says that these four "messengers" will kill "the third part of men," while *Revelation 9:18* says that the fire, smoke and brimstone which issued out of the horses' or comets' mouths will kill "the third part of men." This directly associates the heavenly "messengers" with comets that issue fire, smoke and brimstone out of their mouths or vents. In other words, we are told in two different ways that the comets of the Sixth Trumpet will kill "the third part of men." There is no scriptural precedent for demonic warfare killing great numbers of men. However, there is a history of cataclysmic events from the Old Testament that provides precedent for the events described in the *Book of Revelation*. This will be covered in Chapters 4 and 5. (For a more detailed explanation about the translation of the words "angel" and "messenger" and the retranslation of *Revelation 9:14* see **NOTE TWO**.)

How could four incoming comets match up with the army of 200 million called for in this description from *Revelation*? Astronomers have learned that some comets are consolidated or more densely packed than others. If a loosely consolidated comet or cometary meteor (large comet fragment) enters the atmosphere and then explodes in the atmosphere (instead of going on to hit land or sea), it may completely break up and disintegrate on its journey through the atmosphere. This would result in a bombardment of comet debris of various sizes and infinite numbers, even 200 million pieces of debris, which might be referred to as "mini comets," as well as vast amounts of dust that blacken the sky.

Considering that a comet is a giant dust laden iceball covered with a rocky crust ("dirty snowballs"), the disintegration of a comet can result in fragments that consist of pieces of rock and/or ice. The break up of a comet can generate hailstones that are much greater in size, composition, and density than the hailstones produced by atmospheric conditions. Cometary hailstones may exceed 100 pounds each. Hailstones and rocks would come in burning or in flames due to the combustible elements and compounds they contain (methane, the principle compound of natural gas, sulfur, carbon monoxide, and ammonia) being ignited by the very high temperatures produced by friction with the atmosphere. How can frozen water burn with flames? Well, we have all seen pictures of gasoline floating on top of the water burning. In a similar way the tremendous heat produced when a very large stone of cometary ice comes in at thousands of miles per hour can cause the combustible materials contained in its frozen waters to come to the surface and ignite and burn.* In ancient Near Eastern literature, and in the

---

*   William Corliss, *The Unexplained: A Sourcebook of Strange Phenomena*. (Glen Arm, MD: William R. Corliss, 1974) tells how the *American Journal of Science* in 1835 reported that a faint yellow and slightly oily substance called "inflammable snow" fell on the Russian town of Wolokalamsk in 1832 (p. G1-10). This inflammable snow was 1 to 2 inches thick and covered an area of around 8 to 10 acres. When studied, this soft snow "burnt with a blue flame without smoke," and chemical analysis showed that it contained carbon, hydrogen and oxygen. It seems that the snow came from a comet and the substance that burnt with a smokeless blue flame was methane (CH_4), since methane burns with a smokeless blue flame and methane is found in the frozen waters of a comet. Corliss also tells how the Journal *Nature* reported on the fall of large hailstones that exploded with a sharp report when they hit the ground on January 11, 1912. "As the hail fell, the fragments sprang up from the ground and flew

Bible, these fragments have been referred to as fiery stones, coals of fire, heavy hail, and great hailstones mixed with fire and brimstone. (Brimstone is the ancient term for sulfur. Comets, asteroids, and meteorites often contain relatively high amounts of sulfur, which has a distinctive "old eggy smell" when hot.) Ranging in size from inches to yards across, thousands to hundreds of millions of comet fragments, with a fire burning before them and flames trailing behind them, would cause the land to quake as they came streaming in like an invincible 200 million man army.

> And there fell upon men a great hail out of heaven, every stone about the weight of a talent (100 pounds)
> . . .
>
> *Revelation 16:21 (Seventh Vial)*

While the Scriptures of the Sixth Trumpet convey information that requires a telescope or a spacecraft fly by to see, the Scriptures of the Fifth Trumpet convey information that requires a microscope to see. The Fifth Trumpet (*Revelation 9:1-12*) presents imagery that supports current theories that comets potentially carry bacteria that could cause devastating diseases among mankind. We can easily recognize that the Fifth Trumpet describes cosmic impact, as it begins by telling of a star falling to Earth that opens a deep pit (crater), raises a cloud of smoke and dust, and fills the air with enough smoke and dust to darken the Sun (*Revelation 9:1-2*).* It is difficult to deny that in all directions, looking like a mass of 'popping corn' on a large scale" (p. G2-214). These hailstones may have also come from a comet, since their exploding and jumping suggests methane gas derived from a comet suddenly being released.

*   Most translations refer to a "bottomless pit," based on the Greek word *abyss* (#12 in *Strong's Concordance*) which primarily means 'deep.'

STARS THAT FALL FROM HEAVEN

this is a description of cosmic impact (reasonable scientists agree). There is no contextual basis for saying that the star falling to the Earth is not a literal star, or that it symbolically represents a person, or a demonic angel, or satan himself, as most popular prophecy authors claim.* The denial of the "common sense meaning" of this passage by prophecy authors is quite a surprise, especially since these same authors primarily advocate a literal interpretation of Scripture in their books but fail to consider the ancient meanings and use of words in their books.

One popular author even has a "Golden Rule of Biblical Interpretation" and quotes Dr. David L. Cooper, who wrote, "When the plain sense of Scripture makes common sense, seek no other sense, but take every word at its primary, literal meaning unless the facts of the immediate context clearly indicates otherwise." [33]

The Fifth Trumpet is consistent with the **pattern** of cosmic impact that underlies all Seven Trumpets and Seven Vials. This passage says that the smoke and dust contains locusts that hurt and torment men for a period of time. We know they are not literal or real locusts because the Fifth Trumpet (*Revelation 9:4*) specifically says that these "locusts" do not harm the grass, nor any green thing, nor tree. Nor can we ignore that after Egypt's plague of locusts during the Exodus, *Exodus 10:14* says, there would never be such a plague of locusts ever again. The symbol of "locusts" is used because locusts amass in great numbers, and like locusts, what is in the smoke and dust is carried wherever the wind takes them. Microscopes and Keeping theology from coloring the translation, a better translation would be a "deep pit." A "deep pit" perfectly describes an impact crater. The NIV refers to the "abyss."

*   Interpreting passages in *Revelation* is not a matter of starting with a popular or prevailing "system" of preconceived beliefs about the end times and then making the Biblical data fit this "system."

microbiology and their relationship to that which *Revelation* says has shapes like horses, heads with crowns, faces like men, hair like women, teeth like those of lions, and breastplates of iron (*Revelation 9:7-9*), will be discussed in Chapter 7.

The concept of comets carrying bacteria that can cause pestilence is a theory that was championed by Nobel Prize winning Physicist Sir Fred Hoyle. It is a phenomena mentioned on more than one occasion in the Bible. Additionally, the possibility of a comet carrying bacteria is currently being studied by NASA. It is worth noting that there were accounts of an unusual illness occurring after a confirmed cosmic impact that occurred in 1908. The possibility that cometary bacteria can cause devastating disease is also taken seriously at NASA. All space missions guard carefully against bacteria from space being introduced to planet Earth.

So, neither the Fifth nor the Sixth Trumpet involves demonic warfare at all. Instead the Fifth and Sixth Trumpets add to the escalating level of misery and death to be brought by comets. Beyond possible bacteria, comets do in fact contain harmful chemicals. For example, approximately one percent of a comet's mass consists of highly toxic cyanide (CN), and hydrogen cyanide (HCN). [34] A large impact hitting one of the planet's main watersheds could result in many lakes and rivers being poisoned. Some scientists believe that hydrogen cyanide (HCN) brought by comets played a major role in the widespread extinction of land animals that took place when cosmic impacts brought an end to the dinosaurs. [35]

> And the third angel sounded, and there fell a great star (comet) from heaven burning as it were a lamp, and it fell upon the third part of the rivers, and upon

the fountains of waters; And the name of the star is called *Wormwood*: and the third part of the waters became wormwood; and many men died of the waters, because they were made bitter.

*Revelation 8:10-11 (*Third Trumpet*)*

Large cometary impacts also produce harmful chemicals after they enter the atmosphere. A large impact produces incredibly high temperatures in the atmosphere that causes the sulfur carried by comets to become highly corrosive sulfuric acid, and also some of the nitrogen in the air (the atmosphere is 79% nitrogen) to burn and become nitrous acid (HNO_2) and nitric acid (HNO_3). Lightning discharges and nuclear explosions also do this, but they don't produce the vast quantities and high concentrations of these compounds that large cosmic impacts could create. The nitrogen compounds produced by an impact are acidic, very toxic, and very bitter. They are compared to wormwood in the Biblical passage quoted above because wormwood was an herb known during Biblical times for its very bitter taste.

Besides ingesting poison waters, breathing nitrogen oxides (NO_x) causes lung damage, lung edema, and death. Contact with nitrous acid (HNO_2) is also dangerous because this substance is corrosive, mutagenic, teratogenic, toxic, and universally carcinogenic.[36] The different nitrogen compounds produced in the atmosphere would quickly be caught up in the water vapor of clouds and produce rain as corrosive as battery acid. People continuously exposed to the acid rain or the cosmic dust in the atmosphere that would rain down for weeks to months after an impact would most likely suffer acute deaths, some only chronic rashes and sores. This acid rain would also make lakes and rivers so acidic that most aquatic life in them would die. Large quantities of acid rain hitting the ocean would devastate all forms of sea life down to a depth of tens of meters. [37]

> And the first went, and poured out his vial upon the Earth; and there fell a noisome and grievous sore upon the men which had the mark of the beast, and upon them which worshipped his image.
>
> *Revelation 16:2 (First Vial)*

> And the second angel poured out his vial upon the sea; and it became as the blood of a dead man; and every living soul died in the sea.
>
> *Revelation 16:3 (Second Vial)*

Astronomers were unable to fully appreciate the threat from the sky until they identified an impact that occurred in the last century. This impact event took place in a very remote part of the world in 1908. For years it was thought that a large earthquake had occurred somewhere in Siberia that year. Then, almost twenty years later, a scientific expedition to the remote area discovered the truth about what had actually taken place in Siberia. Since then, there have been over thirty scientific expeditions to the area.

Scientists now recognize that a comet exploded in mid air over the isolated Tunguska River Basin in Siberia on June 30, 1908 at 7:15 A.M. This mid air explosion flattened trees over 1,000 square miles, an area more than half the size of Rhode Island. It was caused by a relatively small 100 meter wide comet fragment which exploded after entering the Earth's atmosphere. Astronomers believe that this small comet fragment resulted in a thirty megaton explosion, an explosion comparable to the biggest of nuclear bombs, an explosion big enough to completely destroy a small state like Delaware or a large sprawling city like Los Angeles.

Tunguska was the first cosmic impact of significant magnitude in modern times, and scientists spoke with over 100 still living eyewitnesses to this cataclysmic event. These people directly experienced and understood how to communicate the reality of what happens during an impact. Both Tungus tribesman and Russian fur traders said they saw a fiery star with a fiery tail cut across the sky and explode high in the air with a loud crack. The explosion was heard over 500 miles away. A Siberian newspaper reporting on the event said that,

> As the luminous body approached the ground, it seemed to be crushed to dust, and in its place a vast cloud of black smoke formed . . . All the buildings shook . . . all the inhabitants of the village ran into the street in terror. Old women wept; everyone thought that the end of the world was upon them. [38]

People in a village 200 miles away from the impact site said they saw a huge mushroom shaped cloud of black smoke rise up and a forked tongue of flame shoot up through it. Others said the whole northern sky appeared to be covered with fire. Some people said they heard bursts of thunder and a strange underground roaring sound as the ground shook. Fierce winds were felt in towns over 300 miles away. One man at the Vanavara trading post about forty miles from the impact site told how his ears were scorched, and another at the same post said his shirt was almost fused to his chest by the fierce heat. Several Tungus herdsmen said they lost entire herds of reindeer and their dogs. In all, it seems that over 1,000 reindeer near the focus of the impact were burnt to ashes. People from a village 125 miles from the impact area told how the ground shook strongly while those forty miles away reported actually being thrown around. One man broke his arm, and another bit off his tongue. A family in a tent twenty-five miles from the impact area was thrown into the air. Several of them were knocked out, only to awaken to see that everything was shrouded in smoke and that they were surrounded by trees that were blown over and burning. The local herdsmen "believed the blast was a visitation by the (thunder) god Ogdy, who had cursed the area by smashing trees and killing animals." [39]

There are no survivor accounts closer to the impact site. Obviously, all living creatures in the area closer to the explosion were annihilated just like ground zero in thermonuclear blasts. A seismograph at a meteorological station in Irkutsk over 500 miles away picked up strong tremors that lasted for more than an hour. A seismograph in St. Petersburg 2,400 miles away also picked up tremors. Even seismographs in the U.S. and Java picked up tremors. The conductor of a Siberian Railway train travelling 400 miles away had to bring the train to a stop after seeing the tracks appear to ripple like water waves. Barographs around the world recorded an atmospheric pressure wave that circled the globe three times. Soot darkened the skies over Siberia for weeks, and Europeans saw a strange glow in the night sky for weeks from moonlight reflecting off the dust thrown up into the atmosphere.

Dramatic pictures of uprooted, flattened and stripped trees to a distance of twenty miles from the impact center were taken by the Russian scientific expedition to the area in 1927. The expedition did not find a large crater, only a shallow, but extensive depression filled with a number of small craters six to thirty feet in diameter. Today little evidence of the event remains except for a vast quantity of tiny spherules, fused material formed during the impact, and soil samples that contain high concentrations of cosmic dust. [40] Some of the reports on the site mention that people who lived near the impact site later developed symptoms of a strange illness, as did some of the members of the first expeditions who came to investigate the site.

We cannot reassure ourselves by thinking cosmic impacts are catastrophes of our distant past or a calamity for Earth's very distant future. There is evidence of another "Tunguska type" mid air explosion that took place on August 13, 1930 in a remote part of Brazil. [41] Also, astronomers have traced the 100 meter Tunguska projectile of ice and rock back to Comet Encke and a stream of related comet debris that may have been responsible for two bombardments of the Moon (the Earth's companion) during the last millenium, one of which was a 100,000 megaton event. [42] Most recently, millions watched twenty-one separate fragments of Comet Shoemaker-Levy 9 plough into Jupiter between July 16 and 22, 1994. These comet fragments ranged in size from one quarter of a mile to 2-1/2 miles in diameter. The impacts looked like tiny pebbles hitting a large ball from our distant point of view.

To illustrate the enormity of such a "fragment" impact, a two mile diameter fragment would produce a ten million megaton explosion compared to the "mere" thirty megaton explosion at Tunguska. This would be big enough to engulf the entire Earth in flames. [43] Add to all of this a number of harrowing (cosmic) "close calls" for Earth during the last decade that the media has reported.

Again, the main point in noting these recent impacts and near impacts is to show that cosmic activity actually does affect the Earth. It is very real and ongoing. They occur continuously, persistently. That the Earth will certainly experience a series of cometary impacts over a short period of time as prophesied in *Revelation* is a very real possibility supported by current scientific knowledge.

While most comets are only a few miles in diameter, some can be quite large. Remember Comet Halley? Comet Halley is the size and shape of the island of Manhattan, about ten miles long and five miles wide. There are even larger comets, which astronomers called "giant comets," that are over forty miles in diameter. Scarier still, the Hubble Telescope has recorded the movements of comets that are over 100 miles in diameter at the edge of the solar system. If one of these monstrous masses of ice and rock were to enter the inner solar system and eventually strike the Earth, it could cause the entire planet to temporarily reel to and fro on its axis (of rotation). This amount of energy that would be released on impact is something that no fault generated nor nuclear explosion generated earthquake could possibly do. [44] The magnitude of shaking that would ensue from such an impact would cause all the great cities of the world to fall, mountains to move, and islands to disappear.

> And the stars (comets) of heaven fell unto the Earth, even as a fig tree casteth her untimely figs, when she is shaken of a mighty wind. And the heaven departed as

a scroll when it is rolled together, and every mountain and island were moved out of their place.

Revelation 6:13-14 (Sixth Seal)

One monstrous mass of ice and rock has already entered the solar system and caused astronomers some concern. In 1977 a 120 mile wide comet named Chiron was found to have a chaotic and unstable orbit between Saturn and Uranus. Computer simulation indicates that gradual disturbances of Chiron's orbit by the larger planets could cause Chiron, with a mass ten thousand times the mass of the comet (or comets) that killed the dinosaurs, to wander into an Earth crossing orbit in 100,000 years.

Yet, if there are more than 1,000 Chiron type comets (called Centaurs) moving about, then one could move into a threatening Earth crossing orbit in about 1,000 years. [45] Furthermore, the experts consulted by *Time-Life* felt that a meteorite that was only six miles wide slamming into the ocean "would transform cool, blue Earth into a flaming crucible" for a short period of time. [46] If a six mile wide impactor could temporarily turn the Earth into a "flaming crucible," what would a 120 mile wide Chiron type comet hitting the Earth do? Could a comet with a mass ten thousand times greater than that of the comet (or comets) that killed the dinosaurs bring what the Bible in effect calls the **"Baptism of Fire"** (*Matthew 3:10-12*)? Such a collision would be so powerful that it would cause the Earth to truly reel on its axis. It would briefly envelop the Earth in flames and produce temperatures that would sterilize the Earth and its atmosphere while boiling the oceans away. Several of the explosions on Jupiter from Comet Shoemaker-Levy 9 slamming into it in July of 1994 were big enough to engulf the entire Earth in flames. A collision with a Chiron type comet would be like the collision that many astronomers believe resulted in the formation of the Moon, when a small planet sized body struck the Earth and removed material that became the Moon. [47] In this event, the world as we know it would end.

> The Earth is broken up, the Earth is split asunder, the Earth is thoroughly shaken, The Earth reels like a drunkard, it sways like a drunkard, it sways like a hut in the wind.
>
> *Isaiah 24:19-20 (NIV)*

> But the day of the Lord will come as a thief in the night in which the heavens shall pass away with great noise and the elements shall melt with fervent heat, the Earth also and the works that are therein shall be burned up.
>
> *II Peter 3:10*

Astronomers now realize that we live in a cosmic shooting gallery of comets, asteroids and meteors. Their concern has prompted a call for intensive efforts to watch for objects on a collision course with the Earth and for programs that can shoot down or deflect incoming objects off their threatening course. In his book *The Prophet and the Astronomer: A Scientific Journey to the End of Time* astronomy and physics professor Dr. Marcelo Gleiser has noted that prophecies of fiery objects falling from heaven has become a "legitimate branch of astronomy" and astronomers have become modern day "prophets of doom."[48] In this regard, consider what the following astronomers have said:

> In just the last couple of decades . . . our cultural evolution has enabled us to become aware of the nature of the threat that doomed the dinosaurs, which could doom us as well. Our planet is a target in the cosmic shooting gallery of high speed asteroids and comets . . . the greatest hazard of all is that civilization could be entirely destroyed any day by the unexpected impact of an asteroid or comet.
>
> Clark Chapman and David Morrison, *Astronomy and Cosmic Catastrophes* [49]

> . . unless we take this issue seriously now, it is unlikely that civilization, and probably our species, will have a long term future. How do we drive home the reality of this danger to national governments? Is it even possible?
>
> Gerrit Verschuur, University of Maryland Astronomer [50]

> Encounters of the Earth with these swarms (of comet fragments) will produce bursts of bombardment . . . with associated Armageddon like effects on the

> ground, analogous to nuclear war....
>
> M. E. Bailey, S.V.M. Clube and W. M. Napier, *The Origin Of Comets*[51]

> ... life on Earth might not survive direct hits from a cometary bombardment.
>
> Donald K. Yeomans, *Comets: A Chronological History* [52]

*   *   *   *

As we learn more about comets, their origin, composition, behavior, and what happens during and after the various types of impacts, our perceptions and understanding of what is described in *Revelation* changes. Armed with this information about comets, the theory that the disasters described in the *Book of Revelation* will be caused by nuclear warfare and/or demonic warfare becomes harder to defend.

Remember that geology and geophysics show that nuclear war and/or fault generated earthquakes cannot produce the level of seismic activity called for in *Revelation's* prophecies. Even volcanic explosions during Earth's initial formation were not of sufficient energy to do so. Likely, only cosmically generated impacts and explosions can produce this level of seismic activity and world wide destruction. Modern science and geo-technical data allow us to draw the same conclusions as the astronomers and planetary scientists: that the catastrophes described in the *Book of Revelation* are caused by a succession of cometary impacts!

In other words, comets will bring the fire and shaking and destruction that are to characterize the end times – not nuclear war, and not demonic creatures. *Revelation* makes it quite clear that the God of the Bible will use comets, the left over building blocks of creation of the solar system, to express His wrath and bring judgment during the end times.

Cometary impacts and their effects precisely fit the descriptions of what the Seven Trumpets and the Seven Vials and similar passages of *Revelation* say will happen during the end times. This cometary explanation resolves too many of *Revelation*'s mysteries to dismiss it as just another opinion about what the Bible says will happen during the end times. Since *II Timothy 3:16* says that Scripture is to be used to interpret Scripture, what are the Scriptures that clarify and confirm that comets are indeed to be the cause of *Revelation*'s disasters?

The next chapter will provide Scriptures that confirm that comets are to be the cause of *Revelation*'s disasters. The next chapter will do this by presenting a witness that is difficult to discredit -- the witness of the God of the Bible based on what He has to say in His word about comets and His use of comets. We will see that the Bible contains amazingly accurate and important scientific information about comets, long before scientists came to recognize this information. We will also see what the Bible tell us about ancient mankind and their response to cosmic events. Understanding this will help us relate to what is prophesied for the end times.

We will learn how the information about comets in the Bible has remained hidden from everyone and what is the key to seeing what so many others have read and studied, and yet missed.

THREE

The Lord of Hosts

Ask me of things to come . . . and concerning the work of my hands command ye me. I have made the Earth, and created man upon it. I, even my hands have stretched out the heavens, and all their **host** (*'stars'* in *Septuagint* and *'starry hosts'* in NIV) have I commanded.

Isaiah 45:11-12*

. . . ***the Lord of hosts*** mustereth the **host** of the battle. They come from a far (place), from the end of heaven, even the Lord, and the weapons of his indignation ('wrath' in NIV) to destroy the whole land. Howl ye, for the day of the Lord is at hand; it shall come as a destruction from the Almighty.

Isaiah 13:4-6

---

*   The *Septuagint is* the oldest and most important Greek translation of the Hebrew scriptures (Old Testament). It was made by 70 Jewish scholars in Alexandria, Egypt around 250 B.C.

Three of the oldest religions of today, Judaism, Christianity, and Islam, share one man in common as the Father of their faith – Abraham. The Bible says that God called Abraham from his family home in the ancient city of Ur to go to a new land. Thus began Abraham's calling as the "father of many nations." Beginning with Abraham, the Bible tells how God began to demonstrate His power over heaven and Earth to mankind in order to establish a name for Himself as the one and only true God. Today, it is difficult for us to fully grasp the importance of what the God of the Bible was revealing about Himself and the cosmos, until we understand what the people of Abraham's time and the prevailing cultures believed about the gods they worshipped.

The Bible speaks of the ancient Sumerians, Akkadians, Babylonians, and Assyrians, yet reading the scriptures alone does not tell us how these different peoples were connected to one another. Fortunately, within the last century archeologists have learned a great deal about the religious beliefs of the Sumerian, Akkadian, Babylonian, and Assyrian cultures from the translation of thousands of records these peoples wrote on clay tablets. Most importantly, these writings reveal that the Akkadian, Babylonian, and Assyrian cultures derived their religious beliefs and many of their practices or rites from the earlier **Sumerians**. Most of the important ancient Near Eastern myths, hymns, prayers, and lamentations were written in both Sumerian and Akkadian, the languages of the Babylonians and the Assyrians.

Among historical works, the Bible is the only ancient document that has provided an accurate account of the religious beliefs and practices begun by the ancient Sumerians. It presents the reader with an account that is wholly consistent with these ancient peoples' own records. Translation of these ancient texts confirm what the Bible says about the pertinent religious beliefs and practices of the Sumerians and the later ancient Near Eastern cultures that adopted the Sumerian religion, such as the Babylonians. Indeed, until the early 1900's, aside from the Bible, there was no evidence that a Sumerian people, language, and religion had ever existed! "The very name Sumer had been erased from the mind and memory of mankind for more than two thousand years." [1] During the nineteenth century, the Bible was only understood as a collection of legends. Even the great Babylonian and Assyrian cultures were held in suspect as myths until their palaces were discovered several decades later. By the 20th century, the excavation of Ur, (*Genesis 11:28, 31*), Erech (*Genesis 10:10*), Lagash, Nippur, and other Sumerian cities would establish that a major civilization older than the Egyptians, the Babylonians, and the Assyrians arose in Mesopotamia, just as recorded in the Bible. *(See Illustration B)*

In terms of the historical record and the validation of the Bible, the importance of these Sumerian cuneiform (written) records can not be overestimated by science or theology. Archeologically, their discovery remains significant because they continue to alter our understanding of who did what first in mankind's great achievements. Up until the twentieth century, Egypt was thought to be the cradle of civilization of the ancient world. But now the historical record gives this distinction to Sumer. The Sumerian civilization is older than that of the Egyptians, and the Sumerians built stepped temple towers (*ziggurats*) that were the forerunners of the stepped pyramids made by the Egyptians. The Sumerians were the first sedentary civilization to domesticate both plants and animals and practice large scale irrigation. As a farm culture arose, it was the

Sumerians who first invented writing and mathematics to help with their burgeoning agri-business. It was the Sumerian Ur-Namma (ca. 2050 BC), not the Babylonian Hammurabi (ca. 1750 BC) who gave us the first law code in recorded history, and the Sumerians, not the Greeks, who gave us the constellations of the Zodiac and the practice of astronomy to help tell them when to till, sow, plant, and reap with the seasons.

Yet in reading many Bible prophecy books by current authors, it is clear that most do not know that the Sumerians preceded the Babylonians or that the religion of the Babylonians began with the Sumerians. As a result, the correct understanding of how this religion relates to what will take place during the end times remains unknown. Few prophecy books even acknowledge that the word "babylon" is an Akkadian word that means a "gate of god," and that it has nothing to do with "babel" and the confusion of languages. It was the Sumerians who first built temple towers or "babylons" to worship their gods. While the current understanding of Babylonian religion in these poorly researched prophecy books is that it was based on their worship of Nimrod, Semiramis, and Tammuz,[2] this belief is incorrect and conflicts with the multitude of written records the Babylonians left that show what they did believe. From these written records, archeologists know a great deal about the Babylonian religion and their worship of the Sumerian goddess Inanna whom they called Ishtar. *(See Illustration C)*

In essence, Sumerian religion involved the worship of a pantheon of gods and goddesses who were led by a woman, the goddess Inanna (literally meaning the Queen of Heaven). Inanna became the patron goddess of the world's first empire, the Akkadian Empire of Sumer and Akkad. This empire united people of different nations and languages. What bound these different peoples together was the worship of the goddess Inanna, the "Queen of Heaven" (*Jeremiah 7:18* and *Jeremiah 44:17-19* and *25*), the rival to the God of the Bible. The false religion of a woman begun by the Sumerians was carried forward in time by virtue of her "holy marriage" (*hierosgamos*) to a succession of empires.

In the following pages, a number of Sumerian, Akkadian, Babylonian, and Assyrian writings will be quoted. These constitute ***primary sources of evidence*** (versus secondary and tertiary sources) as to what these ancient peoples believed, as opposed to the currently held but erroneous ideas that originated with a nineteenth century pastor, Rev. Alexander Hislop, who wrote a book entitled, *The Two Babylons: The Papal Worship of Nimrod and His Wife*. Reverend Hislop wrote this book long before the bulk of these ancient writings were found and translated. [3] For more information, see **NOTE THREE**.

Once again, ancient Sumerian culture and religion were adopted by the later Akkadians, Babylonians, and Assyrians. Sumerians believed that the heavenly gods created people on Earth solely for the purpose of mankind serving the will of these gods. They believed that all mankind was at the mercy of these gods, directing all affairs in heaven and on Earth. Sumerians also originated the notion of the personal god, a kind of divine father, who would act as their guardian and intercessor in the "assembly" of all of the gods. The Sumerians' belief in an assembly of sky gods who were associated with the objects of heaven, such as the Sun, the Moon, the planets, and the comets, became a belief common to all the nations and peoples of the ancient Near East. The late dean of Sumerian scholars, Dr. Samuel Noah Kramer at the University of Pennsylvania wrote in *Inanna - Queen of Heaven and Earth*, that "Sumerian religious faith and practice were based on a cosmology and theology evolved by their thinkers and sages in the early third millennium BC (when comet activity was at a peak) that became the basic ***creed and dogma of the entire Near East.***"[4]

Remember Abraham? He was from a Semitic people who migrated into the land of the Sumerians (*Genesis 11:2*), a fertile plain between the great rivers Tigris and Euphrates, a land known as Mesopotamia (meaning "between the rivers"). These Semitic people were mostly bands of Amorites who had gone from being desert nomads or wanderers to sedentary city dwellers. They built the city of Akkad (spelled Accad in *Genesis 10:10*) and came to be called Akkadians. While the Akkadians kept their own language, they had otherwise fully adopted the Sumerian culture and religion. Over time, the Sumerians were gradually absorbed by the rapidly growing Semitic/Akkadian population. Approximately a hundred years later, the descendants of the Akkadians living in the south of the region came to be known as the Babylonians and those living in the north came to be known as the Assyrians. While the Babylonians and Assyrians both spoke their native "Akkadian" language, they continued following the Sumerian religion and using the Sumerian language as their religious or sacred language, similar to the Catholic Church's use of Latin (still – the official language of the Holy See). The Babylonians even based the curriculum of their schools and academies on the Sumerian language and literature. The late Samuel Noah Kramer wrote:

> . . . it was a Semitic people – the Amorites - who put an end to the Sumerians as a political, ethnic, and linguistic entity . . . the Amorites (who came to

live in the area), commonly known as Babylonians because their capital was the city of Babylon, ***took over Sumerian culture and civilization lock, stock, and barrel. Except for the language, the Babylonian education system, religion, mythology, and literature are almost identical with the Sumerian*** ... And since these Babylonians, in turn, exercised no little influence on their less cultured neighbors, particularly the Assyrians, Hittites, Hurrians, and Canaanites, they, as much as the Sumerians themselves, helped to plant the ***Sumerian cultural seed*** everywhere in the ancient Near East. [5]

Years after Noah's Flood and following the fall of the Tower of Babel, Abraham (Abram) was called from the Sumerian city of Ur (ca. 1996. BC, Third Dynasty of Ur). Ur, the capital of Sumer (*Genesis 11:28, 31*) was then enjoying a period of renaissance or "Golden Age" as a great urban center.* Since Ur was a port city on the Euphrates River not far from the Persian Gulf, it was also a center of commerce and trade. Ships from the Gulf brought diorite, alabaster, gold, copper, ivory, and hardwoods from Egypt, Ethiopia, and India to Ur. During Abraham's lifetime, Ur was also a center of ancient Near Eastern religion. The holy seat of the royal high priestess was in Ur. The city was built around a temple tower called a *ziggurat*. The temple was dedicated to the worship of their Moon god and goddess, parents of Inanna, the Queen of Heaven. These people believed the

---

*   During later Biblical times, Ur, came to be known as "Ur of the Chaldees." "*Sumerian History, Culture and Literature*," Samuel Noah Kramer, translator. Samuel Noah Kramer & Diane Wolkstein, *Inanna - Queen of Heaven and Earth* (New York: Harper & Row, 1983), p. 117.

Queen of Heaven ruled over all the gods of heaven, which the Bible refers to as the "host of heaven" (*Deuteronomy 4:19, 17:3 and II Kings 23:5*).

The Sumerians, their cultural descendants, the Akkadians, and the subsequent Babylonians and Assyrians all worshiped a pantheon or assembly of gods. These people believed their gods guided and controlled the cosmos. They believed these gods were spiritual beings who mainly resided *within* the objects of heaven. These gods were seen daily as the Sun, the Moon, the planets, and occasionally observed as comets. The most important of these gods were cometary because of their powerful effects on Earth.

It seems that these gods were associated with comets because fearful humans had to explain the horrific comet activity that took place during the formative stages of these civilizations. Just as ancient man believed that the Sun and Moon were animated by invisible spiritual beings, comets were also believed to be occupied, animated, and empowered by invisible spiritual beings. The prolific art these people have left behind shows that cometary gods were given human characteristics and human forms replete with wings in recognition of their cometary aspect and ability to fly across the heavens.

Around Abraham's time, the Sumerian assembly of gods, or "host of heaven," was predominately led by a single deity, the mother goddess, known as Inanna. Remember that in the Sumerian language Inanna's name literally means "Queen of Heaven," the highest and most important god for the inhabitants of the Earth. How the other lesser gods had delegated all their power to Inanna so that she ruled heaven is described in a bilingual (Sumerian and Akkadian) work, the "*Elevation of Inanna.*" [6] Babylonian and Assyrian believers called this all powerful ***mother goddess*** Ishtar. To the Canaanites of the

Holy Land, she was called Ashtaroth (*I Samuel 7:3, 31:10, I Kings 11:5* and *II Kings 23:13*). Inanna, Ishtar, and Ashtaroth are just different names from different cultures for the same mother goddess, the same Queen of Heaven. She was also the goddess of light, love and life, who held control over fertility. (Later in this book it is revealed how the "Queen of Heaven," who the Sumerians and Akkadians also called a "harlot," "the hierodule of heaven" and the "vulva of heaven," relates to the "whore of Babylon" referred to in *Revelation 17*! The key elements of her religious system being tied to the political system and carried forward in time by a succession of powerful empires will also be discussed.)

While Inanna was cometary, she also appeared in heaven as the planet Venus, sometimes called the morning star and sometimes called the evening star.* In Sumerian writings, such as the *Self-Laudatory Hymn of Inanna and Her Omnipotence* and the *Hymnal Prayer of Enheduanna: The Adoration of Inanna in Ur*, Inanna is celebrated as the "Life-giving Woman," the "*Queen of Heaven*," the "Supreme One," the "Great Queen of Queens," who received not only "*queenship*" but also "*lordship*." They tell how she is "all devouring in power," "who rain[s] flaming fire over the land," and says "Heaven is mine, Earth is mine . . . Is there one god who can vie with me?" She is the woman who "seized the **Kingship of heaven** . . . (and) has changed altogether the holy rites." [7] Inanna was also celebrated for slaying Kur in battle and turning this monster into a mountain. The cosmic version of Kur was a fire spitting, winged dragon who threw small and large stones from heaven (like the Babylonian's Tiamat), menacing the people of Sumer. [8] In a collection of Sumerian hymns

---

* Later in time when cometary activity died down, a planetary persona was added to the cometary persona of all of the ancient Near Eastern sky gods in a process some call "planetization."

to Inanna, the "queen of both heaven and Earth," we read:

> Proud **Queen** of the Earth Gods, **Supreme** among the **Heaven Gods** ... you make the **heavens tremble** and the **Earth quake** ... you throw your ***firebrands*** across the Earth ... The gods of heaven and Earth kneel before you ... Hail to **Inanna**, Great Lady of Heaven! ... you destroy the wicked ... On the high ***roofs*** of the dwellings, On the ***platforms*** of the city, They (the people of Sumer) make ***offerings*** to her: Piles of ***incense*** ... They ***pour dark beer*** for her. They ***pour*** light beer for her... They prepare gug - ***bread*** in date syrup for her. ***Flour, flour*** in honey, beer at dawn. They ***pour wine*** and honey for her at sunrise. The dogs and the people of Sumer go to her with food and drink. They feed Inanna in the pure and clean place. [9]

The Queen of Heaven was also called the "Queen of Heaven and Earth," and believed to be the single personage who controlled both heaven and Earth. In the ancient Near East it was believed that the Queen of Heaven was the mediator between man and the gods. Offerings to all the different gods were usually made through her or under her auspices. Not until later in Babylonian times did she share some of her supreme powers with her consort, Marduk who had replaced her original consort, Dumuzi, also known as Tammuz, to the Babylonians and the Hebrews. At this time, in a proliferation of written works, the powers and deeds of certain gods were being ascribed to other gods depending on whose religious denomination was in vogue at the time. [10] Belief in the old gods remain as relevant today, as women in the Near East sit outside weeping and wailing for the Queen of Heaven's consort, Tammuz on the day he must leave to spend part of the year in the underworld (*Ezekiel 8:14*). Importantly, "*Tammuz*" is the name of a month in the Jewish calendar.

The Bible specifically refers to the "Queen of Heaven" in *Jeremiah 7:18* and *Jeremiah 44:17-19*, and *25*, and correctly tells how cakes were made for her and drink offerings were poured out.

> The children gather wood, and the fathers kindle the fire, and the women knead their dough, to make cakes to the **Queen of Heaven**, and to pour out drink offerings unto other gods, that they may provoke me (the God of the Bible) to anger. *(See Illustration C)*
> *Jeremiah 7:18*

The Bible refers to the assembly of sky gods or "host of heaven" which the Queen of Heaven led in *Deuteronomy 4:19, 17:3, II Kings 23:5, Jeremiah 8:2, 19:13, Zephaniah 1:5,* and *Acts 7:42,* and correctly tells how incense was burned on house rooftops in worship.

> And the houses of Jerusalem, and the houses of the kings of Judah, shall be defiled as the place of Tophet, because of all the houses upon whose roofs they have burned incense unto all the **host of heaven**, and have poured out drink offerings unto other gods.
> *Jeremiah 19:13*

> I (God) will also stretch out mine hand upon Judah

and all the inhabitants of Jerusalem . . . and them that worship the *host of heaven* upon the housetops.

Zephaniah 1:4-5

Then God turned; and gave them up to worship the *host of heaven.*

Acts 7:42

Throughout the ancient Near East, cultural response to cometary appearances and bombardments was grounded in fear (*Jeremiah 10:2*). Hoping to save their lives, crops, and cities from cometary destruction, people sought the goodwill of the cometary gods. Recall the above quotes from *Jeremiah 7:18, 19:13,* and *Zephaniah 1:45* where the Bible tells how cake was prepared, incense burned, and drink offerings poured out on house rooftops in prayerful worship to the Queen of Heaven and the host of heaven.

For protection from cometary destruction, compare this to earlier quotes from a collection of Sumerian hymns that told how "gug bread" and "flour in honey" were prepared, incense burned, and "wine and honey," "light beer" and "dark beer" poured out on the "high roofs of the dwellings" to the Queen of Heaven. Can there be any doubt that the Bible has correctly recorded key elements of the Sumerian/Babylonian religion? Temples and altars were built to these cometary gods, anthropomorphized (human like) statues were made of them, and sacrifices and offerings were presented in an attempt to appease them and minimize the destruction they brought.

Some modern astronomers have backtracked the orbits of

certain comets and related asteroids and meteor streams (such as Comet Encke and the Taurid meteor stream) to this time period. [11] They discovered there was much greater cometary activity at that time than at any other time during the subsequent 3,000 years.* Therefore, the ancient Near Eastern peoples' preoccupation with comets and destruction from the sky seems more than justified. Because they feared destruction from the sky, it should be no surprise to readers that their gods were mostly cometary in nature and that the ancients were "terrified by the signs of heaven" that these cometary gods sent

---

*   In *The Origin of Comets* reflect on the relatively high frequency of cosmic impacts they and other astronomers have found, especially during the second and third millennia BC. These astronomers note that "most investigations of the (ancient) omen literature have been conducted by historians and scientists working in relative ignorance of modern astronomical developments and of the ways in which disintegrating comets might affect the earth. These specialists have generally adopted a somewhat dismissive attitude toward the physical associations between the sky and the earth implied by Babylonian fears and beliefs." (M. E. Bailey, S. V. M. Clube & W. M. Napier, *The Origin of Comets*, New York: Pergamon Press, 1990: 19.) Ancient texts can now be read in a new way, because 20th century astronomers have finally acknowledged the physical connection between comets and meteors, where meteors are now seen to be the products of the break-up and disintegration of comets. They have recognized that the ongoing decay of comets can result in clouds and streams of comet debris consisting of massive chunks of ice ("heavy hail"), boulders, rocks, and dust. References to: 1) "small and large stones" in the context of fire, 2) earthquakes, 3) the land being torn up and uprooted as by a pickaxe, 4) hurricane winds, 5) flooding, and 6) great destruction are now correctly recognized as references of bombardment by cometary debris. Small and large meteorites and giant-sized chunks of cometary ice weighing hundreds to tens of thousands of pounds each are all described in various ancient texts, including the Bible. For example, Psalm 78:47 in the Amplified translation referring to one of the plagues of the Exodus says, "He destroyed their vines with hail (ice) and their sycamore trees with . . . ***great chunks of ice.***"

(*Jeremiah 10:2 NAS* and *Acts 19:35*).*

Ancient texts tell how the Sumerian cometary gods Inanna, the Queen of Heaven; Ishkur, the Fiery Wild Bull of Heaven; and Ninurta/Anzu, the Winged Lion of Heaven, devastated the land and caused the people to live in fear. These destructive cometary gods would periodically "rain flaming fire over the land" and "rain down on it small stones and large stones" in what seems to be relatively small cometary events. Sometimes, "pits were opened and into them fell hundreds [of] young men." However, on occasion when these gods were truly angered, they would make an all out attack on a city and a major comet impact would take place. *(See Illustration B)*

These people didn't merely imagine that comets brought death and destruction. Along with the recent evidence from orbital backtracking by computers that accounts for much greater cometary activity during these ancient times, geological evidence arising from Earth impacts proves that they actually saw the death and destruction brought by these comets. **Direct physical evidence** for a major impact exists in the two mile wide Amarah Crater of southern Iraq that also dates to this time period. Recall that Chapter 1 related the recent discovery of this crater by geoscientists. The Amarah Crater provides modern researchers clear physical evidence that an enormous comet impact, equivalent to the detonation of dozens of

---

* *Acts 19:34-35* makes reference to "the great goddess Diana, and of the image which fell down from Jupiter," which in the Amplified Bible is rendered "the great Artemis, and of the sacred stone (image of her) that fell from the sky." The image of this goddess was a meteorite and the temple of the goddess Diana (Roman identification) or Artemis (Greek identification) at Ephesus was one of the Seven Wonders of the Ancient world. Also, note that the Kabah, the sacred shrine of Islam at Mecca contains the "black stone" which is believed to be a meteorite.

modern nuclear bombs, took place in the ancient Near East.

Yet, long before the modern discovery of the Amarah crater, the description of such a horrific comet impact was recorded by ancients in literature and referred to in the Bible. Sumerians and Babylonians called an event of this magnitude a "great storm of heaven" or "evil storm" a storm that "attacks the land and devours it" with water "like the flood storm it destroys cities." A "great storm of heaven" would come like "a [cometary] harrow [a farm tool for uprooting the soil] coming from above" to strike a city like a "[cometary] hoe" and make "the land tremble and quake." Such an evil storm involved a "weapon [that] made all cower before it." A "great storm of heaven" was accompanied by "powerful winds," "hailstones and flames," with fires burning "in *front* of the storm," followed by "scorching heat," where in the end "dust was piled high."

In these descriptions are all the modern observational elements for establishing evidence of a large cosmic impact. Characterized first by a brilliant flash of light and a thermal wave, then ground motion from a large earthquake resulting from the impact, followed by a powerful air blast wave, and finally the raining down of dust and debris that was thrown high up into the Earth's atmosphere. For more information about the cometary nature of the Sumerian/Babylonian sky gods, see **APPENDIX A, "Sumerian/Babylonian Cometary Gods."**

After studying the ancient literature, astronomers Clube and Napier reached the same conclusion that some archeologists had previously found evident. In their book *The Cosmic Serpent* they wrote,

The conclusion is reached that *comets were for the most part treated as celestial deities* in prehistoric times . . . *being regarded as gods they [comets] were the subject of worship rather than objective analysis.* As we have seen, newly formed comets and meteor streams were at some not so remote period very prevalent in the sky and man would have been obliged to formulate some sort of picture of what was going on. As it happens, comets were not seen as objects subject to the control of deterministic physical law, but as benign or malignant beings with minds of their own, and as such they were not incapable of influencing the lives of men on Earth below. Indeed, they inspired great *terror* since man was conscious of the disasters they caused. *It may well be therefore that the polytheistic origins of many modern religions relate to primitive beliefs about comets.* [12]

Today it may seem primitive to associate the belief in a god or gods with the physical control of comets, but this association was central to the religious beliefs of ancient Near Eastern cultures. Ironically, this connection between actual control of the comets in the heavens and belief in god is now vitally important to our understanding of the "events" recorded in the Bible. A better understanding of the Sumerians and the ancient religion the Babylonians learned from them, allows us to now see the importance of some of the things the God of the Bible said and demonstrated to mankind during the time of the Old Testament and how this relates to future plans and responses.

When we understand that the God of the Bible is seeking to establish Himself as the one and only true God who controls both heaven and Earth, then it becomes easy to recognize that certain events during Abraham's lifetime are neither arbitrary nor random. There is a prevailing rationale behind events such as: 1) the covenant between Abraham and God being confirmed by causing a "burning lamp," (a meteorite) to pass between the two halves of the animal sacrifice (*Genesis 15:8-18); 2)* the rain of "brimstone and fire" out of heaven (*Genesis 19:24*) sent to destroy Sodom and Gomorrah; and 3) the promise to Abraham for an heir that would be fulfilled years later when Abraham and his wife, Sarah, were "old and well stricken in age" (*Genesis 15:10-14*). The God of the Bible asked them "Is anything too hard for the Lord?"

In reality with Abraham, the God of the Bible addressed the foundations of Sumerian and later Babylonian religious beliefs – that their gods controlled the activities of heaven and Earth, specifically cometary phenomena; and that their gods controlled human fertility. In the action of a burning meteorite passing between the sacrifice, the brimstone and fire from heaven destroying Sodom and Gomorrah, and the restored fertility of Sarah, the God of the Bible taught Abraham about "Himself," the true God, versus the gods of Abraham's native land. Beyond Inanna, the queen of the nation's gods, the Sumerians also had the notion of a personal god, a kind of ***divine father*** who looked out for the individual. The God of the Bible revealed Himself to Abraham as his personal god, a divine father who was also above all other gods, whether national or personal. The God of the Bible demonstrated his power to Abraham by making things happen that were never done by the gods previously known to Abraham had.

When the God of the Bible made promises and a covenant

with Abraham, he asked "whereby shall I know" [that what God said would come to pass] *(Genesis 15:8)*. God then confirmed His covenant with Abraham by doing something no other god had ever done – by causing a "burning lamp," a meteorite to pass between the two halves of the animal sacrifice that Abraham had prepared for God *(Genesis 15:8-18)*.* God **demonstrated** that He and not the Queen of Heaven was controlling the objects of heaven, the host or comets of heaven and that He ruled over heaven and Earth. In *Isaiah 66:1 (NAS, NIV)* God says, "Heaven is my throne and the Earth is my footstool."

In a Sumerian hymn dating to Abraham's time, the Queen of Heaven says, "Heaven is mine, Earth is mine" and she tells how heaven rests on her head as a crown and the Earth is tied on her foot as a sandal." But, Abraham asserts that the Lord is "the most high God, the possessor of heaven and Earth" *(Genesis 14:22)*. [13] That the God of the Bible caused a meteorite of exactly the right size and speed to arrive at a specific place and time, then by its trajectory pass between the sacrifice parts was an astonishing personal signature for a cometary god.†

---

*   In the Ancient Near East a covenant between two parties sometimes was confirmed by the practice of "cutting a covenant." In "cutting a covenant," animals were sacrificed and divided into two halves. Then the parties of the covenant would pass between the divided pieces and say something like "may it so be done unto me if I break my covenant with you." So when the God of the Bible caused a meteorite or comet fragment to pass between the two halves of the sacrifice, He showed His commitment to the promises He made to Abraham. *(Genesis 15:9-21 and Jeremiah 34:18)*. "Covenant," Vol. 5, *Encyclopedia Judaica*, (Jerusalem: Keter Publishing House, Ltd.: 1112).

†   This incident with the "burning lamp" passing between the sacrifice brings to mind occasions when fire from the God of the Bible consumed the sacrifice during the time of Elijah *(I Kings 18:38)*, the time of Moses *(Leviticus 9:22)* and the time of David *(I Chronicles 21:26)*. While these

When the God of the Bible later "rained upon Sodom and Gomorrah, brimstone and fire ... out of heaven" as his chosen method of destruction (*Genesis 19:24*), God again **demonstrated** to Abraham that He, not the Queen of Heaven, ruled over heaven and Earth. Now compare this rain of fire to a Sumerian prayer for the Queen of Heaven from around Abraham's time, called "*The Adoration of Inanna in Ur.*" The prayer says, "Supreme one, who are the Inanna of heaven and Earth, who rain(s) flaming fire over the land." [14] When Sodom and Gomorrah were destroyed with fire from the sky, God showed Abraham that it was He and not the Queen of Heaven who rains fire over the land.

The God of the Bible's omniscience was shown by His telling Abraham that He was going to destroy Sodom and Gomorrah, before it actually happened. In *Genesis 18:17* the Lord asks "Shall I hide from Abraham that thing which I do." Abraham was even allowed to intercede in an attempt to save these doomed cities. Abraham's intercession was not in the form of food and drink offerings but by conversing with his God face to face. And although the cities would be destroyed, God would spare Abraham's nephew, Lot, and his family.

When God promised and then gave Abraham and Sarah a son, long after their child bearing years had passed (*Genesis 15:10-14*), He **demonstrated** to Abraham that it was He, not the Queen of Heaven, who controlled fertility and gave life. In two related Sumerian hymns from around Abraham's time, Inanna, the Queen of Heaven, declares herself to be the "life giving wild cow (of heaven)," the "life giving goddess," and the "life giving woman . . . who multiplies all

particular events were probably not cometary, they do at least represent other occasions when the God of the Bible demonstrated His power to mankind.

living creatures and peoples," and asks "I, the Queen of Heaven am I! Is there one god who can vie with me!" [15]

Given God's meteorite that passed between the sacrifice, the fire that rained on Sodom and Gomorrah, and the gift of a son, Abraham accepts that it was He, the Most High God, who controls the things he had erroneously credited to Inanna (Ishtar) and the host of heaven. [16] God in the Bible reveals these things to Abraham knowing that Abraham would then teach them to his children and their descendants so that over time all the nations of the Earth would be blessed through Abraham (*Genesis 18:18-19*) and his acceptance of the one true God.

In the ancient Near East the word "host" could refer to "an army or great multitude of persons or things." In the Old Testament the God of the Bible is referred to as "the Lord of hosts," or "the God of hosts," or "the Lord God of hosts" over 285 times. For ancient Near Eastern cultures these titles all refer to a god's absolute sovereign power over both the cometary host of heaven and the spiritual host (spiritual beings) of heaven. In these cultures, both were one and the same. Sixteen out of nineteen times the Bible uses the expression "host of heaven" in an astronomical sense to denote the physical objects of heaven. For example, in *Deuteronomy 4:19* the God of the Bible warns Israel about worshipping "the sun, and the moon, and the stars (including comets) even all the host of heaven." In *Deuteronomy 17:3* He warns about serving other gods "either the sun or moon, or any of the host of heaven ('stars of the sky' NIV) and in *II Kings 23:3* He tells about "those who burned incense to Baal (a title meaning 'Lord'- Jezebel, a Phoenician introduced Baal worship to Israel), to the Sun, and to the Moon, and to the planets, and to all the host of heaven." When the Bible refers to the God of the Bible as "the Lord of hosts" in a passage that pertains to the objects of heaven, according to ancient Near Eastern usage, it is in effect referring to the God of the Bible as the "***Lord of Comets.***" Again and again, the God of the Bible represents Himself as being the one who is in charge of the heavens, including the comets, which are part of the "host of heaven." Recognizing that the word "host" can refer to "comets" is the key for unlocking the modern scientific meaning underlying many scriptures.

> I have made the Earth, and created man upon it: I, even my hands, have stretched out the heavens, and all their **host have I commanded.**
> *Isaiah 45:12*

The God of the Bible, the Lord of Hosts, the Lord of Comets specifically says that it is He who created comets and He alone who **commands** the host of comets (*Isaiah 45:12, 40:26, 13:3-5, Daniel 4:35, Psalm 103:20-21, 104:4,* and *148:8*), as the "King of Heaven (*Daniel 4:37*). Modern astronomers sometimes refer to comets as "dirty snowballs" for audiences to grasp their character. Recognizing that comets can be a form of "snow" or "ice" or "hail" helps readers to discover scriptures' scientific content. This is shown when it says that God "reserved" the comets as "the storehouses of the snow" and "the storehouses of the hail" (the Oort Cloud of comets at the edge of the solar system) for times of trouble and days of battle and war (*Job 38:22-23 NIV*) as His "army of heaven" (*Daniel 4:35*). (See Illust. A)

The Bible reveals that He calls all of these comets by name (*Isaiah 40:26*), brings them forth as His mighty ones (*Isaiah 13:3*), as His weapons of wrath (*Isaiah 13:5 NIV*), as His messengers and ministers of wind and flaming fire *(Psalm 104:4 NAS)*, to do whatsoever He commands them to do upon the face of the Earth *(Job 37:12)*. The Bible makes it clear that He calls all of these comets forth in accordance with the ordinances of heaven and Earth *(Job 38:33, Jeremiah 31:35, and 33:25)*, the physical laws, which include the movements of the objects in heaven that it says He set before the foundation of the world.

Concerning who created the heavens, *Nehemiah 9:6* says "Thou, even thou, art Lord alone; thou hast made heaven (the solar system), the heaven of heavens (the Oort Cloud, a spherical reservoir of comets at the end of the solar system), with all their **host** (in this case comets)." * In *Isaiah 40:25-26* the God of the Bible asks:

---

*   In the Bible when the word "heaven" is used in an astronomical sense, the "solar system" is being referred to; and when the phrase "heaven of heavens" is used in an astronomical sense, the "Oort Cloud of comets" is being referred to. Note that *Deuteronomy 10:14* makes reference to both "heaven" and the "heaven of heavens." *Psalm 19:1-6* indicates the astronomical definition of the Bible's use of the word "heaven" for the "solar system" when it talks about "heaven" and tells how God "set a tabernacle (dwelling) for the sun . . . and nothing is hid from the heat thereof," since the objects in the solar system are touched by the sun's heat, as opposed to the absolute cold of space. *Psalm 148:4* indicates the astronomical definition of the Bible's use of the word "heaven of heavens" for the "Oort Cloud of comets," when in a grammatical doublet it says, "Praise him ye heaven of heavens and ye waters that be above the heavens," since the only objects in the Oort Cloud are the frozen waters tied up in the billions of comets found there.

     It is interesting to note that the Hebrew word for "heaven" or "heavens" is *shamayim* (#8064 in *Strong's Concordance)*; and that the Hebrew word for "water" or "waters" is *mayim* (#4325 in *Strong's Concordance)*. The Hebrew word *sham* (#8033 in *Strong's Concordance)* cam be translated as "there" or thither." Thus, the Hebrew word for "heaven" or "heavens", *sham-mayim* can be construed of as "there water."

THE LORD OF HOSTS

> To whom then will ye liken me, or shall I be equal? saith the Holy One. Lift up your eyes on high, and behold who hath **created** these things, that bringeth out their **host** by number ('starry host one by one'- NIV, comets): he **calleth** them (the comets) all by names by the greatness of his might for that he is strong in power; not one (comet) faileth ('fails to appear' in Tanakh).
> *Isaiah 40:25-26*

In *Isaiah 13:3-7*, God speaks of His control of the comets to come during the end times says:

> I have **commanded** my sanctified ones (comets), I have also **called** my **mighty ones** (warriors' NAS - comets) for mine anger … the **Lord of hosts** mustereth the **host of the battle** (army of comets). They came from a far country (place), from the end of heaven (the Oort Cloud, the spherical reservoir of comets at the end of the solar system), even the Lord and the **weapons of his indignation** ('**wrath**' in NIV - comets) to destroy the whole land. Howl ye; for the day of the Lord is at hand; it shall come as a destruction from the Almighty. Therefore shall all hands be faint, and every man's heart shall melt.
> *Isaiah 13:3-7**

---

*   Note that as *Isaiah 13:5 NIV* refers to comets as "the weapons of his (God's) wrath"; *Psalm 78:48-49 NIV* tell how during the Exodus God sent hailstones and thunderbolts and "unleashed against them (the Egyptians) his hot anger, his **wrath**, indignation and hostility – a band

Psalm 103:20-21, 104:4, and *148:8* also tell how the God of the Bible controls comets, and uses them as His "messengers" and "ministers." Referring to comets as "messengers" was a common practice in the ancient Near East.

> (*Psalm 103:20-21*) "Bless the Lord, ye his ***messengers*** (#4397 *maw-lak - **comets***), that excel in strength (*'**mighty ones**'* in NIV - which relates to the 'mighty ones' of *Isaiah 13:3-7* who come from the 'end of heaven to destroy the whole land'), that do his ***commandments***, hearkening unto the voice of his word, Bless ye the Lord, all ye his ***hosts*** (including comets), ye ***ministers*** of his that ***do his pleasure*** ('his will' - NIV) . . . . (*Psalm 104:4*) Who maketh his ***messengers*** (#4397 *maw-lak - **comets***) tempests (#7307 *ruwach* - cometary); his ***ministers*** a flaming fire (cometary) . . . . (*Psalm 148:8*) Fire and hail, snow and vapours, stormy wind (all cometary) ***fulfilling his word*** ('***do his bidding***' - NIV or '***execute*** His commands - *Tanakh*).
> *Psalm 103:20-21,104:4,* and *148:8*[*]

Note that *Isaiah 40:25* says that the God of the Bible calls the "host" or comets "all by names." Indeed, the name of a comet is given in (group) of destroying angels (messengers)" where in the ancient Near East comets were sometimes referred to as "messengers." Both *Isaiah 13:5 NIV* and *Psalm 78:49 NIV* refer to comets as the weapons of God's wrath.

*   Recall that in Chapter 2 and **NOTE TWO** it was explained how the Hebrew word that is translated as "angels" can also be translated as "messengers" depending on context. Also the Hebrew word that is translated as "spirit" can also be translated as "tempest" or "stormy wind."

the Bible. *Revelation 9:11* says that the **name** of the "star" that falls to Earth opening a bottomless or very deep pit, is *Abbadon* in Hebrew and *Apollyon* in Greek. In both Hebrew and Greek this name appropriately means "destroyer."

The author is by no means the first in scientific fields to recognize that the God of the Bible uses comets for His purposes. In keeping good company, it is interesting that the father of modern physics, Sir Isaac Newton (1642-1727), whose theory of gravitation permitted the calculation of the movement of the objects of heaven, also believed that comets represented one of the ways that God operates in the world.

In the book *The Prophet and the Astronomer*, Physics and Astronomy Professor Marcelo Gleiser of Dartmouth wrote that Newton believed that God used comets as His tools or instruments; what the Bible calls "instruments of indignation" or "weapons of wrath" (*Isaiah 13:5 NAS or NIV*). Professor Gleiser wrote, "In Newton's scheme of the world, history was punctuated by catastrophes promoted by collisions with comets through the agency of God, in what might be called a causal theology." [17] Dr. Gleiser also noted that, "In his (Sir Isaac Newton's) view, the scientist's search for a quantitative description of natural phenomena was part of a grander quest, that of deciphering God's plan, or mind: the scientist was a decoder of God's writing." [18]

Many people are familiar with the idea or concept of a gentle, loving, and merciful God. However, there are a number of Biblical scriptures that provide a threatening view of this same God. It is difficult for modern people to envision, yet to people of Old Testament times this broader (good/bad) view emphasizes and respects the God of the Bible's power over heaven and Earth. In this view, the God of the Bible, like the Sumerian/Babylonian gods, causes comets to come, bringing death and destruction upon Earth. Indeed in scripture the God of the Bible is sometimes even described or personified as a Sumerian/Babylonian type cometary god. The Bible tells how God, "covers (Himself) with light as with a garment" (*Psalm 104:2*), and "maketh the (cometary) cloud His chariot" (*Psalm 104:3)*, and "did fly, yea, he did fly upon the wings of the wind"(*Psalm 18:10)*. "He bowed the heavens and came down, with thick darkness under His feet" (*Psalm 18:9 NAS*), and "there went up smoke out of His nostrils, and fire out of His mouth, burning coals blazed out of it" (*Psalm 18:8 NIV*). "He does according to His will in the host of heaven" (*Daniel 4:35 NAS*). "The mountains quake at Him, and the hills melt, and the Earth is burned at His presence" (*Nahum 1:5*). "Bow thy heavens, O Lord and come down: touch the mountains and they shall smoke. Cast forth lightning and scatter them: shoot out thine arrows and destroy them" (*Psalm 144:5-6*). "The heathen raged, the kingdoms were moved: He uttered His voice, the Earth melted. The Lord of hosts is with us" (*Psalm 46:6-7).*

II Samuel 22:8-16 (which is repeated in *Psalm 18:7-15*) gives an extended description of the God of the Bible as a cometary god, not unlike the ancient cometary god Ishkur, the fiery wild bull of heaven, the blast of whose ***nostrils*** does great damage. Another example is the cometary god Ninurta who when in human form is described as riding a chariot of seven hitched winds across the heavens. II *Samuel 22:8-16* uses the same words and personification in the same way to describe the same types of cometary phenomena that were described in the Sumerian/Babylonian literature. II *Samuel 22:8-16 (Psalm 18:7-15) says:*

> Then the Earth shook and trembled; the foundations of heaven moved and shook, because he was wroth. There went up a smoke out of his nostrils, and fire out of his mouth devoured: coals were kindled by it. He bowed the heavens also, and came down; and darkness was under his feet. And ***He rode upon a cherub***, and did fly: and he was seen upon the wings of the wind. He made darkness his secret place . . . At the brightness that was before Him, His thick (cometary) cloud passed hail stones and coals of fire. The Lord also thundered in the heavens, and the Highest gave His voice: hail stones and coals of fire. Yea He sent out His arrows, and scattered them; and He shot out lightnings ('to flash forth'-meteorites) and discomfited them. Then the channels of the waters (seabeds) were seen (tsunami withdrawl) and the foundations of the world were discovered (exposed by deep craters) at thy rebuke, O Lord, at the ***blast of the breadth of thy nostrils***
>
> ll Samuel 22:8-16 (Psalm 18:7-15)

There can be little doubt that the catastrophes of the Bible and the catastrophes prophesied to come involve the God of the Bible, the Lord of Hosts, the Lord of Comets, using comets as His "mighty warriors" (*Isaiah 13:3 NAS*) and the "weapons of His wrath" (*Isaiah 13:5 NIV*) to "worketh signs and wonders in heaven and in Earth" (*Daniel 6:27*). In a sense *Psalm 11:6 NAS* sums up the God of the Bible's use of cometary material to punish the wicked when it says, "Upon the wicked He will rain coals of fire; Fire and brimstone and burning wind will be the portion of their cup."

Considering the negative influence of cometary gods on the people of the ancient Near East, it is no surprise that in *Jeremiah 10:2* the God of the Bible tells the nation of Israel not to be "terrified" at the signs of heaven brought by comets like the heathen nations of the Near East were terrified. *Jeremiah 10:2* says "Thus saith the Lord, Learn not the way of the heathen, and be not dismayed at (***terrified by*** - NAS, NIV) the ***signs of heaven***; for the heathen are dismayed at (***terrified by*** - NAS, NIV) them."[*] For example, *Deuteronomy 4:34* says that the "signs and ***wonders***" sent by the Lord caused "great terrors" in the nation of Egypt during the Exodus.[†] In Chapter 5 we will see how many of the signs and wonders of the Exodus were caused by cometary activity.

When comparing the ancient literature's accounts of cosmic disaster to these Bible scriptures, the idea that the God of the Bible used comets amongst a people who worshipped cometary gods is difficult to deny. In light of these scriptures, their message is clear: the God of the Bible uses comets and the terrifying signs and wonders they bring so that mankind would know that there is only one God with power, ruling over heaven and Earth.

---

[*] *Jeremiah 10:2 NAS* says "Do not learn the way of the ***nations***. And do not be ***terrified by*** the signs of the heavens. Although the ***nations are terrified by them***."

[†] *Deuteronomy 6:22* in the Tanakh (Jewish Publication Society Translation) also refers to the events of the Exodus and characterizes them as involving "***destructive signs and portents***...." Interestingly, the Hebrew word *mo-faith* (#4159 in *Strong's Concordance*) translated as "wonder" or "portent" comes from the Hebrew word *yaw-faw* (#3302 in *Strong's Concordance*), a root word meaning "to be bright." An active comet or incoming comet fragment would "be bright." This helps make it clear that the "signs" that the Bible says caused "terror" were in fact caused by destructive cometary activity.

To support this, *Psalm 77:14 NIV* says "You are the God who performs miracles (wonders); you display your power among the peoples." In the signs and wonders brought by comets, we see the God of the Bible displaying His mighty power to mankind on the biggest stage possible – Heaven – where all can see for themselves. Again and again, the Bible tells how God used the terrifying signs and wonders to let mankind know of His mighty power. (Also see *Exodus 7:3* and *5, 10:1-2, 14:18, 19:4-5* and *8, Deuteronomy 4:35, Joshua 2:10-11, 9:9, Isaiah 63:12,* and *Jeremiah 32:20-21*).

Furthermore, consider the Seventh plague of the Exodus (*Exodus 9:13-28*) which involved "very heavy hail" (NAS) and "fire mingled with the hail" as brought by comet debris. In *Exodus 9:14 and 16* God instructs Moses to tell Pharaoh "For I will at this time send all my plagues upon thine heart and upon thy servants, and upon thy people; that thou mayest *know* that there is none like me in all the Earth . . . to show in thee my *power*, and that my *name* may be declared throughout all the Earth." During Biblical times the Hebrew word *shem* (#8034 in *Strong's Concordance*) which is translated as "name" could also be used to convey a meaning of "revealed person" or "authority." In essence, the miraculous signs and wonders of the Exodus were revelations to mankind about who God is and His absolute power over heaven and Earth.

A closer examination of many of the miraculous disasters and battles of the Old Testament reveals the use of comets and cometary material. For example, this was the case when "the Lord fought for Israel" against a number of nations (multi-national force) in the Valley of Gibeon, by casting down "great stones," and by 'hailstones" from heaven (*Joshua 10:5, 11* and *14).* Another example, comes from *Psalm 83:9,* and *14-18 (NIV)* which says, "Do to them ... as you did to Sisera ('They fought from heaven; the *stars* [cometary] in their courses [paths] fought against Sisera' *Judges 4:7* and *5:20*) .... As fire consumes the forest or a flame sets the mountains ablaze, so pursue them with your tempest (whirlwind - cometary), and terrify them with your storm (cometary). Cover their faces with shame so that men will *seek your name*, O' Lord .... may they perish in disgrace. Let them know that you, whose *name* is the Lord - that you alone are the *Most High* over all the Earth."

Not all kings were like Egypt's Pharaoh of the Exodus. The Bible tells of two kings who recognized the power of the God of the Hebrew people as a result of the signs and wonders from the sky. In *Daniel 6:25-27* we read of an interesting decree made by King Darius of the ancient Persian Empire (550-331 B. C.). *Daniel 6:25-27* says,

> "Then King Darius wrote unto all people, nations, and languages that dwell in all the Earth . . . I make a decree, That in every dominion of my kingdom men tremble and *fear* before the God of Daniel: for he is the *living God* . . . he worketh **signs and wonders in heaven and in Earth.**"
>
> *Daniel 6:25-27*

Daniel 4:1-3, 35 and *37* tell how King Nebuchadnezzar of the Neo-Babylonian Empire (612-536 B.C.) made two similar decrees. The decrees were about Israel's God being the one who works signs and wonders in heaven, as the Lord of Comets, and the one who is able to work His will among the host of heaven, as the true King of Heaven.*

---

* The Neo-Babylonian Empire (612-536 B.C.) officially worshipped a pantheon of cometary gods, where Ishtar was the Queen of Heaven, and her consort Marduk, was the King of Heaven. By the time of the Neo-

Daniel 4:1-3, 35 and *37* entail portions of two separate decrees made by Nebuchadnezzar, the most successful and most celebrated of the Babylonian kings. The ancient King proclaims,

> "Nebuchadnezzar, The King, unto all people, nations and languages that dwell in all the Earth; Peace by multiplied unto you . . . How great are his ('the most High God' NAS) *signs*! And how mighty are *his wonders*! His kingdom is an everlasting Kingdom, and his dominion is from generation to generation . . . he doeth according to his will in the **army of heaven** (*'host of heaven'* - NAS and Tanakh) . . . Now I, Nebuchadnezzar praise and extol and honor the **King of Heaven**, all whose works are truth, and his ways (are right - Tanakh): and those that walk in pride he is able to abase."
>
> *Daniel 4:1-3, 35* and *37*

Should the God of the Bible use comets during the end times, as indicated in the Seven Trumpets and Seven Vials of the *Book of Revelation*, which He says He has "prepared for an hour, and a day and a month and a year" (*Revelation 9:15*), He will in essence be acting in the same way that He did during the time of the Exodus. In *Micah 7:15-17 (NIV)* it says, "As in the days when you came out of Egypt, I will show them my *wonders. Nations* will see and be ashamed . . . they will lay their hands on their mouths . . . They will come trembling out of their dens, they will turn in *fear* to the Lord our God and will be afraid of you" (also see *Isaiah 2:19-21 and Revelation* Babylonian Empire, Ishtar had a planetary expression as the planet Venus in addition to her original and divine expression as a comet.

6:15-17). This use of extraterrestrial comets during the end times to judge nations and again save the Israelite people via catastrophic events including the Seven Trumpets and Seven Vials of the *Book of Revelation* is consistent with *Ecclesiastes 1:9 NAS* which says "that which has been done is that which will be done," and *Ecclesiastes 3:15 NAS* which says "that which will be has already been," and *Malachi 3:6* where the God of the Bible says ". . . I change not. . . ." * The repetition of these events ensures end times impacts by comets.

The God of the Bible's plan to use cosmic activity (comets) during the end times is clearly stated in *Ezekiel 38-39*. *Ezekiel 38-39* speaks of a great end times battle where the God of the Bible will once again use comets to fight for the nation of Israel. In this battle the God of the Bible will fight for and save the Jews who represent the surviving remnant of the nation of Israel. As in the days of Moses and Joshua, God will again use comets to judge Israel's enemies. A number of astronomers see the catastrophic events of the Exodus occurring as a result of cometary activity with the hail and fire that fell during the Seventh Plague [*Exodus 9:18-25*] being the most obvious. During the time of Joshua, the great stones and hailstones that fell from heaven and slew the enemy at Gibeon also speaks of cometary activity (*Joshua 10:10-14*).

In this great end times battle the God of the Bible will again use comets to reveal Himself and His power to many nations. *Ezekiel 38:19-23 and 39:21* give the key elements of comet impact (fire and shaking) that the God of the Bible will send upon the enemy, and then they tell how the God of the Bible will reveal Himself and His

---

*   *Ecclesiastes 1:9* and *3:15* in the *Tanakh* say "Only that shall happen Which has happened, Only that occur Which has occurred; There is nothing new Beneath the Sun!" and ". . . And what is to occur occurred long since."

power to mankind. *Ezekiel 38:19-23 and 39:21* say:

> "Surely in that day there shall be a great ***shaking*** in the land of Israel . . . all the men that are upon the face of the Earth shall shake at my presence, and the ***mountains shall be thrown down*** and the steep places shall fall and every wall shall fall to the ground . . . I will rain upon him, and upon his bands ('troops' - NAS, NIV), and upon the many people ('nations' - NIV) that are with him . . . ***great hailstones, fire, and brimstone (from comets)***. Thus, will I ***magnify myself, and sanctify myself; and I will be known in the eyes of many nations and they shall know that*** I am the Lord . . . (*Ezekiel 39:21*) And I will set my glory among the heathen, and all the heathen shall see my ***judgment*** that I have executed."
>
> *Ezekiel 38:19-23 and 39:21*

In *Joel 2:30-31* we read of signs and wonders to be brought by comets during the end times. *Joel 2:30-31* says, "And I will shew ***wonders*** in the heavens and in the Earth, blood and fire and pillars ('columns' in NAS) of smoke. The sun shall be turned into darkness and the moon into blood (both as a result of the dust raised by cometary impacts) before the great and terrible day of the Lord come." A scientifically accurate description of cometary impact effects is clear and unambiguous.

There are a number of scriptures in the Gospels of the New Testament that tell of the terrifying signs and wonders to be brought by comets during the end times. In *Luke 21:25-26* Jesus speaks of the end times in a prophecy given from the Mount of Olives, called the Olivet Prophecy, He says, "And there shall be *signs* in the sun and in the moon and in the stars (comets); and upon the Earth distress of nations with perplexity; the sea and waves roaring. Men's hearts failing them for *fear* and for looking after those things which are coming on the Earth: for the powers of heaven ('the heavenly bodies' - NIV) shall be shaken."

Even more descriptive and physically consistent effects of cometary impacts are given in *Matthew 24:29* which adds that "the stars shall fall from heaven" in reference to comets falling to the Earth. The dust from such impacts would darken the Sun and the Moon, as referred to in *Luke 21:25-26* and *Joel 2:30* quoted above. In *Matthew 24:29 (Mark 13:24-25)* Jesus says, "in those days shall the sun be darkened, and the moon shall not give her light, and the stars (comets) shall *fall* from heaven and the powers of the heavens ('the heavenly bodies' - NIV) shall be shaken."

Taking *Matthew 24:29* and *Luke 21:25-26* together, and remembering that in Greek the word for "star," the word *aster* can refer to comets, we can see that Jesus in his Olivet prophecy is saying that:

> (1) Stars, specifically comets shall fall from heaven to the Earth (*Matthew 24:29*); where (2) the comets that strike the seas will cause roaring waves (*Luke 21:25-26*); and (3) the comets that strike the land will cause dust to be raised into the atmosphere so that the Sun and the Moon will be darkened (*Matthew 24:29*); (4) so as to produce *signs* in the Sun and in the Moon, and in the stars or comets

(Note that *Joel 2:30* which was quoted above can now be seen to be scientifically correct when it refers to the dust raised from these comet impacts during this same event as "pillars of smoke"); and, (5) so that there will be distress of nations, and men's **hearts will fail from fear** of seeing and being in such a ***terrifying*** and devastating comet bombardment of the Earth (*Matthew 24:29* and *Luke 21:25-26*).

*   *   *   *

There is a hazard in believing end times explanations that do not reflect past Biblical catastrophic events and the numerous scriptures that convey accurate scientific information about comets. The Bible presents a story that does not include nuclear wars and horrific attacks by never before seen demons. That cometary impacts took place in the past and now threaten our future is undeniable.

In this chapter we learned about the religious beliefs of the contemporary cultures of the Bible's Hebrew people – the ancient Sumerians, Akkadians, Babylonians, and Assyrians. As we looked at some of their literature, it is clear that they experienced and wrote about cosmic disasters, and that they believed these disasters were caused by their cometary gods. Ironically, the stories of these ancient cultures were the truth. In this chapter, we saw that the Bible's scriptures ascribe cometary attributes to the God of the Bible, and there are a number of scriptures that proclaim the God of the Bible as the true ruler of heaven and Earth, the "King of Heaven" who, in essence, controls the comets of heaven that bring disaster.

The next two chapters look at the disastrous events described in the Old Testament and show how many of them were caused by cometary activity. We will learn how the events of the Old Testament support what the God of the Bible says regarding His use of comets as His weapons of wrath. A clear pattern for disaster as brought by the God of the Bible will emerge. Chapter 8 will also show how the God of the Bible has dealt with the emergence of a powerful empire in the past. This will provide insight as to how these past events will relate to the emergence of the empire of the antichrist in the future.

> Oh that thou wouldest **rend the heavens**, that thou wouldest come down, that the mountains might flow down at thou presence, as when the melting fire burneth, the fire causeth the waters to boil, to make thy *name known* to thine adversaries, that the *nations* may tremble at thy presence!
>
> *Isaiah 64:1-2*

FOUR

Scientific Explanations for Old Testament Catastrophes
Part 1: Noah's Flood and Sodom and Gomorrah

> The medieval commentators used the most advanced knowledge of their day to understand the Torah. But they did not have the tools we have today. They did not have anthropology, archeology, comparative religion, linguistics, a true grasp of the texture of history.
>
> Chaim Potok
> *In the Beginning* [1]

> Biblical and geological catastrophism are, after all, inextricably linked ... this clearly justifies an urgent reappraisal of the ancient tales of celestial catastrophe. ... Our distant ancestors, it seems have been telling us in simple language that celestial catastrophe has struck, probably more than once, but the message has been lost through the ravages of time.
>
> Victor Clube and Bill Napier
> *The Cosmic Winter* [2]

> ... the point is that we may learn something if we dare examine in more detail the contents of ancient tales that seem to pertain to catastrophic events.
>
> Dr. Gerrit Verschuur
> *Impact - The Threat of Comets and Asteroids* [3]

New astronomical knowledge gained in the last forty years, and especially the wealth of information that has been discovered over the last ten years, has provided new insight into the miraculous catastrophic events that are described in the Bible. Considering what is now known about cometary activity, it is easy to recognize that comets caused many of these catastrophes. A number of astronomers who specialize in the study of comets now believe that comet activity caused several of the Bible's Old Testament catastrophes including Noah's Flood (approximately 3,000 BC), the destruction of Sodom and Gomorrah (approximately 2000 BC), many of the extraordinary events of the Exodus (approximately 1400 BC), and Joshua's great victory in the Valley of Gibeon (approximately 1360 BC). [4] Let us consider the role of comets in these specific events as well as in Deborah and Barak's Great Victory and the "blast" that killed 185,000 Assyrians during Hezekiah's time (*II Kings 19:35*). These catastrophic events reflect the God of the Bible working through nature at specific times for specific reasons to bring disaster within the physical laws of nature. The Bible refers to these physical laws as the "ordinances of heaven and Earth" (*Job 38:33, Jeremiah 31:35*, and *33:25*).

While many people are excited about finding scientific evidence that confirms the reality of Biblical events, there are a few who wonder if scientific explanations will demystify God. Bruce Feiler, in his book entitled *Walking the Bible* asks, "If the Biblical stories can be explained entirely by natural causes, what does that do to the supernatural? . . . Where does that leave God?"[5] Despite these unwarranted fears, scientific explanations for the incredible past catastrophes in the Bible that involve cometary activity will give added credibility to the Bible, since comets are referred to in the Bible as the "weapons of his (God's) wrath." In addition, we gain insight into the incredible end times catastrophes prophesied in the Bible's *Book of Revelation*. The last chapter spoke of cometary activity as a way for God to demonstrate His power to mankind and establish a name for Himself as the one and only true God, the "Lord of Hosts." The perfect timing of each of these past events serves as the awesome fingerprint of God stamped on each catastrophe. Did the God of the Bible, who in *Isaiah 46:9-10* is credited with "Declaring the end from the beginning," program these catastrophes to occur at just the right time in His original plan of Creation?

The Bible provides enough information that a scientific

explanation can be discerned for each of the catastrophic events listed above. In its account of each of these events, there are definitive clues that cometary activity was involved. Let us look at Noah's Flood.

The Flood

> For at the time when the Holy One, blessed be He, wanted to bring a flood upon the world, He took two stars from **Khima** ('the stored aways' as a reference to the Oort Cloud of comets that encloses our solar system, based on a root that means "to store away," #3598 and #3558 in *Strong's Concordance*) and brought a flood upon the world.*
>
> <div align="right">Rabbi bar Nachmani
(3rd century CE)
<i>Babylonian Talmud</i>†</div>

---

*   In the last chapter it was explained how *Job 38:22-23 NIV* tells how comets are *stored away* in "the storehouses of the snow" and "the storehouses of the hail" (the Oort Cloud of comets at the end of the solar system) which are "reserved" for times of trouble, and days of battle and war.

†   From the *Babylonian Talmud*, Sedar Zerafim, Tractate Berakoth Chapter IX Folio 59a translated by Maurice Simon, edited by I. Epstein, Soncino Press, London, 1948. Note that the celebrated medieval sage, Rashi said that in the quoted sentence, the word *Khima* meant a *star with a tail*, that is, a comet; and Samuel said *Khima* was a place of "about a hundred stars" in effect, a "storehouse." Israeli born anthropologist Benny Peiser of Liverpool John Moores University says "When God decided to bring about the Flood, He took two stars from *Khima* [a place of hundreds of stars], put [threw] them on Earth, and brought about the Flood."

Waves caused by ocean impacts may be the most serious problem produced by impacting asteroids (or comets)... tsunamis are probably the most deadly manifestation of asteroid (and comet) impacts apart from the very large... superkillers....

>Jack Hills and Patrick Goda
>*Astronomical Journal*, March, 1993

Now that we can examine the evidence with greater objectivity, it is abundantly clear that, although the continents have not been covered by water during the time that humankind has lived on the Earth, there have nevertheless, been some large scale catastrophic floods.

>Trevor Palmer, Dean of Faculty of Science and Mathematics, Nottingham Trent University [6]

It may come as a surprise just how many Americans believe the Bible story of Noah's ark. In 2005 an ABC New Poll showed that 60 percent of Americans believe in the Bible's story of Noah's Ark and the Flood. Among mainline Protestants, 73 percent believed in the Flood story; among evangelical Protestants, 87 percent believed in the Flood story; and among Roman Catholics, 44 percent believed in the Flood story. Most interesting of all, the poll found that among those who said they had "no religion" 29 percent still believed in the

Flood story.

The Bible's *Book of Genesis* records a "flood of waters" (*Genesis 7:6, 7* and *10*) and says on a certain day "all the *fountains of the great deep* (were) broken up and the *windows of heaven* were *opened* (*Genesis 7:11*)." The story then tells how it rained upon Earth forty days and forty nights. The waters increased (and bore up the ark), and all the high hills and the mountains were covered. Later it tells how the fountains and the windows were stopped and the water abated with the Ark coming to rest months afterward upon the mountains of Ararat in what is now Turkey.

Through the years, many have tried to analyze the Biblical account and explain what caused the Flood. If we could be positive about the cause, it would help in determining if the Flood was a regional or world wide catastrophe. This determination would let us tie the Flood to an actual historical event, which in turn would support the Biblical account. Short of finding Noah's Ark, if we knew the cause, then we would know what type of geological evidence should be sought to help prove this catastrophic event.

When reexamining the Bible's account of the Flood in the light of recent astronomical knowledge, it becomes clear from the nature of the events described in scripture that cometary activity caused the Flood. Indeed, a number of astronomers and geologists have written about a comet impact being the cause of the Flood. For example, discussing the Biblical Flood, Dr. Benny Peiser of Johns Moores University in Liverpool, England told the *Jewish Chronicle*, "Before we did not know of any phenomenon that could trigger such a massive flood disaster . . . Now, recent research on ocean impacts shows that a large body hitting one of the world's oceans could trigger the kind of massive tsunamis capable of wiping out coastal cities." [7]

Astronomers Clube and Napier in the *The Cosmic Winter*, "see a clear astronomical association" between the break up of a "giant comet" and "the Biblical Flood at the start of the third (millenium BC)."[8] Duncan Steel, an astronomer of the Anglo-Australian Observatory in his book *Rogue Asteroids and Doomsday Comets* writes that the connection between the Flood and cometary impact "may well be right."[9] Astronomer Gerrit Verschuur in his book *Impact* mentions a "challenging article in an Austrian geological journal that took the Deluge impact hypothesis a whole lot further."[10] In this 1992 article geologists Edith and Alexander Tollman of the Institute of Geology at Vienna University present a series of geological facts to support their claim "that the Noachin Deluge was the consequence of a cometary impact."* Most important, there is now direct physical evidence to support the belief that the Biblical Flood at the start of the third millenium BC was caused by cometary impact. In 2006 an impact crater dating to this period was found on the seafloor of the Indian Ocean. This crater is in all likelihood the "smoking gun" for the ocean impact that triggered the tsunami that caused the Flood. This very important discovery will be discussed in more detail later in the chapter.

Surprisingly, this new explanation about the cause of the Flood is supported by a better translation of three words in the Bible's

* *http://www.unibg.it/convegni Now-scenarios*, also *http://www.p-kom.com/misti/ext_event.html*. Also Allan O. Kelly a geochemist and Frank Dachille, an amateur astronomer in their book *Target: Earth The Role of Large Meteors in Earth Science* noting the connections between a comet collision and the Deluge wrote, "The literary treasures of so many peoples of the Earth, treasures which were understandably preserved through ... religion, are full of direct and indirect references to a collision-flood." *Target: Earth The Role of Large Meteors in Earth Science*, Allan Kelly and Frank Dachille, Carlsbad, 1953, as quoted in *Impact*, Gerrit I. Verschuur, Oxford University Press, Oxford, 1996, p. 91.

Flood story. The correct translation of these three words shows that the Bible in two separate phrases specifically connects cometary activity with the Flood.

So, what are these three words which, when correctly translated, explicitly reveal what caused the Flood? The words are *"fountains," "windows"* and *"opened."* Genesis 7:11 referring to the Flood says,

> In the six hundredth year of Noah's life, in the second month, the seventeenth day of the month, the same day were all the *fountains* (#4599 in *Strong's Concordance*) of the great deep broken up, and the **windows** (#699 in *Strong's Concordance*) of heaven were **opened** (#6605 in *Strong's Concordance*).
>
> *Genesis 7:11* – traditional translation

This traditional translation of the original Hebrew seems to call for a flood caused by fountains under the deep breaking up and gushing water that suddenly raised the sea level, and for the windows of heaven opening to release torrents of water in the form of rainfall which inundates the Earth and covers the high hills and mountains. However, based on deep drilling and seismic studies, today's geoscientists know there aren't any "fountains" or "springs" under the sea that could cause the sea to suddenly rise. Further, based on extensive telescopic study of the heavens, astronomers know there are no watery windows or floodgates in heaven that could inundate the Earth, not even metaphorically. Is something else being described in the Bible? This is not that the Bible is wrong, but that *the traditional translation does not convey what the original Hebrew said.* Unaware of modern science and the aftermath of cometary impacts, it seems that translators from hundreds of years ago resorted to a descriptive but misleading choice of words ("fountains of the deep" and "windows or floodgates of heaven") to explain the Flood. It is as if these translators wanted to 'help' the Bible narrative on this subject; however, as we shall see, the Bible doesn't need a translator's assistance, but rather diligence in determining what is really being conveyed.

In addition to the traditional mistranslation of these three words, there is also a problem with the traditional interpretation of a key passage about how high the flood waters rose. The Bible says, "And the waters prevailed . . . and all the high hills that were under the whole heaven were covered, fifteen cubits upward (22-1/2 feet or about two stories high) did the waters prevail, and the mountains were covered" (*Genesis 7:19-20*). There seems to be an enigma here. How could such a small rise in water cover all the high hills and mountains? This enigma or problem is easily resolved once we consider that the Flood was caused by a tsunami. In a tsunami the height of the seismic sea wave going over the land is one thing, while the depth of the flood that follows is another thing. The depth of the standing flood waters left behind by a tsunami are very small compared to the height of the waves that brought these waters in. During the 2004 Indonesian tsunami, the flood waters rose several feet, but the two destructive waves were as high as thirty feet in some places. In a flood caused by a tsunami, the tsunami waves reach heights the flood waters never reach, especially in the case of a mega tsunami caused by a comet or asteroid impact. Therefore these Bible verses seem to be conveying that the height of the floodwater that rose up was fifteen cubits, but the hills and mountains were temporarily covered by the towering waves of a mega tsunami and related sloshing of the Persian Gulf within it basin. As noted in Chapter 2, some experts believe that a big cosmic impact can "give rise to a ring of seismic seawaves, or tsunumi nearly as high as the Rocky Mountains."[11]

Traditional interpretations have understood this Bible verse to mean that the flood waters rose fifteen cubits *above* the hills and mountains. Since some of the mountains of Ararat are over 16,000 feet in height, this would of course require a flood that was not just regional, but a flood that was worldwide in scope, that is, a planetary flood. It is important to realize that for flood waters to cover the Himalayas, it would require more than five times the amount of water presently in the Earth's oceans and glaciers! Another problem with this interpretation is that there is no evidence for such a flood in the geological record. (While there is a worldwide geological layer with fossils that relate to the demise of the dinosaurs, there is no worldwide layer with fossils that relate to a planetary flood that took place during Biblical times.)

Geologist David A. Young, an evangelical "defender of Biblical inspiration," agrees with other geologists when he says, "Geology provides no evidence whatever for a universal flood."[12] Young, who has studied the Biblical narrative, notes how the concept of a universal flood is also contrary to evidence from paleontology, biogeography, anthropology, and archeology.[13] For example, archeologists have found hundreds of ancient human habitation sites from coastal areas around the world that would have been destroyed if a universal flood had in fact occurred. Young, like other scientists who have studied the Flood, sees it as a *regional* catastrophe rather than a *worldwide* catastrophe, a catastrophe that affected all the "land" rather than all the "Earth."

When the translation of *Genesis 7* and *8* talk about events occurring "upon the Earth" it must be understood that the Hebrew word *eh-rets* (#776 in *Strong's Concordance*) that has been translated as "Earth" can also be translated as "land." (In the Old Testament *eh-rets* is translated as "land" approximately 1,644 times, and as "Earth" approximately 657 times.) Young says "the modern evangelical church is extremely sensitive about open discussion of scientific issues that bear on *Genesis 1-11* . . . anyone who does try to explore the issues is in ecclesiastical jeopardy. The prevailing atmosphere of fears tends to squelch attempts to deal with these issues." [14] *To refuse to reexamine the scriptures in light of the advancements in archeology, linguistics, history, astronomy, and geology is to shut the door to understanding and appreciating just how amazing the information in the Bible really is.*

So exactly what is the original text saying about the Flood? How should the Hebrew words traditionally translated as "fountains," "windows" and "opened" be translated; and how should the passage about how high the waters rose and the mountains being covered be interpreted? All translations are interpretations of what the particular translator (or translators) thinks the original text said. Common sense dictates that the better the translator understands what is being conveyed, the more accurate the translation. Again, recognizing that scientific knowledge was very limited when the traditional translation was made, the words in question must be reexamined from the standpoint of current knowledge. This reexamination is **not** to force scriptures to align with current scientific knowledge, but to determine the best meaning for what is being conveyed in the scriptures.

First, *Genesis 7:11*, after giving information about the Flood date, reads "The same day were all the *fountains* (*springs* – NIV) of the great deep (the primeval ocean) broken up."* However, the Hebrew word *mah-yaw-naw* (#4599 in *Strong's Concordance*), that is commonly translated as "*fountains*," or "*springs*" could be translated as "sources" in the sense of origination.† This phrase in *Genesis 7:11* would then read, "the same day were all the *sources* of the great deep (the primeval ocean) broken up." That this is a clever reference to comets became apparent when scientists recently discovered that the waters locked up in the ice of comets were in fact the "*sources*" of the great deep, that is, the primeval ocean. Not only does the retranslation provide a logical explanation for what is being described, but it is also a logical explanation for the cause of the Flood. Here is the first way the Bible says the Flood was caused by comets, that is, by comets breaking up and impacting in the ocean which would have produced an impact driven tsunami, a so-called " tidal wave" (or "harbor wave"). It is not the water that the comets brought, but the tsunami waves raised up by the comet ocean impacts that caused the Flood.

Indeed, after entering the Earth's atmosphere, comets are a "source" that can be "broken up" as this phrase indicates. Further, as noted scientists have now compared chemical fingerprint evidence for comets and ocean water, and have determined that the waters locked up in the ice of comets (that impacted the Earth after its formation)

---

* The Hebrew word *teh-home* (#8415 in *Strong's Concordance*) that is translated as "deep" is a loan word borrowed from the Sumerian/Akkadian culture where it is a reference to the primeval ocean or sea.

† Two basic references, *Strong's Concordance*, and *The New Brown-Driver-Briggs-Gesenius Hebrew-English Lexicon* list "source" as one of the possible (figurative) meanings for this word.

were in fact the "*sources*" of the primeval ocean. In particular a new class of comets called "Main Belt Comets" most closely match the isotope chemistry of the Earth's water. [15]

Scientists believe that the Earth and the Moon were hit by many tens of thousands of icy comets or rocky asteroids not long after the Earth was formed. (This produced most of the massive craters we see on the Moon today.) The latter stage of this bombardment is known as the "Late Heavy Bombardment." Based on measuring the amount of the metal iridium in ancient rocks from Greenland, and based on the comparison of the water in "Main Belt Comets' with the water in the Earth's oceans, many scientists now believe that the tens of thousands of impactors of the "Late Heavy Bombardment" were in fact comets. On August 5, 2009 *National Geographic News* reported that these scientists believe that, "By the time of the Late Heavy Bombardment things had cooled down [on planet Earth], allowing meltwater from the flurry of comets to become the world's first seas."[16] *National Geographic News* told how Uffe Jorgensen of the Niels Bohr Institute in Denmark led a group of scientists who studied iridium from rocks dating to the "Late Heavy Bombardment." His team calculated that the ice from the comets coming in "thawed to create a global ocean more than a half a mile deep." [17] Jorgensen and a co-author in a paper to be published in the fall of 2009 in the Journal *Icarus* write: "We may sip a piece of the impactors every time we drink a glass of water." [18] Jorgensen also said, "If it [The Late Heavy Bombardment] had not happened, there would have been no water on Earth, and no life." [19]

The translation of the first phrase of *Genesis 7:11* correctly says that comets are the source of the great primeval ocean, while telling us that the breaking up of comets caused Noah's Flood.

Comets breaking up in the atmosphere as they approached the Earth would result in massive ocean impacts. These cometary ocean impacts would produce towering mega tsunamis, towering walls of water that race across the land that would and cover all the high hills and even the mountains. There is now clear cut evidence of how a tsunami can quickly run up and cover a mountain. On July 10, 1958 eyewitnesses told how a landslide at the head of Lituya Bay in Alaska's Glacier Bay National Park generated a 150 to 450 foot high wave that raced across the bay and ran 1,719 feet up the nearby mountain and stripped away trees. There may have been several different tsunami events where each single tsunami event can involve a series of waves of varying height.* In other words, when the "sources of the deep," that is comets were broken up, there may have been multiple impacts that produced multiple tsunami events. When all the mega tsunami waves stopped racing across the land, the waters brought in by these gigantic sea waves would have kept the land flooded for months. Add to this influx water from much smaller secondary tsunami waves resulting from impact earthquake aftershocks.†

In recent years most scientists have come to believe that comets have not only been "the sources of the great deep" (*Genesis 7:11*) and supplied much of the water in the Earth's oceans, but that comets are also responsible for all of the water that is found in the inner solar system. In light of scientists' evolving belief that comets are the main source of all of the water found in the inner solar system,

---

*   A set of tsunami waves is called a "train."

†   These secondary tsunami waves would not be mega-tsunamis because as explained in **NOTE ONE** only large impacts can produce earthquakes of 10 or more in magnitude, the magnitude needed to produce mega-tsunamis. Even after an impact that produces a magnitude 14 earthquake, the Earth could not store enough energy to produce an aftershock of enough magnitude to produce a mega-tsunami.

it is astounding that the Bible identifies comets as the "sources of the primeval ocean" more than 2,000 years ago. Few Bible readers today know that astronomers and planetary scientists believe that comets supplied much of the water in the Earth's oceans in the first billion years after the Earth's formation and cooling. [20]

The second indication that *Genesis 7:11* says the Flood was caused by comets comes from the unusual second phrase of *Genesis 7:11* which reads, "the windows of heaven were opened." This strange phrase readily stands out as a mistranslation. At face value, it conveys a picture that high in the sky are windows that if opened, would literally drown the Earth with rain, a notion as realistic as the child's sentiment that the sound of thunder is actually God moving furniture in heaven. This mistranslated phrase has led to the erroneous belief that the cause of the Flood came from the forty days and forty nights of rainfall, rather than from an impact driven tsunami.* However, analysis and retranslation of the words traditionally translated as "windows" and "opened" reveals a different picture. (Note that *Genesis 8:2* repeats the key words used in *Genesis 7:11*.)

In *Genesis 7:11* the Hebrew word *ar-oob-baw* (#699 in *Strong's Concordance*), that has been translated as "windows," or "floodgates" (NIV) should be translated as "lurk" or "lie in wait"† to

---

* Note that according to the Almanac, the maximum amount of rainfall recorded for a 24-hour period was 73 inches at Cilaos on Reunion Island in the Indian Ocean on March 16, 1952. If 73 inches or approximately 6 feet of rain were to fall for forty days and forty nights, and one assumes no ground absorption and drainage, this would only give 240 feet of standing water, hardly enough to cover 16,000 foot high mountains. Then there is the meteorological problem of what would charge the clouds with water to produce such a great amount of rain day after day for 40 days.

† The Hebrew word *ar-oob-baw* (#699) is the particle passive of the Hebrew word *ar-rab* (#693) which means "to lurk." In the NAS word #693

convey the idea of something being restrained while it waits to be released. In addition, the Hebrew word *paw-thakh* (#6605 in *Strong's Concordance*), that is translated as "opened," should be translated as "loosed."* So, instead of this phrase being "the windows of heaven were opened," this phrase from *Genesis 7:11* would now read "those that lie in wait in heaven were loosed."

What lies in wait in heaven? Based on other passages in the Bible, "those that lie in wait in heaven" is a reference to comets and the phrase " were loosed" is a reference to the *comets* of heaven being loosed or released from where they reside. This is consistent with the retranslation of the phrase that proceeds this which, as corrected, reads "the same day were all the *sources* (*comets*) of the great primeval ocean broken up." Technically those that lie in wait in heaven being loosed would entail the comets that lie or reside in the Oort Cloud of comets that surrounds our solar system being loosed to travel from the far edge of the solar system to the inner solar system and ultimately hit the Earth. *(See Illustration A)*

Although astronomers did not discover the Oort Cloud of comets until 1950, *Job 38:22-23* (*NIV*) tells of the snow and hail (the frozen waters) which are *comets* that are stored and reserved "for times of trouble, for days of battle and war." Further, in *Isaiah 24:18-19* there is a phrase very much like the phrase of *Genesis 7:11*, that makes it clear that "comets," not rain showers are being referred to in both phrases. The phrase in *Isaiah* says, "the windows (floodgates - NIV) from on high are open," and then it says "and the foundations of the Earth do shake. The Earth is utterly broke down, the Earth

---

is translated as "lurk" or "lie or lay in wait" 21 out of 43 times.

* In the KJV the Hebrew word *paw-thakh* (#6605 in *Strong's Concordance*) is translated "loose or lossed" 11 times and in particular in *Job 38:31*, which like *Genesis 7:11* also seems to have an astronomical context.

is clean dissolved, the Earth is moved exceedingly." It is clear that "rainfall" does not cause planetary shaking and moving. However, from the corrected translation that calls for "those that lie in wait in heaven being loosed," that is, the comets of heaven being loosed, we can certainly imagine how comets hitting the Earth and triggering 13.0 magnitude earthquakes would cause the foundations of the Earth to "shake" and the Earth to "move exceedingly" on its axis of rotation. Even the 9.0 magnitude Indonesian Earthquake in 2004 had some effect on the entire planet and its axis of rotation. For example, Dr. Charles Ammon, a professor of Geosciences at Penn State University said, "Globally, this earthquake was large enough to basically vibrate the whole planet as much as half an inch . . . Everywhere we had instruments, we could see motions. [21]

In a later chapter we will see that the phrase "those that lie in wait in heaven being loosed" relates to the concept of "the brutish ones being loosed" which appears in the Bible. [22] In the ancient Near East the issue of who was in control of the comets, and who can "bind" or "loose" them was an important theme.[*] Just like the first phrase of *Genesis 7:11*, when correctly translated this second phrase of *Genesis 7:11* also has a clear cometary context. It is extraordinary that here in one verse the Bible says, in two different ways, that comets caused the Flood, exactly as a number of modern astronomers have come to believe. *Genesis 7:11* says comets **that lie in wait in heaven were loosed** or released, and upon entering our atmosphere these comets, these **sources of the great primeval ocean were broken up**. Now the forty days and forty nights of rain of *Genesis 7:12* can be

---

[*] Did the "King of Heaven," the "Lord of Hosts" "doeth according to his will in the army of heaven or host of heaven" as *Daniel 4:35* and *37* states or did the Sumerian/Akkadian/Babylonian/ Assyrian "Queen of Heaven" (*Jeremiah 7:8*, *Jeremiah 44:17-19* and *25*) rule?

seen *not* as the cause of the Flood, but as a natural atmospheric response to or consequence of this cometary activity and tsunamis which put a tremendous amount of moisture into the atmosphere. If anyone did the math, they would quickly see that forty days and forty nights of heavy rain would not produce a worldwide flood. The new translation of *Genesis 7:11* would now read:

> In the six hundredth year of Noah's life, in the second month, the seventeenth day of the month, the same day were all the **sources** (#4599 in *Strong's Concordance*) of the great primeval ocean (**comets**) broken up, and those **that lie in wait** (#699 in *Strong's Concordance*) in heaven (**comets**) were **loosed** (#6605 in *Strong's Concordance*).

Genesis 7:11 – retranslation

Note how this new translation of *Genesis 7:11* is consistent with the passage from the *Babylonian Talmud* that was quoted at the beginning of this section on the Flood. In this passage Rabbi bar Nachmani, the Director of the Academy at Pumbedita in the third century AD, whose "astute dialectical abilities earned him a reputation as an 'uprooter of mountains'" wrote, "For at the time when the Holy One, blessed be He, wanted to bring a flood upon the world, He took two stars (comets) from **Khima** ('the stored aways' #3598 and #3558 in *Strong's Concordance*), and brought a Flood upon the world." The phrase "those that lie in wait in heaven" from *Genesis 7:11* directly relates to "the stored aways" of this Talmudic passage; just as *Job 38:22-23 NIV* tells how comets are stored away and "reserved" in "the storehouses of the snow" and "the storehouses of the hail." Also the "loosed" of *Genesis 7:11* directly relates to the phrases "brought a flood upon the world" and "threw them on Earth" of this Talmudic passage. In order for Rabbi Nachmani, Rashi, the celebrated eleventh century Talmudic commentator, and other Jewish sages to write and accept the sentence from the *Talmud* I quoted here, it seems that they translated *Genesis 7:11* in much the same way as done here.

This clearer translation of *Genesis 7:11* shows the Bible's brilliance in conveying scientific information in a simple manner. Notice that this clearer translation of *Genesis 7:11* provides information about the cause of the Flood in a "*doublet*" format. A "*doublet*" is a common type of sentence structure in Hebrew prose in which information or a relationship is stated in two different ways for emphasis.* First, we are told that **comets**, which are correctly identified as the sources of the oceans' water, were broken up and caused the Flood. Then in the second half of the *doublet*, we are told that comets that lie in wait in heaven were turned loose and caused the Flood. Amazingly, the Bible communicated this significant scientific information over 2,000 years before man learned that comets are mainly composed of frozen water, that comets were the sources of the Earth's water, that comets lie in wait in heaven in a storehouse we now call the Oort Cloud, and that a comet impacting a very large body of water can cause a tsunami and a flood.† For

---

* Regarding "***doublet***s," *Matthew 18:16* and *Deuteronomy 19:15* say that by two or three witnesses shall a matter be established. This Biblical principle is typed (illustrated) by Joseph telling Pharaoh that Pharaoh's dreams of cattle and of ears of corn were "one," and Joseph telling Pharaoh that his dream was doubled "because the thing is established by God" (*Genesis 41:25* and *32*).

† Dr. Trevor Palmer of Nottingham Trent University, United Kingdom, one of the world's foremost experts on catastrophism read an excerpt

more scriptural support for the translation of these two phrases of *Genesis 7:11* see **NOTE FOUR**.

More importantly, the Bible specifically describing the Flood as caused by comets is consistent with the other physical events it says took place during the Flood. Based on the Bible's Flood narrative it is now abundantly clear that the Flood was caused by comet bombardment. We can picture how a comet *broke up* while approaching the Earth, with fragments hitting the ocean in the area of the Arabian Sea or the Indian Ocean. These impacts would have caused sudden shifting in the seafloor and earthquakes of cataclysmic size. In turn these earthquakes would have produced a series of giant seismic sea waves (tsunami) that went out in all directions. One set of tsunami waves would have washed across part of low lying Saudi Arabia and funneled up the Persian Gulf. As the waves approached land and shallow water, they would have rose up into giant walls of water. By the time they reached land near the confluence of the Tigris and Euphrates Rivers in Southern Iraq, some waves would have been a few miles wide and perhaps a mile or so high. Moving inland at over 400 miles per hour and hemmed in on the east by Iran's (north-south trending) Zagros Mountains, these waves would have rushed unchecked to the north across the plain of the Tigris and Euphrates (in what geologists call a structural trough) until they slammed into the mountains of Ararat in Turkey. Much smaller secondary waves caused by aftershocks could have kept the waters sloshing back and forth between the Gulf and the mountains for months.* *(See D & F)* of my writing on the retranslation of *Genesis 7:11* which I sent him and in an email to me (1/09/04) he wrote, "Whilst it would be going too far to say your retranslation provides proof of a major cometary-related inundation event in relatively recent times, it nevertheless gives additional support to such a scenario."

* Dr. Gerrit Verschuur in *Impact* (p. 154) notes how calculations

During the Flood, the Bible says that the waters only rose 15 cubits (22-1/2 feet) (*Genesis 7:19-20*), but the towering walls of water from the tsunami waves would have covered all the high hills and the mountains. During the 9.2 magnitude Indonesian Earthquake and Tsunami in 2004, the waters only rose several feet but a tsunami with waves over thirty feet in height was set in motion, which ultimately caused over 200,000 deaths. This brought the public's attention to the devastation that even a small tsunami can bring. An earthquake caused by a big comet impact could trigger an earthquake 100,000 times more powerful than the earthquake that produced the devastating 2004 Indonesia tsunami. Try to imagine a Flood brought by a series of tsunamis with waves nearly as high as mountains.

How could the Ark have survived such a tsunami? In 1868 a huge tsunami hit the Chilean port of Arica. According to Lieutenant L. G. Billings, an eye witness, when the massive wave hit the U.S.S. Wateree, a single stacked, two-masted, wooden side wheeler, and engulfed and buried it under a mass of sand and water.[23] After staying underwater for what seemed a suffocating eternity, it pushed its way to the surface. Then the ship was carried along at great speed until it suddenly went motionless at the foot of the coastal range of the Andes several miles inland. Nearby lay a vertical cliff and the wrecks of two bigger ships. How then did the Wateree survive? "Our survival was certainly due to the construction of the ship, her shape,

---

by Jack Hills and Patricia Goda of the National Laboratory in New Mexico show that if a six-mile wide comet were to strike the ocean, the deep water wave that would result from the impact would be about 1.8 miles high 600 miles from the point of impact, which could translate into a tsunami that is over 60 miles high at it's inception. Verschuur says, "With a deep-water wave that large, it is moot as to whether it would notice the shoreline and would, instead, wash over a good fraction of the continent and break up against the Rocky Mountains!"

and her fitting out, which allowed the water to pour off the deck almost as quickly as if she had been a raft." [24] Further, in the opinion of the eye witness, "We can only assume that the wave caught her just right; that instead of suffering instant capsizement like the other ships in the bay, she lifted to the wave's thrust and was borne by it."[25] In the Bible we read ". . . and the waters increased and bore up the ark" (*Genesis 7:17*).

Does the ancient Near Eastern literature support the Bible's view that comets caused Noah's Flood or offer any added insight about the events preceding this massive regional flood? According to Sumerian and Babylonian cuneiform accounts, a devastating flood was an actual historical event. The Sumerians and the Babylonians (regional neighbors to the Bible's Noah) each have their own accounts of the Flood (Sumerian "Ziusudra," and Babylonian "Utnapishtim" or "Atrahasis"), a flood that covered the mountains on the first day and only lasted six or seven days as opposed to the forty days called for in the Bible (*Genesis 7:17*). In the Babylonian account of the Flood, as given in the "*Epic of Gilgamesh*," Ut-Napishtim (the equivalent of the Biblical Noah) tells Gilgamesh how "their heart led the great gods to produce the Flood."

Because most of the Sumerian/Babylonian gods were originally cometary in nature, we are in effect being told that cometary activity caused the Flood. Indeed, the events that are then described by Ut-Napishtim all speak of comet activity. First, we are told that before the Flood began, the cometary god Ninurta, who is associated with the phenomena of slinging fiery stones (from a cometary cloud) came forth and "a black (cometary) cloud rose up from the horizon." Then posts are torn out by what is presumably a fierce wind. Next, the story says that "bewilderment" arose as "torches," came in "setting the land ablaze with their glare." (Tablet XI, lines 8, 14, and 97-113). [26] The "torches" that set the "land ablaze" before the Flood began seem to be references to impacts from fragments of a comet that is breaking up as it is coming into the atmosphere. Next a comet impact that raises a massive dust cloud is described when the cometary god Adad, the "Wild Bull of Heaven" is referred to, and we are told that *"the wide land was shattered like a pot"* and that Adad turned "to blackness all that had been light" (lines 105-107). (Another translation says that Adad "turned daylight into darkness, when he smashed the land like a cup.") [27] Then "For one day the south storm blew, gathering speed as it blew, submerging the mountains. Overtaking the people like a battle (lines 108 –110)." Only the towering waves of a powerful tsunami could cover the mountains on the first day of a flood. This is a description of a comet impacting the ocean that produced a towering seismic sea wave. Dr. Sharad Master, a geologist at the University of Witwatersrand considers this same passage from the *Epic of Gilgamesh* and agrees, asking, "Could this be a reference to a bolide impact (a meteoroid that explodes audibly as it passes through the atmosphere) which triggered a tsunami?" [28] This Babylonian account of the Flood is comparable to the Biblical account of the Flood (*Genesis 7:11*), since it also speaks of a comet or comets coming in and breaking up as the cause of the Flood which quickly covered all the high hills and the mountains with water.

The Flood is a vital part of the Sumerians' historical and literary tradition. For example, the Sumerian "King List" refers to the Flood, and says, "The Flood then swept over the land. After the Flood had swept over the land ... Kingship descended from heaven a second time," and it identifies the city states of Kish, Erech, and Ur as the first of the new Sumerian dynasties.[29] This has led archeologists to date the Flood to the start of the third millennium BC. In fact, there seems to be some geological evidence preserved at these sites that a flood occurred during this time period.

It is interesting to note that the archeological date for the Flood is consistent with the great amount of cometary activity some astronomers believe happened at the start of the third millennium BC. This is based on the orbital backtracking of certain comets and Earth crossing asteroids traveling in related orbital streams.[30] There could be another surprising clearer translation – the word "flood." The narratives of the seventh and eighth chapters of *Genesis* refer to a "flood," "a flood of waters" and the "waters of a flood." The Hebrew word that is translated as "flood" (*mab-bool*, #3999 in *Strong's Concordance*) in these chapters, a word that only occurs in these chapters and *Psalm 29:10*, is of uncertain derivation and meaning. It could also be translated as "cataclysm" or "catastrophe." In the *Septuagint*, the third century BC Greek translation of the Old Testament by Jewish scholars, this word is always translated as **kataklysmos**. Further, when this event is referred to in the New Testament (*Matthew 24:28-29, Luke 17:27,* and *II Peter 2:5*) the Greek word that is translated as "flood" is the same word **kataklysmos** (#2627 in *Strong's Concordance*). So, the seventh and eighth chapters of the book of *Genesis* actually refer to a "cataclysm" or "catastrophe" (namely as when comets come in and break up), a catastrophe that caused the "catastrophe of waters" or tsunami we know popularly as Noah's Flood.

While the clearer translation of *Genesis 7:11* makes it likely that the Flood was caused by an impact driven tsunami, physical evidence is still needed to prove that such an event actually took place. Unfortunately, evidence of impact generated tsunamis can be hard to find. Dr. Lewis in *Rain of Iron and Ice* notes "all ocean impacts throw up massive tidal waves that can devastate coastal regions without leaving any distinctive signature that says 'this was an impact event.'" [31] Ocean impacts must be large enough or occur in waters shallow enough to leave a telltale crater in the ocean floor. If the impact that caused the 120 mile diameter Chicxulub Crater in the Gulf of Mexico north of the Yucatan Peninsula 65 million years ago had occurred in very deep waters, it may never have been found.

Evidence for a tsunami caused by an ocean impact can also come from the marine sediments and debris that it deposits inland near the high water mark. For example, on the Hawaiian Island of Lanai, there is a tsunami deposit (tsunamite) 1,000 feet above sea level. [32] The impact that blasted out the Chicxulub Crater in the Gulf of Mexico left tsunami debris containing tektites and grains of shocked quartz as far inland as the Brazos River in Texas and possibly Kansas City.

Incredibly, there is now evidence for a tsunami caused by a large ocean impact around the time the Bible gives for Noah's Flood! **Both** the crater and some of the tsunami deposits left by this large ocean impact have been found. On November 14, 2006, the *New York Times* announced the discovery on the front page of their weekly science section in a feature entitled, "Ancient Crash, Epic Wave – Did catastrophe fall from above in 2807 BC?" [33] The *Times* reported on the discovery of a very large ocean impact that left an eighteen mile wide crater at the bottom of the Indian Ocean. This horrendous impact caused a sudden shift in the sea floor and an earthquake that was tens of thousands of times more powerful than the earthquake that shook San Francisco in 1906. This sea floor impact rebound and shift with accompanying earthquake in turn triggered a tsunami where massive seismic sea waves went out in expanding concentric rings from the impact site. *(See Illustration D)*

This Indian Ocean impact crater was named the Burckle Crater. It was discovered in 2005 by Dr. Dallas Abbott, a research scientist at Columbia's Lamont-Doherty Observatory in New York. The *Times* told how Dr. Abbott, a geologist, and a self proclaimed "band of misfits" came together in 2004 to form the "Holocene Impact Working Group." The members of this international research group are recognized experts in geology, geophysics, geomorphology, tsunamis, tree rings, soil science and archeology. While some astronomers still have doubts that large comets or asteroids have hit the Earth in the last 10,000 years, a time period called the Holocene epoch, the *New York Times* reported that members of the Holocene Impact Working Group say that

> astronomers simply have not known how or where to look for evidence of such impacts along the world's shorelines and in the deep ocean. Scientists in the working group say the evidence for such impacts during the last 10,000 years, is strong enough to overturn current estimates of how often the Earth suffers a violent impact . . . Instead of once in 500,000 to one million years, as astronomers now calculate, catastrophic impacts could happen every 1,000 years.[34]

Note that the recently found Amarah Crater in Iraq, discussed in

Chapter 1, and Burckle Crater in the Indian Ocean as discussed both represent Holocene Impacts.

The Holocene Impact Work Group's breakthrough for locating ocean impact craters came when they began using satellite images of the Earth to search for deposits that have been carried onto the land by impact generated mega tsunamis. Such deposits typically occur in a series of massive wedge shaped formations tens to hundreds of miles long and upwards of 1,000 feet high called "chevrons." These chevron shaped deposits are very important, because their apex point in the direction from which the mega tsunami originated, which is the same as the direction of the ocean impact crater. Sites with chevrons have been identified all around the world's coasts, including the Caribbean, Australia, Africa, India, Vietnam, the North Sea, and even Long Island, New York. [35]

To find the exact location of the impact crater that produced the tsunami deposits Dr. Abbott used sea surface altimetry data collected by satellites that scanned the ocean's surface and precisely measured its height. Underwater features such as mountains, volcanoes, canyons, and very large craters can affect the Earth's gravitational field and cause minute differences in the height of the surface of the sea. To determine if a circular depression found along the line indicated by a chevron is an impact crater, sink hole, or volcanic crater, core samples must be taken of the sediments on the seafloor in the area. Cores containing products characteristic of a cosmic impact, such as impact glass and impact spherules amidst high levels of metals that speak of extraterrestrial origins would provide conclusive evidence that a circular depression is an impact crater.

The effectiveness of this altimetry technique was

demonstrated in 2005. That year, Dr. Ted Bryant, a geomorphologist at the University of Wollongong in New South Wales, Australia, and a member of the Holocene Impact Working Group, identified two chevron shaped deposits four miles inland from the Gulf of Carpentaria in northern Australia. Both chevrons pointed north and Dr. Abbot was asked to find the craters that relate to them. Dr. Abbott used satellite altimetry to hunt for the craters. After several days she located two craters in the shallow waters north of the two chevrons. A check of deep sea sediment cores previously taken in the area by the Australian Geological Survey showed that they contained materials only produced by cosmic impacts (including marine microfossils fused to spherules of melted glass). Based on the dating of sediments in the area of the craters and the dating of the sediments in the chevrons, Abbot and Bryant were able to determine that both the chevrons and the craters are of the same approximate age, which they estimated to be about 1,200 years old. [36]

This dramatic demonstration of the connection between ocean impact craters and chevrons deposited by tsunamis set the stage for the discovery of the Burckle Crater. As noted earlier, the Burckle Crater was created by a very large ocean impact that occurred around the time of the Biblical Flood. The impact that left this large undersea crater also generated a mega tsunami with waves that went out in all directions, similar to a stone hitting the water and sending out a series of expanding concentric rings. Some of the sediments carried by these tsunami waves resulted in the chevron shaped land deposits found around the Indian Ocean.

Months after the initial discovery of the underwater craters near Australia, Dr. Abbott, Dr. Bryant and Dr. Gusiakov of the Novosibirsk Tsunami Laboratory in Russia and other scientists traveled to the southern end of Madagascar (the very large island that lies off of Africa's east coast) to take samples from four massive chevrons lying several miles back from the coastline. They wanted to confirm that these chevrons were indeed deposits from a tsunami, a mega-tsunami generated by a large cosmic body that struck near the middle of the Indian Ocean. The chevrons showed that a very large tsunami had dumped massive deposits inland over a distance of 28 miles. [37] These Madagascar chevrons and other chevrons around the Indian Ocean found in Western Australia and India all point back to a crater site in the middle of the Indian Ocean (30.86 South latitude and 61.36 East longitude). [38] A year earlier Dr. Abbott used satellite images from Google Earth to find all of these chevrons. She then used satellite altimetry of the Indian Ocean's surface to search the area of these chevrons pointed to and discovered the 18 mile in diameter crater that is now called Burckle Crater.

Burckle Crater lies about 900 miles southeast of the Madagascar chevrons, below 12,500 feet of water. While samples have not yet been taken directly from Burckle Crater, three cores previously taken from sediments on the sea floor in the area around the crater contained impact glass, impact spherules, and high levels of nickel, all of which are products of cosmic impact. [39] High nickel content is one of the ways scientists recognize rocks of cosmic origin, since nickel is extremely rare in terrestrial rocks. For example, nickel is over 130 times more abundant in carbonaceous meteorites than it is in the earth's crustal rocks! One of the three core samples taken near Burckle Crater even contained tiny drops of pure nickel (Ni). Since pure Ni melts at 1453° C (2,647° F), these drops of nickel offered unequivocal evidence that an explosive impact had taken place, "where the shock waves from the explosion would have heated the air to more than 3,000 degrees Fahrenheit" and drops of pure nickel were separated out. [40]

As stated, Burckle Crater has not yet been directly dated, however, based on the ages of the sediments found in the three deep sea cores taken near the crater, Dr. Abbott estimates that it is approximately 4,500 to 5,000 years old. [41] The work of one of the members of the Holocene Impact Working Group supports Abbott's estimate. Dr. Marie-Agnès Courty, a soil scientist at the European Center for Prehistorian Research in Tautavel, France has been studying soil deposits from around the world for evidence of dust from cosmic impacts and she believes that there was a major impact around 4,800 years ago.[42] There are also a number of other lines of physical evidence that call for a large scale catastrophic event of some sort around 4,500 to 5,000 years ago. This is around the time the Bible gives for the Flood!

Hopefully, the National Science Foundation will send an oceanographic research ship to take a closer look at Burckle Crater and take samples from it. If the radiocarbon dating (C-14) of shells collected from several of the chevron shaped tsunami deposits match the probable 3,000 BC age of Burckle Crater, this will provide powerful evidence for establishing the reality of the Biblical Flood. It would connect a story told in the Bible to a real historical event, a regional event triggered by a comet impact in the Indian Ocean.

If the dating is confirmed, this would be evidence that the Scriptures pertaining to the Flood are correct. When words within the passages are correctly translated, we find a mountain high tsunami caused by comet bombardment being the cause of the Flood, instead of underground fountains gushing forth and forty days of rainy weather. Recall that when these Scriptures are correctly translated and interpreted, they tell first of comets being "loosed" and "broken up," then describe how the waters of a towering tsunami quickly covered all the high hills and mountains, and how the towering sea waves left the land flooded as they withdrew back towards the sea.

When Dr. Abbott and her team went to study the Madagascar chevrons, they quickly found that the chevron deposits contained deep water shells mixed and fused with melted glass and shattered types of rock typically formed during cosmic impacts. Later, when the samples were examined under a scanning electron microscope by Dee Breger, the director of microscopy at Drexel University in Philadelphia, she recognized a metallic fingerprint of sorts. What she found were splashes of iron, nickel, and chrome in the proportions that characterize impact ejecta. Ms. Breger also saw that these metals were fused to micro fossils from the ocean floor. [43] In their report on the trip to Madagascar, the Group wrote, "Because extraterrestrial material contains a very high nickel content, the Fe-Ni-Cr metal is evidence that these marine fossils and concoidally (curved) fractured grains are impact ejecta." [44] Recall, that Dr. Abbott also found a very high nickel content in the core samples taken in the area near Burckle Crater. The very high nickel content in both the crater deposits and the tsunami deposits reflect the relationship between the crater and the ensuing aftermath tsunami.

The Fenambosy Chevron was the largest of the four chevrons discovered on Madagascar. It was over 600 feet high, covering an area twice the size of Manhattan. [45] It seems that it takes a tsunami of 1,000 feet high or more to leave deposits that are this thick. While large tsunamis can also be caused by volcanoes, earthquakes, and landslides, they cannot produce deposits of the size, extent, and composition as tsunamis caused by cosmic impact. Dr. Gusiakov of the Novosibirsk Tsunami Laboratory in Russia said, "The measured run up heights and inland penetration . . . are far beyond the range produced by the largest historically known tectonic tsunamis (seismic and volcanoes)." [46] Thus, the international expedition to Madagascar was able to quickly establish that the chevrons represented deposits from an impact generated mega tsunami. Dr. Bryant, the geomorphologist on the team, told the *New York Times,* "We are not talking about any tsunami you have ever seen . . . Aceh (the 2004 Indonesian tsunami) was a dimple. No tsunami in the modern world could have made these features. End of the world movies do not capture the size of these waves." [47]

Immediately after the impact in the Indian Ocean seismic sea waves would have gone out from the Burckle Crater at over 450 miles per hour.* Anywhere these waves approached landfall the shallow water would have caused these waves to pile on top of each other and rise up into powerful walls of water as high as mountains. Depending on the topography of the land where the massive waves struck, they could have rolled far inland. One need only look at a map to see that a tsunami propagated in the middle of the Indian Ocean would not only strike Madagascar, Australia and India, places where related chevrons have been found, but also the Middle East.

In less than five hours, these transoceanic waves would have traveled due north from the Indian Ocean to the Arabian Sea and across part of Saudi Arabia which is barely above sea level, then up the Persian Gulf and into the mouths of the Tigris and Euphrates Rivers before slamming into the mountains of Turkey. The low lying lands of the Tigris and Euphrates river valleys, would have acted as a

---

*   The very deep Burckle Crater would involve major vertical displacement of the sea floor which is necessary to produce big tsunami waves.

funnel of sorts.* The sound would be deafening as the first towering walls of water moved across the land like high speed freight trains, passing over the land of Noah, and the Sumerians and Akkadians until the waters of these powerful waves were broken up by the high mountains of Ararat in Turkey.† After the waters of these towering waves were stopped and fell back, these lands would have been covered with dark silty floodwater tens of feet deep and all humans and animals in the region would have drowned. There would have been a number of smaller secondary seismic sea waves generated by strong aftershocks from the cataclysmic earthquake caused by the ocean impact. These powerful aftershocks could have continued for months before significantly diminishing in size and frequency. For example, during the month following the 1964 Anchorage, Alaska magnitude 8.3 earthquake, "19 aftershocks were reported with magnitude 6 or above." [48]

Based on preliminary findings, Dr. Abbott and her associates estimate that this Burckle Crater impact occurred around 2,500 to 3,000 BC and they suspect it ties to one of the many ancient stories, legends or myths about a flood. The question is which flood story, if any, offers any information that directly ties it to this historic catastrophic event? One of the members of the Holocene Impact Working Group, Dr. Bruce Masse, an environmental archeologist at the Los Alamos National Laboratory in New Mexico thinks he

---

* The 2004 Indonesian Tsunami showed how landmasses can cause tsunami waves to diffract or bend. Diffraction would have let the Burckle Tsunami run directly up the Persian Gulf.

† A tsunami usually involves a series of waves of varying heights coming in, not just a single wave. For example, the Hilo Hawaii tsunami of 1946 which was triggered by a 7.1 magnitude earthquake in the Aleutian Islands, involved eight major waves coming in. The 2004 Indonesian Tsunami involved 4 or 5 waves hitting Bandai Aceh.

knows when a comet fell to produce Burckle Crater. Dr. Masse has analyzed nearly 200 flood stories and myths from around the world. He says many stories and myths describe torrential rains, hurricane force winds, and darkness during the storm. Some even describe a full solar eclipse. Dr. Masse believes that "the Burckle Crater impact event is in the right location to be the source of devastating rains, tsunamis, wind, and associated social upheaval about 2807 BC." [49] More specifically, Dr. Masse thinks the Burckle Crater relates to a solar eclipse that occurred in May 2807 BC. [50]

However, critics do not see the relevance of a flood story that does not give definitive information about the flood in question being caused by a cosmic impact. While a flood story may tell of a solar eclipse, devastating rains, hurricane force winds, darkness, and people dying, it still needs to clearly tie the flood to an impact and towering walls of water washing over the land. Dr. Masse recognizes that "extraordinary proof" is needed to connect a particular flood story to the Burckle Impact and tsunami.

Among the flood stories of the world, only "The Flood" story in the Bible gives specific indications that Noah's flood ties to the Burckle Impact. Assuming that the estimate of Burckle Crater's age is correct, then the Burckle impact is consistent with the date of the Biblical Flood. When the Biblical Flood story is properly translated, it provides "extraordinary proof" that it talks about the Burckle Impact and tsunami. This is because hidden within poorly translated words and phrases, the Bible specifically demonstrates that the Flood was caused by comet bombardment, and then relates a number of events that can only be caused by towering walls of water from an impact driven tsunami washing across the land.

Recalling, the clearer translation discussed earlier, the Biblical

Flood story in *Genesis 7:11* tells how the sources of the great deep, that is, comets were broken up; and how those that "lie in wait in heaven," the comets of the Oort Cloud were turned loose. In this, we are told in two different ways that comets came in and broke up bombarding the Earth. *Genesis 7:19-20* then tells how the waters of towering tsunami waves covered all the high hills and mountains, and how the remaining waters flooded the land and rose up fifteen cubits. In this, we are told the waters of an impact driven mega tsunami, with waves upwards of several miles in height, covered all the high hills and mountains and the waters brought in and left behind by the tsunami then flooded the land so that the waters on the land rose to a height of 15 cubits. Recall from a few pages earlier, it was pointed out that related Sumerian/Akkadian flood stories like the one in "*The Epic of Gilgamesh*" tell how the mountains were submerged on the *first* day. [51]

    Note that when *Genesis 8:3-4* says "the waters returned from off the earth continually" it indicates that the waters driven upon the land by the tsunami began to drain off back into the ocean. In addition, when this same verse says Noah's Ark then came to rest upon the mountains of Ararat months after the Flood began, it indicates that latter tsunami waves drove the Ark up this high. Remember, that in a mega tsunami it is the waters of the seismic waves that reach up high to cover hills and mountains, not the flood waters that remain behind, as the 2004 Indonesian tsunami demonstrated. Do not be confused by *Genesis 8:5* when it says that the flood waters continued to subside and that the tops of the mountains were seen months after the Ark come to rest on Ararat. This does not indicate that the flood waters were high enough to cover the tops of mountains and that they then subsided.

We have learned that forty days of rain cannot cover mountain tops. This concept is physically impossible, and is inconsistent with other verses in the Biblical Flood story. (Remember, *Genesis 7:12* says that the standing waters of the Flood only rose 15 cubits or 22.5 feet.) Rather *Genesis 8:5* indicates that while the flood waters brought in by the tsunami continued to subside, the clouds shrouding the mountain tops during the preceding meteorologically tumultuous and stormy months finally cleared away so that the mountain tops could again be seen. The incoming rushing waters of the towering tsunami waves would cover all the hills and the mountains only temporarily. A mega tsunami from another ocean impact could have driven Noah's ark up into the mountains and left it to rest there months after the first impact took place (*Genesis 8:4*). Noah's Ark being left to rest upon the mountains of Ararat is not unlike the 1868 Chilean tsunami that deposited large ships miles inland at the base of a mountain cliff, as was discussed earlier. Thus, in *Genesis 8:4* we have added information that the Biblical Flood involved multiple impact driven tsunami waves. To properly understand all the passages of the Bible's Flood story, one must first understand what happens when an impact generated mega tsunamis occurs.

Dr. Abbott and her colleagues have collected shells for radiocarbon dating (C-14) from two of the Madagascar chevron shaped tsunami deposits. This will give us a direct dating of the Burckle Impact and the tsunami it generated. We await the results.

To establish even more evidence for the Biblical Flood one would expect to find marine sediments and debris in the mountains of Ararat. Jumbles of broken trees, shells and marine sands should have been swept up into the mountains. In a tsunami deposit "caused by an impact, definitive evidence for an impact would come from glassy tektites, grains of shocked quartz and an overabundance of the elements that characterize the dust from comets and asteroids. In 1996, Dr. Lewis said that no suspected tsunami deposits (tsunamites) have been searched for such evidence, and this still seems to be the case with one notable exception. [52] Neither have lake deposits in properly situated lakes been searched for such evidence. The notable exception to this lack of searching is, of course, Dr. Abbott's recent work on chevron shaped tsunami deposits in Madagascar and Australia.

Unfortunately, over time, some of the evidence from a tsunami could have been washed away. Does this mean that this added evidence of the mega tsunami that caused the Flood can't be found? No! Some of the tsunami materials from the Flood should have been trapped in one of the large lakes in the mountains of Ararat, such as Lake Van, the largest lake in Turkey. Lake Van lies in the second largest interior basin of Turkey. The lake sits at an altitude of 5,250 feet and is ringed by mountains. *(See Illustration F)*

Instead of just searching for the Ark, some of the expeditions to the Ararat area would do well to take core samples from the sedimentary deposits found at the bottom of this 330 foot deep lake. Each year a new layer of material is laid down at the bottom of a lake. These layers, called varves, reflect the history of the lake. One of these layers should contain evidence from the Flood. It should be a relatively thick layer reflecting a sudden increase in the amount of material deposited occurring in the year of the Flood. As best as can be determined, the Bible indicates a date of around 3000 BC for the Flood. Most importantly this layer should contain debris picked up by the rushing wave (including ocean sand, marine shells, and fish teeth). This layer should also contain materials associated with cosmic impact (glassy tektites, grains of shocked quartz and dust with a cosmic chemical signature).

Actually, a core from Lake Van in Turkey has already been taken which shows that a sudden change occurred around 3,000 BC (based on a varve count calendar). This core shows a dramatic increase in quartz content around 3,000 BC. [53] While this sudden increase in quartz content could reflect quartz from ocean sand or shocked quartz from cosmic impact, the core was not specifically studied for evidence of marine sediments nor evidence of cosmic impact. Interestingly, what happened at Lake Van around 3000 BC could correspond to the marine sediment record which shows "a cometary scale dust flux event and aridification, ca. 3500 BC over North Africa and Arabia."[54] It should be noted that this same Lake Van core also shows "an abrupt doubling of the quartz content . . . (and) a tripling in other Van sediments records from 4,200 to 4,000 BP (2200-2000 BC)." [55] As we shall see, this particular abrupt increase in quartz content has been associated with a sudden change in climate that occurred around 2,200 BC, which in turn is associated with a sudden increase in regional dust deposits. Some scientists now attribute this sudden increase in dust to the cosmic impact that produced the two mile wide Amarah Crater in Southern Iraq.

It is interesting to note that while all of the possible direct evidence from the Biblical Flood has not yet been collected and studied, there is physical evidence in addition to the Burckle Crater and the Madagascar chevrons for a large scale catastrophic event of some sort around the time of the Flood. A number of scientists believe that a major catastrophic event took place around 3000 BC, a time where there was also an abrupt climatic downturn. Their beliefs are based on a number of lines of physical evidence. [56] For example, ice cores from Greenland show that 3000-4100 BC is one of the periods over the last 12,000 years where the chemical composition of the atmosphere experienced an abrupt increase in concentrations of sea salt and torrential dust. [57] This date from ice cores is consistent with data from cores taken in eighteen different Northern European peat bogs which show a cosmic component (micro meteorites, micro tektites and the geo-chemical signature of the peat ash) for a number of periods over the last 9,000 years, one of which is 3000 BC. [58]

A third line of physical evidence beyond the Burckle Crater that speaks of a large scale catastrophic event taking place around 3000 BC involves cores taken from a North American lake. "Walter Dean of the U.S. Geological Survey in Denver found three sharp peaks in the amount of dust settled to the bottom of Elk Lake in Minnesota. Dust peaked at about 5,800, 3,800 and 2100 BC, plus or minus 200 years, according to the counting of annual layers in the lake sediment." [59] So cores taken from ice in Greenland, peat bogs in Northern Europe, and a lake in North American all reflect sudden change as a result of a catastrophic event around 3000 BC. As mentioned earlier, based on the orbital backtracking of certain comet and asteroid families, some astronomers believe that a great amount of cometary activity occurred at the start of the third millennium BC. [60]

While not as conclusive as data from ice, peat, lake cores, chevron deposits and craters, there are other data that indicate a catastrophic event took place around 3000 BC. This includes: (a) the water levels at the Dead Sea rose 300 feet around 3000 BC [61]; (b) tree ring data from oak trees in Northern Ireland indicates a major climatic event at 3150 BC; [62] (c) ice core data from Greenland shows a big spike in sulfate concentrations at 3200 BC, and (d) a drop in the oxygen isotope ratio at 3,000 BC. [63] One researcher believes that the data also indicate short term oscillations in sea level at 3200 BC.

Between 1928 and 1934, British Archeologist Sir Leonard Wooley excavated the ancient Sumerian city of Ur and found a nine foot thick layer of alluvial silt, which he dated to around 4,000 BC, and took to be evidence of Noah's Flood. [64] Unfortunately, no other sites were found which showed similar alluvial deposits from this time period. Wooley also found evidence of flooding at the ancient Sumerian cities of Shuruppak (the home of the Sumerian Noah – modern Fara) and Kish, which he dated to around 2750 BC. [65] Again, it is unfortunate that no follow up has been done at other sites to see if there was widespread flooding in Sumer during this time period and no direct dating of these alluvial deposits has been conducted. Also, these alluvial deposits have not been tested for a cosmic component, and they may only represent local flooding of the Euphrates River. Nevertheless, Dr. Sharad Master, the geologist who found the Amarah Impact Crater in Southern Iraq, thinks that, "The 2.6m thick 'Flood' layer at Ur could be a tsunamite." [66]

One note of caution: two Columbia University marine geologists William Ryan and Walter Pitman have recently written a book entitled *Noah's Flood: The New Scientific Discoveries About the Event that Changed History*. It has drawn a lot of attention. [67] This book describes a flood involving the Black Sea, where salt water from the Mediterranean suddenly breached the Bosporus, a narrow strait between these two seas into freshwater of the Black Sea. However, their attempts to connect their theorized Black Sea Flood with the Biblical Flood and the Babylonian Flood are flawed. Ryan and Pitman completely ignore the specific details of the Biblical and Babylonian narratives and their archeological context. Further, the evidence they provide for their theorized flood has been challenged by a number of other geologists who are more familiar with the area. For example, the date for their theorized Black Sea Flood is 5,500 BC, which is over 2,000 years earlier than the date indicated for the Biblical Flood and the Babylonian Flood. In addition, a flood in the Black Sea area is too far away to make an automatic connection to a flood in the Mesopotamian (Biblical/Babylonian) area. Totally different linguistic stocks and cultures are involved and there is no archeological evidence for migration from one area to the other. These marine geologists are clearly in over their heads in regards to the archeology, history, and the literature of the ancient Near East.

Most importantly, an international team of geologists who have researched the Black Sea area dispute Ryan and Pitman's claim that the intrusion of Mediterranean Sea water into the Black Sea was the result of a catastrophic flood. They say several lines of new evidence (including drill cores) show that the intrusion of the Mediterranean Sea into the Black Sea was gradual and that it took place thousands of years earlier than 5,500 BC. [68]

In review, we have seen that the Flood is an important historical event for the ancient literate cultures of Mesopotamia (the Sumerians, the Akkadians, the Babylonians, the Assyrians, and the Hebrews). All of these cultures record the same basic story of a catastrophic Flood that occurred around 3000 BC. All of these cultures directly or indirectly indicate the same physical cause for the Flood – a cometary impact event that produced a tsunami so big that waves came in from the Persia Gulf and washed over the land and mountains. The writings of these cultures tell of this cometary event in different ways, giving us not only a more complete picture of what happened, but also eliminating the concern about whether the

Bible's story of the Flood was copied from others. Note that only the Biblical account says that comets are the sources of the water in the Earth's oceans, an important fact that scientists discovered just two decades ago, and are still learning about.

While direct physical evidence for the Biblical Flood is starting to be collected and studied, there is still another line of evidence that a catastrophic cometary event took place around 3000 BC. This date is consistent with the great amount of cometary activity some astronomers believe occurred around 3000 BC, based on the orbital backtracking of certain comet and Earth crossing asteroid families.[69]

In the quest for direct evidence for the Flood we at least know now what caused the Flood and what types of physical evidence need to be found or studied in greater detail. Detectives can't solve a homicide case by searching in the wrong place for the wrong type of murder weapon. An authoritative radiocarbon date for the Burckle Impact Crater in the Indian Ocean must be obtained. Also, drill cores from Lake Van in the mountains of Ararat could contain evidence of marine tsunami deposits showing a cosmic signature consistent with samples from Burckle Impact Crater and the related chevron shaped deposits. Were this evidence available, the case would be solved.

Finally, it has been hypothesized that the recently found two mile wide Amarah Impact Crater in Southern Iraq represents the "smoking gun" of the Flood impact. After all, it involves an explosion equivalent to that of dozens of modern nuclear bombs that occurred less than 6,000 years ago in what was then a shallow sea. However, this impact did not occur in water deep enough or distant enough ("fetch") to develop the giant seismic sea waves needed to wash across the Mesopotamian plains up to the mountains of Ararat.* But, the Burckle Impact which occurred in water 12,500 feet deep and over 2,000 miles from the Persian Gulf does meet these criteria. In a later chapter we shall see that the Amarah Impact Crater is the "smoking gun" of another Biblical catastrophe.

Sodom and Gomorrah

A number of astronomers believe that the destruction of Sodom and Gomorrah involved comet activity. Dr. John S. Lewis, a professor of Planetary Sciences at the University of Arizona and Co-Director of the NASA Space Engineering Research Center at the University of Arizona, is one scientist who believes that the destruction of Sodom and Gomorrah was caused by cosmic bombardment. [70] *Genesis 19:24* says, "Then the Lord rained upon Sodom and upon Gomorrah brimstone and fire from the Lord out of heaven." "Brimstone (burning sulfur) and fire raining down from heaven" describe the breaking up and disintegration of a comet in the Earth's atmosphere above these ancient cities, since large chunks of rocky and icy material falling from heaven would be seen as fire raining down from heaven. In addition, cometary material is rich in sulfur. Even a small meteor fall can produce a smell of sulfur that is so strong that it is almost suffocating. [71] In Chapter 2 a portion of the *Aeneid* was quoted which told how a meteorite struck and "the place around reeked with the smell of sulfur." [72]

Few Bible commentaries have grasped the true meaning of what is being described in the phrase about fire and brimstone falling

---

*   "Fetch" is the stretch of open water over which the generating wind (or seismic event) is blowing (or operating). Over 100 miles of fetch is needed to develop waves of maximum size.

from the sky. (Volcanic activity cannot be used to explain the fire and brimstone, because there are no volcanoes nor volcanic deposits in the region.) Determined to explain this phrase, some have even said that the destruction of Sodom and Gomorrah was caused by an earthquake which somehow explosively ignited the methane gas and sulfur that was found in the local tar deposits and shot it up into the sky. However, earthquake activity such as this is unprecedented and implausible in terms of geology. For those familiar with astronomy and ancient literature, the expression "fire falling from heaven" is not a cryptic expression nor a literary device that needs some sort of fanciful explanation; it is a simple descriptive phrase. "Fire from heaven" is an accurate description for cosmic material, either a meteor, asteroid, or comet, burning in the atmosphere ("fire") as it comes in to strike the Earth. Consider this 1833 account of the Leonid Meteor Shower which truly frightened people.

> For several hours over the United States there was a continual blaze of thousands and thousands of meteors at a time. One estimate was that over 240,000 meteors fell during that period, so many meteors in the sky at a time that many people were woken from their beds and stared at the sky in panic, believing the *sky to be on fire.* Many feared that it was the end of the world and dreaded what they would see at daybreak. [73]

Large meteors can shine as bright as the planet Venus or the full Moon; they are appropriately called "fireballs." In a Griffith Observatory publication the night of the 1833 Leonid Meteor Shower is referred to as "the night it rained fire," and it says "the most terrifying aspect of the shower was the many brilliant *fireballs* . . . never did rain fall much thicker than the meteors fell towards the Earth , , , for the stars are falling." [74] Also consider that a 14th century Chinese report told how "It rained iron," that is, iron meteorites near the Erh River one day and many people and animals were killed. [75] The rain of fire and brimstone and of iron from heaven speaks of the natural phenomena of cosmic bombardment. (In December of 2005, *The History Channel* aired a documentary about meteors, asteroids, and comets entitled "Fire in the Sky.")

Further indication that a cometary bombardment took place during the destruction of Sodom and Gomorrah comes from *Genesis 19:28*. This verse tells how Abraham "looked toward Sodom and Gomorrah, and toward all the land of the plain, and beheld, and, lo, the smoke of the country went up as **the smoke of a furnace**."* The "smoke of a furnace" speaks of the rising smoldering cloud that appeared after the explosive cometary impact. An earthquake opening a possible fissure would not produce a towering, smoldering cloud nor create the nodules of sulfur encased in ash found in the area. The impact of even a small cometary fragment over Sodom and Gomorrah could release energy equivalent to the explosion of many hydrogen bombs and raise a mushroom cloud just like that in nuclear explosions. Note that the Bible also uses the expression "the **smoke of a furnace**" in *Revelation 9:1-2*, when telling how a star (a luminous heavenly body - a comet) from heaven falls to the Earth, and opens a great pit (impact crater). It says smoke will arise

---

*   *Genesis 19:28* in the *Septuagint* reads " . . . and behold a *flame* went up from the Earth, as the *smoke of a furnace*." *Genesis 19:28* in the NIV reads, "he saw dense smoke rising from the land, like smoke from a furnace."

out of the pit as *the smoke of a great furnace*, and the air will be darkened and the sun will be darkened by the smoke. Without a doubt all of this describes an explosive cometary impact. Remember that *Revelation 9:1-2* was the passage astronomer John S. Lewis said clearly dealt with "impact phenomena." [76]

This Biblical account of fire and destruction raining down from the sky to destroy Sodom and Gomorrah brings to mind the eyewitness accounts of the large comet fragment that broke up in the atmosphere over Tunguska, Siberia in 1908 presented in Chapter 2. Further, the concept of fire from heaven bringing destruction is also found in the literature of the ancient Sumerians, Akkadians, Babylonians, and Assyrians, the contemporaries of the peoples of the Old Testament. In this literature we read about their (cometary) gods throwing down fire and firebrands from heaven to Earth, burning brought by "hailstones and flames," and of the Queen of Heaven and her consorts "who rain flaming fire over the land" in contexts consistent with cosmic impacts. [77]

More importantly, the Bible itself shows that the expression and concept of fire falling from heaven indeed pertains to cosmic material raining down from heaven because of the other times this expression or variations of it are used in the Scriptures in a context relating to cosmic impact. For example, *Exodus 9:23-25* tells how the Lord rained devastating hail and fire during the Exodus; *Isaiah 30:30* tells how the Lord will show indignation and anger with the flames of a devouring fire and hailstones. *Ezekiel 38:19-22* tells how there will be great shaking and the Lord will rain great hailstones, fire, and brimstone on those who invade Israel during the end times. *Revelation 8:7* tells of hail and fire being cast upon the Earth; *Revelation 8:8* tells of a great mountain burning with fire being cast into the sea; and *Revelation 8:10* tells of a great star from heaven burning as if it were a lamp falling upon the rivers. *Revelation 9:1-2* tells of a star falling form heaven to Earth, a great pit being opened, a cloud of smoke being raised and the air being darkened.* Recall Chapter 2 discussed how astronomers have discovered that comets have vents or mouths in their rocky crusts, which issue fire, smoke, and brimstone: and *Revelation 9:17-18* correctly tells how comets have mouths which issue fire, smoke, and brimstone and how these three will kill many men.

The Bible reiterates the cometary nature of the destruction of Sodom and Gomorrah in *Isaiah*. *Isaiah 13* describes what we recognize as cometary events, with comets coming in from "the end of heaven . . . to destroy the whole land," heaven and Earth being shaken and the Sun and the Moon being darkened (*Isaiah 13:5, 10, and 13*). Then verse 19 says these events "shall be as when God overthrew Sodom and Gomorrah." In *Luke 17:26-29*, Jesus likens the day of his return to **both** the day of Noah and the day of the destruction of Sodom and Gomorrah. In terms of catastrophe, the common denominator between all three days will be comet activity. Recall how the God of the Bible began to reveal himself to man by calling Abraham out of the Sumerian/Akkadian culture at Ur, which worshipped cometary gods, and introduced himself as the highest of the gods, the "Lord God of heaven." He left the dramatic personal signature of a cometary god when he "passed a smoking cometary fragment or meteorite, which is described as "a smoking furnace and

---

*   In Chapter 2 we learned that a comet hitting the Earth can cause an earthquake with great shaking; with great hailstones, fire, and brimstone falling from the sky; a great pit or crater being opened; the smoke of a towering mushroom cloud rising up, and enough dust being raised to darken the sky.

a burning lamp" (*Genesis 15:17*), between the pieces of the sacrifice Abraham made to God in accordance with the ancient practice of "cutting a covenant."* When the God of the Bible told Abraham that he was going to destroy Sodom and Gomorrah, it would have been no surprise to Abraham that the God of the Bible did so as a cometary god would, via cometary activity (*Genesis 18:17, 20, 23,* and *19:13, 14, 24,* and *28*).

Now that we know that natural phenomena can cause fire and brimstone to rain from heaven, the question is if there is any archeological evidence to support the account of Sodom and Gomorrah's destruction. Is there any direct physical evidence to support the Bible's story of such a catastrophe or is it just a morality tale regarding the wages of sin?

At the southern end of the Dead Sea in an area characterized by tar pits and oases (*Genesis 14:10*) archeologists have found the ruins of two ancient Bronze Age cities (*Genesis 13:12* and *14:3*). Burnt and reddened bricks have been found. Both cities were destroyed by fire. Abundant potsherds indicate a dense population dating to a period between 2500-2000 BC that ended abruptly around 2000 BC. The discovery of these settlements is consistent with the first century Jewish historian Josephus writing that "traces of the cities are still to be seen." [78]

---

*   In the Ancient Near East a covenant between two parties sometimes was confirmed by the practice of "cutting a covenant." In "cutting a covenant," animals were sacrificed and divided into two halves. Then the parties of the covenant would pass between the divided pieces and say something like "may it so be done unto me if I break my covenant with you." So when the God of the Bible caused a meteorite or comet fragment to pass between the two halves of the sacrifice, He showed His commitment to the promises He made to Abraham. (*Genesis 15:9-21* and *Jeremiah 34:18*). *Encyclopedia Judaica*, Vol. 5, "Covenant," Keter Publishing House, Jerusalem Ltd., Israel, p. 1112.

The most definitive evidence for the destruction of Sodom and Gomorrah would come from physical evidence associated with cosmic impact. Core samples from buried sediments dating to the time these cities were destroyed by fire should contain high concentrations of cosmic dust with very high concentrations of iridium, nickel, and other elements raised at impact. There could also be grains of shocked quartz, whose structure stems from the high pressures of impact or tiny spherules of fused glass like material that stems from the high temperature of impact. Volcanic activity does not produce shocked quartz nor dust with very high concentrations of iridium and nickel.

While no formal scientific testing has been done yet, there is some physical evidence that indeed indicates that a cosmic event took place. Dr. Benny Peiser, an expert on cosmic impact from Johns Moores University in England, reports that deposits of a form of calcite only found in meteorites has been discovered near the sites. Then there is the sulfur found in the area. In gypsum deposits, sulfur occurs in small marble to palm sized nodules or balls, which is not typical of the way sulfur is usually found. The sulfur is tightly compacted and over 95% pure. A glassy ash encloses the sulfur nodules indicating burning and vitrification from great heat. Several different amateur groups have filed reports about this sulfur and posted pictures of these unusual sulfur nodules on the web. Petrographic study of the sulfur could reveal its origins and how it came to form these unique nodules. Could the atmospheric conditions that cause hail to come down in spherical form have caused the sulfur found at Sodom and Gomorrah to have come down in spherical form? It is also interesting to note that the surface of the Dead Sea suddenly dropped by several hundred feet around 2200

BC, and some people have speculated that "the whole southern part of the Dead Sea may be an impact crater that was caused by a cosmic disaster."[79] They made this speculation because the Lisan Peninsula of the Dead Sea almost cuts off the small southern basin of the Dead Sea from the much larger northern basin of the Dead Sea, and the mean depth of the southern basin is only thirty feet, which is much less than 1300 foot mean depth of the northern basin. Until the 19th century it was possible to make a two mile ford across a ridge that went from Lisan to the west shore. Today the southern basin is now separated from the northern basin, and the southern basin has been divided into small ponds. The bottom line is this: if the destruction of Sodom and Gomorrah was caused by cosmic bombardment as the Bible indicates, there should be evidence waiting to be found. In a culture that generally believes that science and faith in the Bible are incompatible, what does it mean if there is scientific evidence to support the Bible's account of what happened to Sodom and Gomorrah?

In the Bible story of Sodom and Gomorrah we have another incident like that of the Flood, where the God of the Bible said destruction was coming and behold, the destruction came in the form of cosmic impact. Since this happened on more than one occasion, it may not be a coincidence. We should give God's description of His "ministers of flaming fire" a closer look for what we can learn about comets and what this means for mankind's future.

*   *   *   *

In this chapter we saw that interpretation of certain words from the Bible are a reflection of the translator's understanding of the ancient Hebrew culture and what is specifically being spoken about in a passage. An understanding of the archeological/historical context and even the scientific context is also important. We saw that within the Flood story when three words in *Genesis 7:11* were examined, the more accurate translation presented here not only conferred a significantly different interpretation to an old familiar story. Instead of mysterious "fountains of the deep" breaking up, we have "sources of the deep" breaking up which are comets. And instead of mysterious "windows of heaven" being opened, we have "those that lie in wait in heaven" being loosed, which are comets. This more accurate translation is not a matter of forcing scripture to fit a scientific mold. This more accurate translation is derived from examining the Hebrew words within the context of other scriptures, and allowing the words to convey their meaning within a context of what we now know to be true about cosmic activity in our solar system and its effects on Earth.

This chapter showed that based on the nature of the events described in scripture, cometary activity caused the Flood. Amazingly there is now physical evidence to support this translation and interpretation. The recent discovery of the Burckle Crater and chevron shaped tsunami deposits provide physical evidence to support the explanation that a massive tsunami caused by a large ocean impact around the time the Bible gives for Noah caused the Flood! In addition, we saw that a number of astronomers and geoscientists are now acknowledging that the descriptions in the Bible of certain catastrophes appear to be descriptions of cosmic activity.

The headline on the 2008 *History Channel* documentary ***Mega Disasters: Comet Catastrophe*** read, "The cause of the Biblical Great Flood may have been a massive three mile wide comet that crashed in the Indian Ocean, 4800 years ago." What would the scientists interviewed in the documentary say after they learn that the Bible *specifically* says that comet impact caused the Great Flood when the scriptures are more accurately translated?

When scripture says that God rained upon Sodom and upon Gomorrah brimstone and fire, and Abraham told how after the destruction of these cities he beheld smoke going up as the smoke of a furnace, it is not difficult to recognize how all of this was caused by a cosmic impact. We learned this included a description of cometary material breaking up and disintegrating in the Earth's atmosphere, with sulfurous rocky and icy material blazing as it fell from heaven to earth.

It is interesting to note that Jesus spoke of a relationship or connection between the day of his return, Noah's Flood, and Sodom and Gomorrah.

FIVE

Scientific Explanations for Old Testament Catastrophes
Part 2: The Exodus, Joshua's Great Victory, Debra and Barak's Victory and the Blast that Killed 185,000 Assyrians

> Our fathers understood not thy wonders in Egypt.
>
> *Psalm 106:7*

The Exodus

In light of what is now known about comets, we are compelled to consider whether many of the extraordinary events of the Exodus were caused by cometary activity. In their book *The Cosmic Serpent*, British comet experts Victor Clube from the Department of Astrophysics at the University of Oxford and William Napier from the Royal Observatory at Edinburgh show that they are far from championing the God of the Bible; nevertheless, after taking a detailed look at the miraculous events of the Exodus, they conclude "the Biblical account of the Exodus itself amounts to a description of an impact associated with a terrible comet." They add that "a reasonable interpretation of the Exodus account is that it took place much as described and the events seen until now as miraculous simply describe an impact with a fragment from a great comet during a close encounter."[1]

Apart from the killing of the firstborn sons of the Egyptians, Clube and Napier have divided the plagues of the Exodus into two groups. The first group reflects natural biological phenomena common in Egypt and includes the plagues of frogs, lice, flies, disease on beasts, and locusts. They maintain that the second group reflects rare, but natural cosmic phenomena resulting from cometary activity and includes the plagues of blood, boils on man and beast, heavy hail, and darkness. Stones of ice and fiery rock, and irritating iron rich dust from disintegration debris traveling in a stream weeks to months ahead of and/or behind a comet would explain what turned the waters of Egypt blood red and killed the fish.* It would explain what caused the boils on man and beast, and caused the fire and "a ***very heavy hail (ice)*** such as has not been seen in Egypt" (*Exodus 9:18* and *24 NAS*). In addition, cometary activity could throw enough dust into the atmosphere to bring about three days of "thick darkness," "darkness which may be felt" (*Exodus*

---

*   Actually a comet within our solar system could produce two periods of cometary activity on the Earth on its orbit around the Sun. One period of activity could be as the comet and its tail passed close by the Earth on its journey ***inward*** to go around the sun; and a second period of activity could be as the comet and its tail again passed close by the Earth on its journey ***outward*** after it rounded the Sun.

10:21-23). The irritating comet dust and the noxious nitrogen and sulfur compounds produced in the atmosphere by cometary activity could certainly have produced the "boils on man and beast". Recall that in Chapter 2 we learned that vast quantities of nitrogen oxides can be produced and fall as acid rain which can be toxic, mutagenic, teratogenic, and carcinogenic. Man and beast would develop a variety of rashes and sores.

Astronomers Clube and Napier, reflecting on the Exodus and all of the new knowledge that has been gained about comet activity, wrote "These events are by now so ***obviously cosmogonic*** that we need not labor the point." [2] In their second book *The Cosmic Winter*, Clube and Napier gave added details about the "comet/impact elements" of the Exodus. [3] The question follows, does the field of astronomy provide us with independent evidence to support comet activity as the cause of the extraordinary events in the book of Exodus? Incredibly, the answer is yes! Scientists seem to have identified the specific comet and the year of these events in Egypt.

Orbital backtracking done by American astronomer Donald Yeomans and Chinese astronomer Tao Kiang showed that Comet Halley had an extremely close encounter with planet Earth on September 7, 1404 BC.[*] This date is within the general time frame for the Biblical Exodus. Based on this orbital backtracking, some

---

[*] D. K. Yeomans and T. Kiang, "The long-term motion of Halley's Comet," *Monthly Notices of the Royal Astronomical Society*, 1981, 197: 633-646. On pages 633 and 642 Yeomans and Kiang write, "The dynamic model used to compute the long-term motion of Comet Halley successfully represented the ancient Chinese observations over nearly two millennium . . . . The interval from 141 to 1404 BC was free from severe perturbations so that the computed times of perihelion passage back to 1404 BC are not likely to be in error by more than a month." Also see Donald K. Yeomans, *Comets*, John Wiley and Sons, 1991, p. 265.

astronomers believe that the Earth most likely passed through Comet Halley's tail during this encounter. They say it was probably a terrifying sight. In *The Cosmic Winter* Clube and Napier say "*It is conceivable therefore that the Exodus story contains the earliest record of Comet Halley.*"[4] Not only is the date of 1404 BC for Comet Halley's extremely close encounter with the Earth within the general time frame for the Biblical Exodus, it agrees with another scientific date pertaining to the Exodus, a radiocarbon date.*

Archeologists have a radiocarbon date for the fall of the ancient city of Jericho of 1400 BC plus or minus forty years, which according to the Bible occurred forty years after the Jews left Egypt. In an article in *Time* entitled "Score One for the Bible - Fresh clues support the Story of Joshua at the walls of Jericho," the details of this radiocarbon dating are discussed. After the plagues of the Exodus, the Bible says the Israelites wandered forty years in the wilderness before they entered the Promised Land and began their conquest by taking the city of Jericho. So, the date for the plagues of the Exodus can be calculated by subtracting forty years from the date of the fall of Jericho. It had long been known that the city's walls fell in a way suggestive of sudden collapse and that Jericho burned after its walls collapsed. However, only recently could we see that these events took place at the right time to match the Biblical account of collapse (*Joshua 6:20* and *24*). Dr. Bryant Wood, an archeologist at the University of Toronto was able to get a radiocarbon date for the fall and burning of the walls. The *Time* article states:

---

*   Note that Greenland ice cores which reflect dust concentrations laid down due to volcanic and/or cometary activity shows a strong signal in the ice dated at 1388 +/- 50 years BC, which agrees well with the Comet Halley date of 1404 BC for the Exodus. See Victor Clube and Bill Napier, *The Cosmic Serpent*, Faber and Faber, London, 1982, p. 237.

A thick layer of soot at the site, which according to radioactive carbon 14 dating was laid down about 1400 B. C., supports the Biblical idea that the city was burned, not simply conquered. Finally, Egyptian amulets found in Jericho graves can be dated to ***around 1400 BC*** as well. Says Wood: 'It looks to me as though the Biblical stories are correct.'[5]

Dr. Wood also showed that the pottery dug up at Jericho in the 1930's was in common use around 1400 BC, a fact that supports the radiocarbon dating of the event.

An archeological date of "around 1400 BC" for the fall of Jericho basically agrees with the Comet Halley date of 1404 BC for the plagues of the Exodus. The correspondence between the radiocarbon dating for the fall of Jericho, and the Comet Halley dating of the Exodus is impressive.* Combining the two objective dating methods,

---

*   To fully understand the magnitude of this correspondence, we need to know that radiocarbon dates are not really exact to a single year, but are subject to a percentage of error so that the actual date lies within a range of years. Direct reference to Dr. Wood's report on his research in the March/April 1990 issue of the *Biblical Archeological Review* shows that the radiocarbon date cited in the *Time* article was subject to an error margin or range of plus or minus 40 years ("Did the Israelites Conquer Jericho? A New Look at the Archeological Evidence," Bryant Wood, *Biblical Archeological Review*, March/April, 1990, p.53.) Based on this radiocarbon date, the walls of Jericho could have fallen at a date of 1360 BC (1400 BC - 40 = 1360 BC). Using this end of the radiocarbon date range is within reason since two of the scarabs found buried at a cemetery at the site bear the name of a Pharaoh (Amenhotep III) who reigned from ca. 1386 -1349 BC If we count forwards and add the 40 years the Israelites wandered in the wilderness to the date for the plagues of the Exodus obtained from the extremely close encounter with Comet Halley, 1404 BC, we get a date of 1364

we can say that Jericho fell between 1364 to 1360 BC. In other words, **the orbital backtracking of the comet caused plagues of the Exodus and radiocarbon dating for the burning of the walls of Jericho are two independent scientific dating techniques that are consistent with the Biblical narrative for these two events.** In his report Dr. Wood said, "When we compare the archeological evidence at Jericho with the Biblical narrative describing the Israelites' destruction of Jericho, we find quite a *remarkable* agreement."[6]   Comparing Dr. Woods' radiocarbon dating of Jericho's fall to the astronomical dating of the plagues of the Exodus supports the *entire* Biblical narrative pertaining to the Israelites leaving Egypt and entering the Promised Land. Further confirmation will come as we take a close look at the specific events of the Exodus.

Not only are the Exodus events "obviously cosmogonic," but the God of the Bible in *Psalm 78:49* specifically says (in ancient Near Eastern terms) that He sent comets as His messengers during the Exodus. *Psalm 78:43-49* talk about the signs and wonders sent by God during the Exodus, with verses 47 and 48 talking about the "hail" ("great chunks of ice" in the *Amplified Version*) and "hot thunderbolts" that fell. *Psalm 78:49 (NAS)* says, "He sent upon them His burning anger, fury, and indignation, and trouble. A band (sending) of destroying angels (messengers-comets)." In the Hebrew *Tanakh* (the Hebrew version of the Old Testament) the last verse of *Psalm 78:49* reads, "a band of deadly messengers." Returning to the work of cometary specialists Clube and Napier, we see that among the events that are "obviously cosmogonic" is the very heavy hail mixed with fire that fell. For the non-scientist however, the waters of Egypt turning to blood and fish dying may not be so "obviously BC (1404 BC minus 40 = 1364 BC) for the fall of Jericho. This date falls well *within* the range of the radiocarbon date for the fall of Jericho.

cosmogonic." The Bible's account of the first plague of the Exodus given in *Exodus 7:20-21* says "all the waters that were in the river were **turned to blood**. And the *fish* that was in the river **died**."

There are several comet related reasons why the waters of Egypt could have turned to blood, or turned red, and caused the fish to die. First, typically the iron rich, dark red colored dust from a comet and the deep reddish brown acid rain could color the waters.[7] A significant amount of atmospheric comet dust and acid rain containing nitrogen compounds would explain how there was even blood "in vessels of wood and in vessels of stone" (*Exodus 7:19*). The fish died because the Nile and the waters of Egypt would have become so acidic and toxic that most aquatic life in them would have died. Second, the dust from a comet or its tail could suddenly raise the temperature of the water and put nutrients (chemical elements which fertilize) in the water which would in turn cause a population explosion ("bloom") of plankton or algae (dinoflagellates) that release red pigments into the water in a phenomena known as a "red tide."

The red tide phenomenon is not confined to oceans as commonly thought. In modern times this phenomenon has been observed to occur on a number of occasions in the brackish fresh waters of Egypt and Israel, especially in the Nile River. These red tides or blooms of plankton can also be toxic enough to kill the fish in a body of water. Fish may be killed by the plankton (dinoflagellates) themselves or by the build up of toxins released by these plankton "that act as nerve poisons causing **red blood** cells to lyse (break open)" so that gills **hemorrhage**, and blood is **literally** released into the water.[8]

As explained above, comet activity could indeed cause the Nile River to turn the color red like blood and cause the fish in the river to die, or perhaps to turn to blood in the literal sense of blood from hemorrhaging fish. It should also be noted that the Hebrew word for blood (*dam* #1818 in *Strong's Concordance*) during Biblical times was sometimes used as a metaphor for "death," not unlike the English metaphors "his blood is on your head" or "his blood is on your hands" when someone has caused another's death. For example, the NAS and the NIV translate the Hebrew word *dam* (#1818) as "death" in *Psalm 94:21* (also see *Romans 3:25*). So *Exodus 7:20-21* could be translated to say "all the waters that were in the river were turned to *death* (*dam*). And the fish that were in the river died."

Translating **dam** as "death" instead of "blood" in *Exodus 7:17-21* is consistent with the fish in the river dying and with the river stinking as these verses also say. Support for this translation and interpretation of *Exodus 7:17-21* comes from *Joel 2:30* which says, "And I will shew wonders in the heaven and in the Earth, blood (death), fire and pillars of smoke," a passage we will see is cometary. An Egyptian hieroglyphic manuscript written about the time of the Exodus, the "*Ipuwer Chronicle*," says that after a catastrophe, "the River is blood. If one drinks of it, one rejects it." [9]

With the waters of Egypt turning to blood and stinking (hydrogen sulfide), we have the basis for the second, third, and fourth plagues of the Exodus. The second plague would involve frogs leaving the stinking and toxic waters of Egypt en mass (*Exodus 8:3-6*). With their regular habitat and food supply poisoned, it would seem that these frogs soon died en mass, with their rotting flesh providing a feast for the lice (NAS and NIV - gnats) of the third plague (*Exodus 8:16-18*). Continuing this biological chain reaction (as begun by the comet activity) would yield the swarms of flies of the fourth plague, as the eggs laid in the rotting flesh of the frogs became larvae and then hatched to produce a plague of flies (*Exodus 8:21-24*).

Cometary dust circulating in the atmosphere could block sunlight and cause sudden atmospheric cooling which would kill plant life in some areas so the search for food could cause locusts to come from afar to seek the plant life in Egypt. The Scriptures say that the locust arrived after an east wind blew all day and all night and later they were removed by a strong wind (*Exodus 10:13-19*). Could comet activity have also set in motion the all night "east" wind, which was strong enough to split the Red Sea (Sea of Reeds) as told of in *Exodus 14:21-22*?

Remember the account of the 100 meter comet fragment from Comet Encke that caused the huge mid air explosion and powerful wind that flattened more than half a million acres of forest in the Tunguska River Basin in Siberia in 1908? A cometary fragment exploding low in the atmosphere, could easily produce a very strong wind, and even cause ground motion like a quake and tsunami. Thus, the after effects of a comet impact could perhaps cause a cascade of forces effecting the waters of the Red Sea to part (*Exodus 14:16* and *21-31*). Even though great earthquakes produce only a tiny fraction of energy compared to that of a of cometary impact, the 1964 Anchorage, Alaska earthquake and related tsunami left the bay at Valdez dry for a period of time. Likewise according to eyewitness Eliza Bryan, the New Madrid, Missouri earthquake of 1811-1812 caused the bed of the Mississippi River to be exposed and "its waters gathered up like a mountain" even running backwards upstream against its natural flow to the Gulf. [10]

The Sumerians and Babylonians especially feared comets for the powerful scorching winds they set in motion; winds that were sometimes referred to as "whirlwinds," "tempests" and "storming winds." If a comet with an exceptionally long, dust laden tail, such as the Great March Comet of 1843, whose tail was about 180 million miles long, passed low through the Earth's atmosphere, it would create horrific winds that could last for days.[11] The cloud or stream of debris travelling with a comet passing low through the Earth's atmosphere could also cause powerful winds, especially if there was an atmospheric impact. Thus, a comet or comet fragment need not strike the Earth to cause powerful winds.

Philo (ca 30 BC - 50 AD) was a Hellenistic Jewish scholar who had access to the ancient world's famed library in Alexandria, Egypt. In Philo's account of the splitting of the Red Sea, he indicates that something more than just a strong wind occurred as evidenced by a huge "dense and black cloud" that appeared. Such a "dense and black cloud" could have been cometary. Philo first tells of a wind of "tremendous violence" (Colson Translation) arising after sunset and says,

> . . . under the influence of which the sea retreated; for as it was accustomed to ebb and flow, on this occasion it was driven back much further towards the shore, and drawn up in a heap as if into a ravine or whirlpool. And no stars were visible, but a ***dense and black cloud (cometary)*** covered the whole of the heaven, so that the night became totally dark, to the consternation of the pursuers. And Moses, at the command of God, smote the sea with his staff. And it (the sea) was broken and divided into two parts.[12]

In Philo's account, the "dense and black cloud" that covered and completely darkened the night sky could have been either the dust laden tail of a comet or a cloud of dust and debris given off by a comet as it traveled along its orbit. Imagine a locomotive train that belches out smoke from its smokestack as it travels along and you have a picture of a comet discharging clouds of debris as it travels around the sun.

The next key question is "Could a strong east wind cause the Red (Reed) Sea to part as readily as an impact generated earthquake and tsunami could cause it to part?' The answer is yes! Two oceanographers recently calculated that a strong wind blowing in the narrow and shallow Gulf of Suez, which is the northwest arm of the Red Sea, could indeed account for this phenomenon. A *New York Times* article, entitled "Red Sea parting may have been wind caused, scientists say" by John Noble Wilford, reports how Doron Nof, a professor of oceanography at Florida University in Tallahassee and Nathan Paldor, a specialist in atmospheric sciences at Hebrew University in Jerusalem and a visiting scholar at the University of Rhode Island's Graduate School of Oceanography at Narragansett, came to this finding. It discussed their research published the week of March 16, 1992 in the *Bulletin of the American Meteorological Society*.[13] Wilford reported that the research of these scientists provided a "*scientific explanation*" for the parting of the Red Sea. The scientists explained that this phenomenon could have been created by winds of 46 miles per hour or more and said "the parted waters could have spilled back into place in only four minutes."[14] Professor Paldor is quoted as saying, "The Gulf of Suez provides an ideal body of water for such a process because of its unique geography."[15] Wilford reports that these oceanographers also said "the Biblical description of the Israelites going 'into the midst of the sea upon the dry ground' could be explained by the presence of a *natural* ridge in the bottom of the Red Sea in the Gulf of Suez. The account of a wall of water on either side, they said, supports the theory that the wind was pushing back the water." [16] We can now see that the strong wind probably caused by comet activity could have caused the Red Sea to part.

In the Bible, God is specifically given credit for exposing the bed of a sea (the Red Sea?) through such comet activity. *Psalm 18:9-15 (II Samuel 22:8-16)* first describes a comet "bowing the heavens," traveling with clouds of "hail stones and coals of fire." Then it tells how God sent out comet fragments or meteorites as his "arrows," so that "the channels (seabeds) of the sea appeared." Similarly, *Nahum 1:3-4* refers to a comet as a "whirlwind" and as a "storm" (ancient terminology which will be fully explained in a later chapter) and says that God "rebuketh the sea and maketh it dry." *Psalm 77:16-20* speaks of Moses and Aaron leading Israel during the time of the Exodus and tells how God's "arrows" (meteorites) went out and the "waters" and the "seas" were afraid, and "the Earth trembled and shook."

The NAS and NIV translations of these verses (*Psalm 77:16-20*) tell how God's "thunder" (sonic booms from meteorites) was heard in the "whirlwind," that is, in the rotation stream or cloud of cometary debris that can be associated with a comet. The Hebrew word *galgal* (#1534 in *Strong's Concordance*), translated as "whirlwind" in *Psalm 77:18 NAS, NIV*, literally means "wheel." Note that the ancient Sumerians, Akkadians, Babylonians, Assyrians, Israelites, and Greeks sometimes perceived of comets as "wheels." When these verses from *Psalm 77* tell of the Earth trembling and shaking in concert with meteorites ("arrows") being sent out and the "seas" being disturbed, we know that a cosmic impact or debris blast is being described. In these verses the word "sea" may refer to the

"Red Sea," and the word "waters" may refer to a later Exodus event, a time when Joshua and Israel entered the Promised Land. At that time, the waters of the Jordan River split and "stood and rose up upon a heap . . . and the people passed over right against Jericho (*Joshua 3:16*)," much like the Mississippi River's "waters gathered up like a mountain" during the 1811-12 New Madrid Earthquake that was discussed earlier.

When the waters of the Jordan River backed up to make a dry passage, the "winds" of a comet may not have been the immediate cause. Instead, a comet fragment or meteorite, as an arrow of the Lord, could have caused a small earthquake, which in turn could have caused a large section of the river bank to fall in and dam up the river for hours. It would have been similar to an event in 1927. In 1927 following a earthquake, a section of one bank of the Jordan River fell into the river not far from the town of Adam and blocked the water for over 21 hours. This enabled people to easily cross over the dry river bed just as the Bible says the Israelites crossed the Red Sea. [17]

Astronomers Clube and Napier, in addition to saying that some of the plagues of the Exodus and the splitting of the Red Sea were caused by comet activity, state that "the Exodus account itself contains an accurate description of an outstandingly brilliant comet." [18] So, beyond telling of plagues and the splitting of the sea caused by a particular comet, the Exodus account also tells of a *second comet* and gives an accurate description of this *second comet*, an "outstandingly brilliant comet." The description of this second comet, and its movements over a period of a few days is found in *Exodus 13:21-22* and *14:19-20*. These verses tell how when the Jews left Egypt "the Lord went before them by day in a pillar of a cloud . . .

and by night in a pillar of fire," but then "the pillar of the cloud went from *before* their face and stood *behind* them."

Clube and Napier believe that the progenitor of Comet Encke, a comet with a low inclination and an orbital period of just 3.3 years (the shortest orbital period of all known comets) is described in the verses of *Exodus 13:21-22* and *14:19-20*. Sumerian, Akkadian, Babylonian, and Assyrian texts, which tell of great battles in the sky, indicate that *two* comets were sometimes seen in the sky at once during the third and second millennium BC! While orbital backtracking indicates that Comet Halley, with a ***high inclination***, and an orbital period of approximately 76 years, brought many of the plagues recorded in Exodus; the orbital characteristics gleaned from *Exodus 13:21-22* and *14:19-20* indicate that a ***second comet***, a comet with a ***low inclination*** and a very short period was ***also*** visible in the sky at the time of the Exodus. This could have been the progenitor of the comet we now know as Comet Encke.

Chinese records going back to 1058 BC indicate that Comet Halley was very bright or active (-7.7 magnitude) at these early dates and plainly visible during the day. Some astronomers believe that the progenitor of Comet Encke was even brighter (-12.0 magnitude) than Comet Halley at that time. These astronomers believe that the progenitor of Comet Encke (proto Encke) was as bright as the Moon when it came close to the Earth, and, with its cloud or coma, was even larger than the full Moon in size. [19]

These two very active comets (Halley and Encke) would have been the dominant objects in the sky both at night and during the day. Comet Encke is the same comet that is associated with the June 30, 1908 Tunguska event, the June 22-26, 1975 and June 25, 1978 bombardments of the Moon, and it could have been the comet that brought the "very heavy hail" (*Exodus 9:18*) of the Seventh Plague of the Exodus. As the progenitor of comet Encke broke up over time to become what is now Comet Encke, it seems that it had some very destructive episodes of disintegration. For example, astronomer Duncan Steel tells how the late Harvard astronomer Fred Whipple in collaboration with Egyptian astronomer Salah El-Din Hamid, used modeling and orbital backtracking and "found there was evidence of an exceptional event, which they saw as a breakup of a proto-Encke's comet around 4,700 years ago . . . (and) evidence of another break up about 1,400 years ago." [20]

Today Comet Encke crosses the Earth's orbital path every 3.3 years and the Earth crosses the very broad stream of debris left by Comet Encke along its orbital path, the Taurid meteor stream, twice per year in June and November. [21] What did men see and what happened on the Earth during proto Encke's major breakup? The point here is that a comet coming close to the Earth need not strike the Earth to show wonders in the sky and to rain fiery cometary debris upon men.

These two active comets, Comet Halley and Comet Encke, could have been involved with the Exodus. Several other ancient cultures record similar unusual incidents. Two very active comets seemed to have put on such a spectacular show that the Sumerians, Akkadians, Babylonians, Assyrians, Egyptians, Minoans, and Greeks of the third and second millennium BC all took notice and thought they were two warring sky gods. The theme of warring sky gods, that were in fact two active comets, is a fundamental thesis of Clube and Napier's studies of ancient catastrophe (*The Cosmic Winter*, 1990, and *The Cosmic Serpent*, 1982). In these books, the authors describe two combating sky gods, one of which is often serpent like or dragon like, like those in the Sumerian myth of Inanna versus Kur and in the Babylonian myth of Marduk versus Tiamat.

Similar serpent like sky gods appear in the Assyrian myth of Ninurta versus Zu, the Egyptian myth of Horus versus Seth, (where in the "*Ipuwer Chronicle - Admonitions of a Sage*," Seth is known as Typhon), in the Greek myth of Zeus versus Typhon, and the Greek myth of Zeus versus Phaethon. The Greek word *Phaethon* (from the word *phaino* - "to shine"), which has been interpreted to mean "blazing star," is a very appropriate description for an active and fast moving comet. The mythical Phaethon drove a fast moving sun chariot across the sky, which was struck by a thunderbolt and sent to the ground trailing ashes and glowing dust as it fell. It is interesting to note that in Egyptian hieroglyphics the symbols for "thunderbolt" and "meteorite" are the same and contain a star." [22]

There are also many similarities between the specific catastrophic events of the Biblical Exodus, the Egyptian Typhon myth from the *Ipuwer Chronicle*, and the Greek Phaethon myth.[23] The Egyptian *Ipuwer Chronicle*, was recorded in a hieroglyphic manuscript that was written about the time of the Exodus (ca. 1400 BC), and, like the Biblical Exodus account (*Exodus 7:17-21*), tells how during a time of catastrophe, the "river" turned to "blood," "stank," and became "loathe" to drink. [24] All three stories, (the Exodus, Ipuwer and Phaethon) tell of the activity of **two separate comets** observed concurrently. The clearest statement that Typhon was, in fact, a comet comes from the ancient Roman author Pliny (23-79 AD) who in his *Natural History* writes, "A terrible **comet** was seen by the people of Ethiopia and Egypt to which **Typhon**, the king of that period, gave his name. It had a *fiery* appearance and was twisted like a coil (that is, serpent like), and it was very grim to behold. It was not really a star so much as what might be called a ***ball of fire***." [25]

As noted, Clube and Napier explain that *Exodus 13:21-22* and *14:19-20* give the description of a brilliant comet. The tail of a brilliant comet with a ***low inclination*** lying near the ecliptic plane, as in the case of Comet Encke, with its head or nucleus ***below*** the horizon at that latitude "would appear to rise vertically up." It would look like a pillar of fire during the day and pillar of smoke during the night. They explain that in the latter part of the ***night*** this vertical tail would appear "as a red band of light," and that would be the pillar of fire. During the early part of the ***day***, "after sunrise the white inner tail would dominate, and there would be a pillar of a cloud (smoke)."[26] They explain that this pillar would change from a position of "before" to a position of "behind" Israel, from the ***eastern*** horizon to the ***western*** horizon, as the comet passed through perihelion, that is, its closest approach.[*]

Based on the description given in *Exodus 13:21-22* and *14:19*, Clube and Napier explain that the appearance and behavior of this "pillar of fire (and smoke)" is what one would expect during a close encounter with a comet. They write, "We have then a plain description of an exceptional ***comet*** in a direct orbit of low inclination and small perihelion distance . . . the progenitor of **Comet Encke** satisfies all the ***orbital*** criteria." [27] With a very short period of just 3.3 years, the progenitor of Comet Encke could have been visible in the sky long enough to be seen as the "pillar of fire" or the "pillar of a cloud" that

---

*   Being seen on one horizon and then on the other horizon is quite normal for an active comet approaching the sun. For example, in A. D. 1066 Comet Halley was first seen as a morning object in the ***east***, and then as an evening object in the ***west***, and in 164 B. C. Babylonian tablets in the British Museum also tell how Comet Halley was first seen in the ***east***, and then in the ***west***. *Comets*, Yeomans, 1991, pp. 364 and 392.

led Israel into the desert (*Exodus 13:21-22, 14:19, Psalm 78:14*, and *105:39, Numbers 14:14*, and *Nehemiah 9:12 and 19*).

Exodus 14:19 identifies the "pillar of the cloud" as an "angel or messenger of God" (#4397 *malawk* - ***cometary messenger***) "which went before the camp of Israel, removed, and went behind them; and the ***pillar of the cloud*** went from before their face, and stood behind them."* This verse is key because it provides us with an undeniable description of cometary phenomena. Comet experts Clube and Napier conclude that based on orbital criteria, the pillar was indeed cometary. Further, *Exodus 14:21* and *22* tell of a "strong east wind" that divided the waters of the sea. As discussed earlier, this is also a cometary event. The events of *Exodus 14:19* and *Exodus 14:21-22* probably involve the activity of two separate comets. We learned that the strong winds created by the comet of *Exodus 14:21-22* could have caused the waters in the northward arm of the Red Sea to split just as the oceanographers suggested would occur with very strong winds. The dust laden tail of a comet passing overhead could have been the final instrument of the split, which brings to mind the "dense and black cloud" of Philos' account of this event.†

* In previous discussions we saw that the Hebrew word *malawk* (#4397 in *Strong's Concordance*), which is usually translated as "angel" can also be translated as "messenger." Since both the "messenger" and the "pillar of the cloud" made the *same* move, from in front to "behind," the "messenger" and the "pillar of the cloud" must be the same thing. In ancient Near Eastern usage, and Biblical usage, the "messenger" of *Exodus 14:19* is best understood as a "cometary messenger."

† Space doesn't permit a full explanation, but there is a seeming enigma in *Exodus 14:20*, which says, "And it came **between** the camp of the Egyptians and the camp of Israel, and it was a cloud and darkness (to them) but it gave light by night (to these); so that the one came not near the other all night." As translated, this verse sets up an impossible astronomical circumstance. More correctly translated and interpreted, the "*it*"

The big question now is: does any direct physical evidence remain to support this belief of the cometary activity of the Exodus? Since there are strong indications that comet activity was involved with the Exodus, there are specific types of physical evidence to look for. As we have learned, cometary activity produces certain types of evidence, including grains of shocked quartz, tiny spherules of fused material called tektites, and deposits of dust that have a unique cosmic chemical signature. For example, dust particles found at the site of 1908 Tunguska, Siberia impact site contained unusually high concentrations of iridium, nickel, and other elements. [28] Thus, it may be possible to find physical evidence for some of the plagues of the Exodus. While the Nile river bed has been scoured and filled many times, and the wind has long since blown away ancient surface deposits, it still may be possible to find buried deposits dating to the time of the Exodus. Dried up ancient lakes, marshes, or caves may still contain such deposits. Core samples from these ancient deposits could reveal sediments dating to the time of the Exodus. These ancient sediments when studied could conceivably contain physical evidence of cometary activity, just as many scientists who are studying the Amarah Crater in Southern Iraq now recognize that the dust deposits associated with the fall of the Akkadian empire have a cometary signature. [29]

---

should apply to the second comet, the comet referred to in *Exodus 14:21-22*, not the comet of *Exodus 14:19*. And the Hebrew word that is translated as "between" (*beyn* or *biyn* #996 or 995 in *Strong's Concordance*) should be translated as "above, or "perceived." So a more accurate translation would read, "And *it* came or passed (*bow* #935 in *Strong's Concordance*) **above** (#996 in *Strong's Concordance*) the camp . . ." as a reference to the second comet's tail.

Joshua's Great Victory

A number of astronomers have come to believe that Joshua's great victory in the Valley of Gibeon involved comet activity. In fact, it is difficult now for just about anyone to arrive at a different explanation for the description of this event. Co-Director of the NASA Space Engineering Research Center at the University of Arizona, astronomer John S. Lewis believes this Biblical event involved cosmic bombardment. He wrote: "Reports of persons being injured or killed by meteorites are uncommon. The account in *Joshua 10:11*, dating from about 1420 B.C., may be the oldest."[30] Joshua 10 tells how Joshua went to fight against the five Amorite kings and their armies gathered at Gibeon for battle. Then *Joshua 10:11-13* says:

> And it came to pass as they (the Amorites) fled from before Israel . . . that the Lord cast down **great stones** from heaven upon them . . . and they died; they were more which died **with hailstones** than they whom the children Israel slew with the sword . . . And the sun stood still and the moon stayed, until the people had avenged themselves upon their enemies . . . So the sun stood still in the midst of heaven and hasted not to go down about a whole day.
>
> Joshua 10:11-13

This seems to be a terrifying, yet accurate description of great or large hailstones coming down from heaven killing people. These verses speak of rocky and icy debris from a close encounter with a comet's dust laden tail, or a "cloud" of comet debris and the bombardment that followed. That the sun stood still a day speaks of this local bombardment being part of a much larger bombardment event that interfered with the rotation of the Earth.

In *Joshua 10:11 NIV* "large hailstones" are referred to, however, the Hebrew word translated as "large" (*gawdol* #1419 in *Strong's Concordance*) can just as well be translated as "great" or "heavy," where "great" or heavy hailstones are a reference to hailstones (debris) from comets. "Great, heavy, or large hailstones" are also referred to in *Exodus 9:18* and *24*, *Ezekiel 38:22*, and *Revelation 16:21*. *Exodus 9:18* and *24* speak of the "very heavy hail" mingled with fire that came as the seventh plague of the Exodus.* *Exodus 9:24* and *25* then tell how this hail killed man and beast that were in the field and broke every tree and how there had never been a hail such as this in the land of Egypt. When *Exodus 9:23-24* tell of heavy hail and fire mingled with the very heavy hail, they describe frozen water that contains combustible materials that are on fire. These added insights about what the Bible calls heavy hailstones or heavy hail make it clear that the very heavy, very great, or very large hailstones of *Joshua 10* were not atmospheric (weather derived) hail, but astronomic (cometary derived) hail.

Ancient Near Eastern accounts of comet debris bombardment causing massive damage and loss of human life have already been noted. From the similarities in descriptions, it becomes obvious that

---

*   Recall in the King James translation, *Exodus 9:18* and *24* refer to "very grievous hail" where the Hebrew word translated as "grievous" (*kaw-bade* #3515 in *Strong's Concordance*) primarily means "heavy," but can be translated as "heavy" or "great" or "large." For example, *Exodus 9:18* in the NAS refers to "very heavy hail."

these ancient accounts firmly link the events spoken of in *Joshua 10:11-13* to bombardment by comet debris. In works such as the *"Epic of Gilgamesh"* and *"Lamentation Over the Destruction of Sumer and Ur,"* (presented in Chapter 2), we read about hailstones and flames, and great stones falling one after another with great thuds, which left the land "shattered like a pot," left people with "crushed heads," and created large pits or craters into which one hundred to two hundred men fell. [31]

In the ancient literature of the Near East there were two different types of "hail" (*barad* #1259), one astronomic (or meteoric) and one atmospheric; just as in the ancient Near East there were two types of "lightning" (*barak* - "a flashing," #1300 in *Strong's Concordance*). These are the "lightning" or "flashing forth" of an atmospheric electrical discharge and the "lightning" or "flashing forth" of a meteorite or a piece of comet debris.[*] Context determines what types of "lightning" or "hail" is being referred to. For example, *Ezekiel 38:22* refers to "great hailstones, fire and brimstone" falling upon men, where the words – "fire and brimstone (burning sulfur)" clearly indicate hailstones or hail not derived from the atmosphere but derived from comet debris.

How great is great, or how large is large, or how heavy is heavy? In *Revelation 16:21* reference is made to "**a great hail out of heaven**, every stone about the weight of a talent" falling upon men. Here is specific information about what the Bible (*Exodus 9:18* and *24*, *Joshua 10:11 NIV*, *Ezekiel 38:22*, and *Revelation 16:21*) calls great, large or heavy hailstones or hail. When *Revelation 16:21* says

---

[*]   Note that in the words "hail" or "lightning" as used in the ancient Near East, the emphasis is on the description of the phenomena, not the science behind the phenomena.

that each stone weighed about the weight of a talent, it means that each stone would be about 100 pounds. Hailstones that weigh 100 pounds would certainly kill men and cattle and break trees as spoken of in *Exodus 9:25*.* A hailstone weighing a hundred pounds would represent approximately 12 gallons of water frozen into a sphere almost 18 inches in diameter. A basketball is a sphere approximately 10.5 inches in diameter, so a hundred pound hailstone would be close in size to a beach ball or a twenty inch television set. Bombardment by great hailstones would represent very, very large balls of ice coming in at the speeds of bullets or faster, smashing whatever they hit and sending out icy shards as they shatter upon impact. When we realize that each stone represents 12 gallons of water we can see why the Bible and the literature of the ancient Near East often associates bombardment by great hailstones with "overflowing waters" (for example, *Isaiah 28:2, 17, 30:30 NAS, Ezekiel 13:11, 13, 38:22*; *Lamentation Over the Destruction of Sumer and Ur*, Kramer, line 224, *Lamentation Over the Destruction of Ur*, Kramer, line 174, and *The Curse of Agade*, Kramer, line 149).

In addition to the bombardment by comet debris that is referred to in *Joshua 10:11-13*, there is reference to the phenomena of the Sun and the Moon standing still. This unusual phenomenon seems to be related to the bombardment by comet debris. The 'sun and the moon standing' still speaks of a comet (probably the same comet that brought the bombardment by comet debris) impacting the Earth at just the right angle and direction and force to temporarily

---

*   The term "great, large, or heavy hailstones or hail," as used in the Bible, is related to the term "small and large stones," the stereotypical term for "hailstones" or "hail" in the Sumerian literature And it should be noted that in Akkadian one of the words used for "stone," the word *abnum*, can also mean "hail."

interrupt or suspend the normal daily rotation of the Earth. This would of course make the Sun and Moon appear to stand still. Astronomers have already acknowledged that a comet impact can indeed affect the Earth's rotation. Astronomers Clube and Napier in their book, *The Cosmic Winter* write, "An oblique impact will change the rotation of the Earth by a very small amount." [32] Of course, bigger impacts could cause bigger, longer interruptions in the Earth's rotation. Based on recent computer simulations, scientists believe that the Earth's current rate of rotation was established by a "giant impact" when a very large comet or planet like body about 600 miles in diameter slammed into the Earth. The molten debris ejected then accumulated to form the Moon. [33] If a 600-mile in diameter impactor established the Earth's rate of rotation, what would a six mile in diameter or sixty mile in diameter impactor do? What if a large impact took place on the opposite side of the world as Joshua fought and the comet debris that fell during the battle came from this distant impact? Is there an impact crater, possibly like Burckle Crater, deep below the sea that dates to Joshua's time yet to be found?

Recalling Chapter 2 we note that laser studies involving reflectors placed on the Moon showed that the Moon is still rocking on its polar axis from an impact recorded by five Canterbury monks during the twelfth century. One of the Apollo missions photographed a "fresh" thirteen mile in diameter crater with bright rays (indicating recent formation) in the very same area where the monks saw the explosion. Laser observations of the Moon have enabled astronomers to discover that over a three year period, the Moon oscillates or wobbles 15 meters about its polar axis. This is consistent with an impact, caused by a comet or asteroid about a mile in diameter that hit with an energy of about 100,000 megatons, an energy about ten times greater than the combined nuclear arsenals of the world, having taken place 800 years ago. [34]

Drs. L. Mansinha and E. Smylie of the University of Western Ontario in Canada have analyzed data collected by the international time service, the Bureau International de l'Huere in Paris on the Earth's rotation and wobble. They have found that temporary changes in the path of the North pole (the north end of the Earth's rotational axis) closely correlate to the occurrence of earthquakes with a magnitude greater than 7.5 in the Richter scale. [35] Not only was the 9.2 magnitude Indonesian earthquake that caused the tsunami in 2004 large enough to vibrate the whole planet, but Enzo Boschi the head of Italy's National Institute said, "the quake even disturbed the Earth's rotation." [36] Richard Gross, a geophysicist with NASA's Jet Propulsion Laboratory in California theorized the quake "caused the planet to spin 3 microseconds . . . faster and to tilt about an inch on its axis." [37]

If earthquakes greater than 7.5 on the Richter scale, like the 1906 San Francisco earthquake, and the 2004 Indonesian earthquake can cause temporary changes in the path of the Earth's axis of rotation and if the impact of a one mile in diameter comet or asteroid can cause a long term fifteen meter oscillation on the Moon, what could the "*oblique impact*" of a six mile in diameter comet, producing shock waves many hundreds of thousands of times more powerful than the San Francisco earthquake, do to the Earth's rotation? It seems possible that such an impact could temporarily interrupt the rotation of the Earth and make the Sun and the Moon appear to stand still as spoken of in *Joshua 10:11-13*.

Another explanation for the "long day" of *Joshua 10* is offered by astronomer John S. Lewis who writes, "The most direct interpretation is that one entire night was about as bright as day. If we recall reports from England on the night after the Tunguska (Siberia, 1908) fall, or imagine the night sky dominated by a brilliant passing comet, the juxtaposition of these two phenomena makes some physical sense." [38] The point here is not exactly how cosmic bombardment brought about a "long day" but that it could have happened through either axis rotation or through a brightening of the night sky. The most important point is that cosmic bombardment was the ultimate cause, and that some astronomers now feel compelled to explain this Biblical event in terms of astrophysics.

Additional support for the idea of a cosmic impact affecting the Earth's axis of rotation, as in the case of Joshua's long day, is found in *Isaiah 24:18-20*. Here the Bible speaks of even more dramatic effects on the Earth's axis of rotation as a result of cosmic impact. Earlier in this chapter, while discussing the role of comets in the Flood (*Genesis 7:11*), the phrase translated, "the windows of heaven were opened" was better translated to read "those that lie in wait in heaven were loosed" (so that upon entering our atmosphere these comets, "the sources of the great deep were broken up"). A similar Hebrew phase in *Isaiah 24:18* which usually is translated "the windows from on high are open" should also be better translated to read "those that lie in wait on high (in heaven) are loosed." Recall that the phrase "those that lie in wait in heaven" is a reference to the snow and hail, that is, the comets that are stored and reserved in heaven "for times of trouble, for days of battle and war," as spoken of in *Job 38:22-23* (NIV). Now *Isaiah 24:18-20* can be seen clearly to speak of comet bombardment and impacts affecting the Earth's axis of rotation, even causing the Earth to rock back and forth on its axis. *Isaiah 24:18-20* (clearer translation) says:

> . . . those that lie in wait on high (in heaven) are loosed and the foundations of the Earth do shake (as a result of impacts). The Earth is utterly broken down, the Earth is clean dissolved (broken up) the Earth is **moved exceedingly**. The Earth shall reel to and fro like a drunkard, and shall be removed like a cottage.
>
> *Isaiah 24:18-20* (retranslation)

Isaiah 13:13 also talks about the Earth's axis of rotation being affected by comet activity. *Isaiah 13:13* says:

> Therefore I will shake the heavens, and the Earth shall **remove out of her place** (shall be shaken from its place - NAS), in the wrath of the Lord of hosts, and in the day of his fierce anger.[*]
>
> *Isaiah 13:13*

In *Isaiah 13:13* we know that it is comet activity that causes the Earth's axis to move because of the verses that come before. For example, *Isaiah 13:3-5* say:

> I have commanded my sanctified ones (set apart ones, that is, **comets** - *Job 38:22-23*), I have also called my mighty ones (**comets**) for mine anger . . . . They come

---

[*] *Job 9:6* also speaks of the God of the Bible shaking "the Earth out of her place."

from a far country (*place*), from the end of heaven even the Lord and the weapons of his indignation (*comets*), to destroy the whole land.*

Isaiah 13:3-5

(At the end of heaven or the solar system is the heaven of heavens, which contains the host of heaven, that is, the comets of heaven. In other words in an astronomical sense the heaven of heavens is the spherical cloud of comets that survived the solar system, which we now call the Oort Cloud of comets - *Nehemiah 9:6, Psalm 148:4,* and *Deuteronomy 4:19*).

Job 38:22-23 tells how the Lord has reserved comets for times of trouble, just as the Lord used comets to bring about the great catastrophes of the Old Testament. We have clear evidence as to why the Lord's "sanctified ones," "mighty ones," and the "weapons of indignation" are direct references to comets.

Just twenty years ago talk about comets hitting the Earth, affecting the Earth's axis of rotation, and the Earth's stability, would have sounded pretty far fetched to scientists. Now, as explained in more detail in Chapter Seven, scientists have come to believe that the Earth was slammed into and rocked by a very large comet or small heavenly body about 600 miles in diameter that tore material

---

*   In *Isaiah 13:5* the Hebrew word *eh-rets* (#776 in *Strong's Concordance*) that is translated as "country" can also be translated as "land," "world," "region," or "place." This retranslation lets us recognize the "*doublet*" (a Hebrew sentence construction where the same thing is said twice, but in slightly different ways) in *Isaiah 13:5*. We now have "They came from a far place (#776), from the end of heaven." While there are no countries at the end of heaven, that is, the end of the solar system, there are billions of comets at the end of the solar system.

out of the Pacific Ocean basin to form our Moon. If such an impact happened once, it could happen again. The Bible describes a real possibility when it talks about such an impact to come in *Isaiah 24*.

Deborah and Barak's Great Victory

Another catastrophic event in the Old Testament that involved comet activity is the account of Deborah and Barak's great victory over the Canaanites, whose captain was named Sisera (*Judges 4:2, 12-15*, and *5:20*). *Judges 5:20* says:

> They fought from heaven; the stars (***comet debris*** or ***meteorites***) in their courses fought against Sisera.
>
> *Judges 5:20*[*]

This reference to the ***stars in their courses (incoming flight paths) fighting against the captain Sisera*** describes comet debris or meteorites from heaven coming in and hitting the enemy.[†] The declaration of

---

[*]   "From the heavens the stars fought, on their courses they fought against Sisera," *Judges 5:20 NIV*. To keep it consistent with *Judges 5:20* a better-translation of *Judges 4:15* says, "And the Lord discomfited Sisera and all his chariots and all his "host with the ***blowing*** (#6310 in *Strong's Concordance* "mouth," "blow," edge") ***destruction*** (#2719 to #2717 in *Strong's Concordance* – "sword," "destroy") before Baraki; so that Sisera lighted down off his chariot and fled away on his feet."

[†]   The Hebrew word *mes-il-law* (#4546 in *Strong's Concordance*) that is translated as "courses" in *Judges 5:20* relates to "thoroughfare" and it can also be translated as "paths," which it is in the *Septuagint*. Pieces of comet debris coming in on their respective ***incoming flight paths*** (#4546) is what is being referred to.

stars fighting against the enemy refers directly to comets and comet debris or meteorites. Remember in the ancient world, the word "star" could refer to any luminous body in the heavens (a comet, asteroid, meteor, planet or the sun). Identification of the specific object being referred to comes from the context in which the word is used. *Judges 5:20* describes stars from heaven in their courses, or incoming flight paths, fighting against the enemy. The sun or the planets in their courses or orbits couldn't fight against the enemy by striking the Earth, but meteorites or comet debris could. 'Meteorites' or 'comet debris' is the indicated translation, just as 'comet debris from heaven' fighting against the enemy is told of in *Joshua 10:11-13* when it says "they were more which died of hailstones than they who the children of Israel slew with the sword." Just as cometary debris from heaven enabled Joshua to win his great victory, comet debris or meteorites from heaven enabled Deborah and Barak to win their great victory. In the Hebrew language Barak's name means "lightning" (*barak* - "a flashing"). As noted earlier in the ancient Near East there were two types of "lightning" or "flashing," the "lightning" or "flashing forth" from an electrical discharge and the "lightning" or "flashing forth" from a meteorite or a piece of comet debris moving across the sky. Thus, in *Judges 5:20* we have "Barak," or "flashing forth" winning a great battle by means of cometary material "flashing forth."

The Blast That Killed 185,000 Assyrians

Another catastrophic event in the Old Testament that involved cometary activity is the "blast" that killed 185,000 invading Assyrians outside the gates of Jerusalem. The scriptures tell how in

701 BC, during the fourteenth year of King Hezekiah's reign, the Assyrians invaded Judah with a great army and camped outside the walled city of Jerusalem poised to take the city (*II Chronicles 32:1-2* and *Isaiah 36:1-2*). Judah's King Hezekiah went into the Temple and declared the day "a day of trouble" (*Isaiah 37:3* and *II Kings 19:3*). Then, via the prophet Isaiah, the God of the Bible tells Hezekiah to not be afraid because "Behold I will send a *blast* upon him, (the King of Assyria, Sennacherib, and his army), and he shall (*get the message*), and shall return to his own land" (*Isaiah 37:7* and *II Kings 19:7*).[*] Then we come to read about how a "Messenger" of the Lord smote or struck down 185,000 Assyrians during the night. *II Kings 19:35-36* (*Isaiah 37:36-37*) says:

> And it came to pass that night, that the angel (*messenger*) of the Lord went out and smote in the camp of the Assyrians an hundred and fourscore and five thousand (185,000); and when they arose early in the morning, behold they were all dead corpses. So

---

[*] The King James says "he shall hear a rumor and shall return"; but the Hebrew word translated as "hear," the Hebrew word *shema* (#8085 in *Strong's Concordance*) can also be translated as "obey, consider, regard, or listen"; and the word translated as "rumor," is the Hebrew word *shem-oo-aw* (#8052 in *Strong's Concordance*) which can also be translated as "notification," to give "he shall regard the notification and shall return"; which in today's jargon would be "he shall get the message and return." This translation is now consistent with verses later in this chapter such as, *Isaiah 37:36-37* (*II Kings 19:35-36*); where we are told that 185,000 men of Sennacherib's army were suddenly killed and Sennacherib returned to Ninevah. In other words, it wasn't a "rumor" that got the King of Assyria, Sennacherib to stop his siege of Jerusalem and return to his land; but the message he got about the danger of attaching Jerusalem, after 185,000 men of his army were suddenly killed.

Sennacherib, King of Assyria departed and went and returned, and dwelt at Ninevah.

II Kings 19:35-36
(Isaiah 37:36-37)

Recall the discussion earlier (in the section on the Exodus) about how the Hebrew word translated "angel" can also mean and be translated as "messenger." Context and similar usage in the Bible must be considered to understand and determine the proper translation.

Clearly in *Isaiah 37:36* (*II Kings 19:35*), the *malawk* (#4397) of the Lord that struck 185,000 Assyrians was a "messenger" not an "angel" and this messenger was cometary. A cometary messenger is consistent with the earlier verse in this chapter (*Isaiah 37:7, II Kings 19:7*) in which the God of the Bible promised to send a "blast." The Hebrew word *ruwach* (#7307 in *Strong's Concordance*) can mean spirit, blast, tempest or wind. So God sent a blast upon Sennacherib that destroyed his army and caused him to return to Assyria.

The God of the Bible's use of a cometary "messenger" (*malawk* #4397) to send a "blast" (*ruwach* #7307) upon Sennacherib is also consistent with *Psalm 104:4 NAS*, which speaking of God says, "He makes the winds or **blasts** (*ruwach* #7307) his messengers (*malawk* #4397), flaming fire his ministers."* Since the same words

---

*   Word #7307 in *Strong Concordance* is the Hebrew word *ruwach*, which can mean "spirit, blast, tempest, or wind." In *Psalm 104:4 NAS* "tempests" and "flaming fire" go together, but "spirits" and "flaming fire" do not go together. Also in *Psalm 104.4* "messengers" and "ministers" go together. So in *Psalm 104:4 NAS* we have a Hebrew "**doublet**," (where the same thing is said in two different ways) involving "tempests" as "messengers, and "flaming fire" as "ministers."

are used in the same ways in *Isaiah 37:7* and *37:36* (*II Kings 19:7* and *19:35*) and *Psalm 104:4 NAS*, we can conclude that in *Isaiah 37* (*II Kings 19*) God used a cometary messenger to produce the blast that caused Sennacherib to return to Assyria. That the *ruwach* (#7307), the "blast" or "tempest" or "wind" referred to in both *Isaiah 37:7* (*II Kings 19:7*) and *Psalm 104:4 NAS* is cometary in origin is also supported by *Psalm 11:6*. *Psalm 11:6* uses the Hebrew word *ruwach* (#7307) when speaking of God. It says:

> Upon the wicked he shall rain snares ('coals of fire' in NAS and 'fiery coals' in NIV), fire, and brimstone (burning sulfur from a comet), and an horrible tempest (*ruwach* #7307; burning wind in NAS, 'scorching wind' in NIV, and 'stormy blast' in the Septuagint), this shall be the portion of their cup.
>
> *Psalm 11:6* *

Psalm 148:8 also uses the Hebrew word *ruwach* (#7307) when speaking of God. It says:

---

*   *Psalm 11:6 KJV* makes reference to a "horrible (#2152) tempest" (#7307), which the NAS translates as "burning wind", the NIV translates as a "scorching wind," and the Hebrew *Tanakh* as a "stormy blast." Since the Hebrew word, *zil-aw-faw* (#2152), which is translated as "horrible," in the KJV can also be translated as "a glow," we can also have reference to a "glowing tempest," "glowing wind" or a "glowing blast." A "glowing blast" seems to best point to the cometary origin of this blast. A comet passing through the atmosphere would appear as a ball of fire and cause the sky to glow as it did during the Tunguska event in 1908. If an impact occurred, the explosion would cause a blindingly brilliant fire ball and scorching wind effects like those that occur after a nuclear explosion.

> Fire and hail *(cometary)*; snow and vapour (*'ice'* in *Septuagint* and 'cloud' in NAS and NIV); stormy (#5591 *schawrah* stormy, tempest, whirlwind) wind (*ruwach* #7307), fulfilling his word ('do his bidding' in NIV and 'executes His command' in the Hebrew *Tanakh*).
>
> *Psalm 148:8*

From *Psalm 104:4 NAS, Psalm 11:6,* and *Psalm 148:8* we see that God uses "tempest like winds or blasts" caused by comet activity as His messengers. We see that these "tempest like winds or blasts" are accompanied by flaming fire, fire, hail, ice, burning coals, and brimstone, all brought by comet activity to "do his bidding and execute his commands." The God of the Bible chooses to "rain" down comets and cometary debris upon the "wicked."* Although this concept may be foreign to many today, in the ancient Near East comets were universally considered to be "messengers" and sometimes the agents of catastrophe from their displeased gods.

Further support that the "blast" sent by God upon Sennacherib (*Isaiah 37:7* and *II Kings 19:7*) is cometary in origin comes from British Astronomers Victor Clube and Bill Napier. In their book *The Cosmic Winter*, Clube and Napier tell of a 'blast from God" described in the ancient Egyptian literature "which left Egypt in state of dire affliction." They determined that this expression "a blast from

---

* Note how the statements in *Psalm 104:4 NAS, Psalm 11:6,* and *Psalm 148:8* about how comets serve God as his ministers, and messengers, by doing his bidding and executing his commands upon the wicked, are consistent with *Isaiah 40:26*, which tells how God calls comets by their names and none fail; *Isaiah 45:12*, which tells how God command comets; and a number of other scriptures that speak of the Lord of Host's command over comets, and the host of heaven. This was discussed in Chapter 3.

God" represented "destruction from the sky." [39] Clube and Napier noted that the ancient Egyptian literature and the Bible used these same words (a "blast from God") in the very same way. They also wrote in their earlier book *The Cosmic Serpent* about Sennacherib's army "allegedly" being destroyed by a "host from heaven," a local catastrophe, which they associate with "the close passage of a great comet." [40]

This blast from a comet could have killed Sennacherib's army in a number of different ways. The strong winds of the blast could have brought death, but something even more deadly may have been involved – poison gas! This blast could have contained poisonous gas and amounted to what we call today chemical warfare. As noted in Chapter 2, a comet passing through the atmosphere would produce deadly nitrous oxide, and since comets contain hydrogen cyanide, they could also bring the deadly gas cyanogen $(CN)_2$. The discovery of cyanogen gas (through spectral analysis) in comets was responsible for alarm in 1910 when the Earth passed through the tail of Halley's comet. [41] Many people in London even sought gas masks and sealed off their windows.

Impact specialist Peter Schultz of Brown University believes that a low angle grazing impact would produce hurricane force winds and **lethal clouds of carbon monoxide,** while ionization of the atmosphere [would] engulf the region with **noxious** NO and NO_2 gases." [42] So here we have another aspect of how a "blast" from a comet can bring poisonous gas.

Added support for the "blast" that killed off Sennacherib's army as told in the Bible comes from a passage in an important Sumerian work that also indicates that cometary activity caused some people to be gassed. The Sumerian work called "*Lamentation*

Over the Destruction of Ur," translated by Samuel Noah Kramer of the University of Pennsylvania, tells how the gods judged and destroyed the city of Ur, letting the city be taken over by Ur's enemies after *first* sending a "great storm of heaven," a "storm" of comet debris that put the city in ruins around 1950 BC. [43] In the context of this "great storm of heaven" or cometary storm that "attacks the land and devours it" and "makes the land tremble and quake," which broke down the city's massive walls and killed most of its inhabitants; it seems some people were gassed so that, "Although they were not drinkers of strong drink, they drooped neck over shoulder . . . standing near the weapons . . . Mothers and fathers who did not leave their houses were overcome by fire; the young lying on their mother's laps" (lines 224-228). [44]

That the "blast" the God of the Bible sent upon Sennacherib (*Isaiah 37:7* and *II Kings 19:7*) caused devastation is consistent with other Biblical passages that speak of devastating "blasts" sent by the God of the Bible. *Exodus 15:8* says, "And with the **blast** of thy nostrils the waters were gathered together . . . ." *Job 4:9* says "By the **blast** of God they perish, and by the **breadth** of his nostrils are they consumed." *Psalm 18:15* (*II Samuel 22:16*) says, "The channels of waters were seen and the foundation of the world were discovered at thy rebuke, O Lord, at the **blast** of the **breadth** of thy nostrils."[*] The "blast" of *Psalm 18:15* is clearly cometary since this verse refers to "the channels of waters" or the seabed being exposed and the "foundation" or the deep layers of the Earth being exposed, (which

---

[*] In *Exodus 15:8* the Hebrew word *ruwach* (#7307) is translated as "blast"; but in *Job 4:9* and *Psalm 18:15* the Hebrew word *nesh-aw-maw* (#5397), which means "blast, wind or spirit" is translated as "blast," while the Hebrew word *ruwach* (#7307) is translated as "breadth"; so it can be seen that in these verses the Hebrew word *ruwach* (#7307) and the Hebrew word *nesh-aw-maw* (#5397) are equivalent to each other.

would occur after big cosmic impacts). This verse is at the center of a number of verses (*Psalm 18:7-15, II Samuel 22:8-16* quoted in Chapter 3) that personify the God of the Bible as a cometary god who bows or traverses the heavens with smoke coming out of His nostrils and fire coming out of His mouth, while sending out hailstones and coals of fire like arrows.

It should now be clear that, according to the Bible's own definition of its words and the patterns of catastrophe in the Bible, that the "blast" and the "messenger" that God sent upon the Assyrians were both cometary. They are one and the same. A "***cometary blast***" was promised and delivered by the "***cometary messenger***" that killed 185,000 Assyrians. In *Isaiah 37:7* (*II Kings 19:7*), the God of the Bible promised to send a "blast" or "wind" or "tempest" (*ruwach* #7307) upon Sennacherib. Then in *Isaiah 37:36* (*II Kings 19:35*), his "messenger" (*malawk* #4397) killed 185,000 Assyrians. These verses, as told, are consistent with the text of *Psalm 104:4 NAS* which when referring to God says, "He makes the ***winds*** (*ruwach* #7307 or ***blast*** or ***tempest***) His ***messengers*** (*malawk* #4397)."

The cometary activity that involved the "messenger" that delivered the "blast," is described as a day of trouble or a day of cometary activity by the Bible. King Hezekiah declared the day that the Assyrians came to make war on Jerusalem was "a day of trouble" (*II Kings 19:3* and *Isaiah 37:3*). This term acts as a label of sorts. The term "a day of trouble" ties the activities of this day to *Job 38:22-23*. *Job 38:22-23 (NIV)* speaks of "times of trouble . . . days of war and battle," and associates such days with cometary activity. These verses also speak of the storehouses of the snow and the storehouses of the hail, which is one of a number of references the Bible makes to the Oort Cloud, that spherical shell like reservoir or storehouse of comets that encloses the solar system. *Job 38:22-23 (NIV)* says:

> Have you entered the storehouses of the snow or seen the storehouses of the hail, which I (have) reserved for ***times of trouble***, for days of war and battle?
>
> *Job 38:22-23 (NIV)*

Job 38:22-23 indicates that the icy comets of the Oort Cloud are stored away and reserved for use during times of trouble, when war and battle are to take place. We have already seen cometary activities play a role in other times of trouble like the seventh plague of the Exodus in the confrontation with Pharaoh (*Exodus 9:18 and 24*), Joshua's victory over the Amorites (*Joshua 10:11-13*), and Deborah and Barak's victory over the Canaanites (*Judges 5: 20*).

*   *   *   *

This account of cometary activity (the "blast," and the "messenger") in Hezekiahs' confrontation with the invading Assyrians, on a day Hezekiah called "a day of trouble," fits the ***pattern*** called for by *Job 38:22-23*. Remember that in times of trouble, war, and battle when the God of the Bible takes action, this action often involves cometary activity of some sort. Since times of catastrophe are also times of trouble, the pattern of comet caused catastrophe called for in *Job 38:22-23* can be seen to be behind all the Biblical stories of catastrophe that have been discussed in the last two chapters. These stories include:

1. Comets that broke up and impacted into the sea to

cause a gargantuan tsunami that brought the waters of Noah's Flood (*Genesis 7:11*).

2. Cometary activity that brought the fire and brimstone that destroyed Sodom and Gomorrah (*Genesis 19:24*).

3. Cometary activity that brought the plagues of blood, boils, great hail mixed with fire, and darkness of the Exodus (*Exodus 7-10*).

4. The comet whose tail represented the pillar of fire and smoke that led Israel out of Egypt during the Exodus and passed between the camps of the Israelites and the Egyptians (*Exodus 13:21* and *14:19*); and the comet whose tail brought the great wind that caused the splitting of the Red Sea during the Exodus, when Pharaoh's army was drowned (*Exodus 14:21-28*).

5. Cometary activity that rained down great stones and heavy hailstones from heaven and caused the Sun and the Moon to stand still, which taken together enabled Joshua to win a great victory over the Amorites (*Joshua 10:11-13*).

6. Cometary activity that brought the debris that fell when the "stars (meteorites) . . . fought from heaven," which enabled Israel led by Deborah and Barak to win a great victory over the Canaanites (*Judges 5:20*).

7. Cometary activity on the "day of trouble" during the reign of Hezekiah that caused a "blast" that killed 185,000 invading Assyrians outside the walls of Jerusalem (*II Kings 19:3, 7, 35-36,* and *Isaiah 37:3, 7, 36-7*).

In Chapter Eight we will see that this pattern, the pattern of icy comets causing catastrophic and dramatic events, also brought about the destruction of the city and temple tower at Babylon as spoken of in *Genesis 11:1-9*. As we have begun to see, the pattern of comets being used as the weapons of God's wrath is also related to the Bible's prophecies of catastrophe that are to come during the great events of the end times.

SIX

The Beginning of the End:
The First Four of the Seven Trumpets and Seven Vials
Of the Book of Revelation

> The central theme is clear and unambiguous: the events described in *Revelation* are of astronomical origin; and describe real physical events, not mere portents or symbols.
>
> John S. Lewis
> *Rain of Iron and Ice* [1]

> Most human activities at the present time are dominated by the same mistaken view of the heavens and it may be some time before we begin to accept that nature has **Armageddon** in store.
>
> S. V. M. Clube
> *The Dynamics of Armageddon* [2]

> And there shall be signs in the Sun, and in the Moon, and in the stars; and upon the Earth distress of nations, with perplexity; the sea and the waves roaring . . . .
>
> *Luke 21:25*

Many people are familiar with the Bible's Exodus story, usually having seen Cecil B. DeMille's 1956 movie version *The Ten Commandments* which depicts a number of terrifying plagues God sent upon ancient Egypt. Even to a generation who loves to watch "disaster" movies, the plagues of the Exodus would have been spectacular to see live. Although some may feel comfortable delegating the Exodus events to myth or fiction, the previous chapter of this book offered scientific explanations for many of the plagues of the Exodus as descriptions of cometary related events. These ancient catastrophes are not just strange stories from the distant past. According to the Bible, the plagues of the Exodus are a clue to Earth's future. The Bible foretells that mankind will experience an even more dramatic demonstration of God's power and an even greater period of natural wonders. In

Micah 7:15-16 (*NIV*) the God of the Bible speaks of Israel's enemies during the end times saying, "As in the days when you came out of Egypt, I will show them (Israel's enemies) my wonders. Nations will see and be ashamed . . . They will lay their hands on their mouths." If the pattern of the natural wonders of the Egyptian Exodus is followed, the coming end times wonders will also involve related natural events caused by cometary activity.

Astronomers have learned a great deal about cometary activity in the last twenty years. Many have recognized and come to acknowledge that some of the events described in the *Book of Revelation* speak of cometary activity. However, as we have seen, cometary activity was not first mentioned or introduced in the *Book of Revelation*. Abraham, for example, came out of a culture (Sumerian/Akkadian in the city state of Ur) that worshipped **cometary** gods. In a particularly telling incident (*Genesis 15:8-18*), the Bible tells how God used cosmic activity to confirm one of His promises to Abraham when He passed a "burning lamp" (meteor) between the animals that had been sacrificed. During this unusual encounter, God told Abraham that his descendants would serve and be afflicted by a nation in a strange land (Egypt) for four hundred years and that He would then judge that nation. In part, the Exodus story tells of this judgment and as shown in Chapter 5, how it involved related natural events caused by cometary activity. Chapter 3 of this book showed how the Bible refers to comets as the weapons of God's wrath. Because comets are considered the weapons of God's wrath and comets caused the disasters of the Exodus, there should be no surprise that the Bible says judgment of the end times will come again by cometary activity.

The descriptions of the end times impact events found in

the *Book of Revelation* are consistent with *Job 38:22-23 (NIV)*. The "storehouses of the snow" and the "storehouses of the hail" are actually the comets of the Oort Cloud the God of the Bible says He has "reserved for times of trouble, for days of battle and war." The Bible indicates that comets will come during the three and a half year great tribulation period of the end times.

In Chapters 4 and 5, we saw that the catastrophic events of the Old Testament were caused by some of the comets reserved for the times of trouble and days of battles and war spoken of in *Job 38:22-23 (NIV)*. For example, the clearer translation of *Genesis 7:11* said that on the day of the Flood "the sources of the great deep (were) broken up (comets), and those that *lie in wait* in heaven (*comets*) were *loosed*." Similarly, the clearer translation of *Isaiah 24:18-20* said that "those that *lie in wait* on high (the comets of *Job 38:22-23*) are *loosed* and the foundations of the Earth do shake (as a result of the impacts)." Based on Noah's Flood, Sodom and Gomorrah, the Exodus, Joshua's great victory over the Amorites, Deborah and Barak's great victory over the Canaanites, and Hezekiah's day of trouble when a blast killed 185,000 Assyrians, Chapters 4 and 5 revealed precedence in the Bible for the comets that are stored and reserved in heaven, the comets of *Job 38:22-23 (NIV)*, being "loosed" (*Genesis 7:11* and *Isaiah 24:18-20*) and coming to Earth during the end times, "the time of Jacob's trouble," (*Jeremiah 30:7*), which will be a "time of trouble . . . a day of battle and war." (*Job 38:22:23*). Actually, *Isaiah 24:18-20* and the verses that follow specifically refer to an important cometary event that is to take place during the end times when "the Lord shall punish . . . the kings of the Earth upon the Earth," and there will be signs involving the Sun and the Moon as a result of the dust raised by impact (*Isaiah 24:21-23*).

According to many passages in the Bible, especially passages in the *Book of Revelation*, the God of the Bible is to send a second and far more deadly and larger-scaled sequence of related plagues than the terrifying plagues that occurred during the Egyptian Exodus. The *Book of Revelation* refers to seven rounds of these end times wonders or plagues as the "Seven Trumpets" and the "Seven Vials" (often translated as "bowls") (*Revelation 8:6-9, 12; 11:15-19;* and *16:1-21*). Each round will consist of one Trumpet and one Vial (Bowl), which occur concurrently and represent two different views of the same disastrous event.* These wonders or plagues are *interrelated* and will involve the Earth undergoing very rare, but natural cosmic impact phenomena over a three and a half year period. During this time, a sequence of seven separate and increasingly severe rounds of bombardment by icy comets coming in from space will occur. Jesus described the last of these bombardments when he referred to the "tribulation" of the end times in *Matthew 24:29* saying "the stars (comets) shall fall from heaven (the solar system)," and that there would be signs involving the Sun and the Moon.†

The "Seven Trumpets" of the *Book of Revelation* tell of the **disastrous effects** of each round of comet bombardment **on the Earth.** While the "Seven Vials" of the *Book of Revelation* tell of the **disastrous effects** of each round of comet bombardment **on mankind.** Thus, the

---

*   While the writing about the Trumpet and the Vial for each event is linear, this doesn't mean that these two different views of the same event occur one after another. Careful analysis of the Trumpets and the Vials shows that the First Trumpet and the First Vial occur **concurrently**; and so on for each of the seven separate rounds of bombardment. In each round the same **cometary messenger** both "sounds" (Trumpets) and "pours" (Vials).

†   As shown in the preceding paragraph, *Isaiah 24:18-23* also speaks of comet bombardment and signs in the Sun and the Moon. Also see *Isaiah 34:4*.

First Trumpet and the First Vial taken together give the full set of effects upon Earth and upon mankind (people) for the first round of comet bombardment. This sequence of seven separate rounds of comet bombardment occurring during a three and a half year period matches what scientists have only recently discovered about cosmic bombardment – that cosmic impacts rather than occurring singly as previously thought, can be bunched in time or "concentrated into brief catastrophic periods in which multiple impacts occur."[3] Recall the 21 separate large fragments (mini comets) from the break-up of Comet Shoemaker-Levy 9 that hit the planet Jupiter over the course of a week in July of 1994.[4] As the world watched on television, this provided the ultimate proof that multiple impacts can and do occur during a short period of time.

What is most important is that the seven rounds of the "Trumpet and Vial" sequence of comet bombardments given in the Bible are specific, detailed and scientifically testable. Remember that these Biblical prophecies contain information about comets that was not known by scientists until long *after* the Bible was written. Although we cannot "test" the events of the future, we can at least test the scientific accuracy of many of the Bible's accompanying details for each round of prophesied comet bombardment against what we already know about the origin, composition, behavior, and impact effects of comets. Statisticians refer to this as pursuing a "correlation statistic." If proven to be accurate, the currently testable components of a prediction or prophecy provide insight into the possibility of the prediction or prophecy itself being accurate. Thus, if the previously unknown testable components of a prediction or prophecy are proven accurate, greater confidence can be placed in the likelihood of the prediction or prophecy coming to pass. For example, one of the Trumpet prophecies describes how fire, smoke and brimstone come out of the comets' mouths. Only recently have astronomers learned that comets have vents or mouths that outgas fire, smoke and brimstone. Does this detail about comets now make the prophecy more believable? What about the other details in this prophecy?

Considering the information on the subject of comets in the Bible explored in the previous chapters, it should be easy to recognize that the Seven Trumpets and the Seven Vials of the *Book of Revelation* tell about catastrophes brought by comets involving different types of impacts. Nevertheless, a number of well meaning Bible scholars have not recognized this in their interpretations of the Seven Trumpets and the Seven Vials. So, before studying the details of the Seven Trumpets and the Seven Vials let us consider some guidelines for interpretation of Bible prophecy as set forth by the Bible itself.

First, the interpretation of the prophecies of the Bible should be based on the Bible's own definitions, terms, rules, and principles of interpretation. This means that the proper definition of words and phrases must be ascertained by determining their meaning in the original Hebrew language. The selected definition should then be consistent with both the usage of these same words and phrases in similar contexts in other Biblical passages and their usage by cultures that were contemporary with the Hebrew culture. Further, interpretations of Biblical prophecy should meet the Bible's own criteria for determining the validity of information. For example, there must be two or three witnesses to establish a matter (that is, other scriptures that support an interpretation); what is prophesied to happen has to have happened before (there is nothing new under the sun – *Ecclesiastes 1:9* and *3:15*) and what is sown affects what is reaped (cause and effect – *Galatians 6:7* and *Hosea 8:7*).

It is standard scientific procedure, when studying any important ancient document to obtain an accurate understanding of the words based on the culture that produced the document and on the ancient historical, geographic, and environmental contexts. This is how anthropologists have come to properly understand what an "informant" from another culture is trying to tell them. Final determination of what is being said in the Bible must come from the Bible itself (*II Timothy 3:16*). Further, the Bible says, "Do not treat prophecies with contempt. Test everything. Hold on to the good" (*I Thessalonians 5:20-21 NIV*). See **NOTE FIVE** for more information about guidelines for evaluating an interpretation of Bible prophecy.

Throughout the centuries there has been a road block made by God Himself to interpretation of the Bible's end times prophesies. The Bible says that certain "words are closed up and sealed till the time of the end" (*Daniel 12:4* and *9*), and that certain prophecies of great catastrophes won't be clearly understood until the last days (*Jeremiah 23:19-20* and *30:23-24 NAS* 'In the last days you will clearly understand it'). On this basis alone, interpretations of certain end times prophecies put forth before the beginning of the end times must also be suspect. Not until the end times will knowledge have increased enough for certain prophecies to be correctly interpreted (*Daniel 12:4*) and understood.

Recent advances in scientific knowledge about comets has made it possible to understand a number of the miraculous and catastrophic events the Bible says took place during Old Testament times. Likewise, this same information about comets brings new understanding about the miraculous and catastrophic events

THE FIRST FOUR TRUMPETS AND VIALS

described in the Trumpet and Vial prophecies. This new information is consistent with a quote from Jesus in *Mark 4:22 NAS* (*Luke 8:17* and *Matthew 10:26*) saying, "For nothing is hidden, except to be revealed; nor has anything been secret, but that it should come to light." We will see that there is both Biblical and scientific precedent for the types of events that the Seven Trumpets and the Seven Vials prophesy to occur during the end times.

With the Bible's criteria for interpretation in mind, we can now look at the Trumpets and Vials of the *Book of Revelation*. The seven interrelated "wonders" or "plagues" call for the Earth to undergo a sequence of seven separate and increasingly severe bombardments by comets over a three and a half year period. In this description of the events of the end times, the Seven Trumpets describe how each round of comet bombardment will affect the Earth. The Seven Vials describe the disastrous effects of these same rounds of comet bombardment on the human race. The Seven Trumpets and the Seven Vials constitute two witnesses to these seven rounds of bombardment. This interrelationship between the Seven Trumpets and the Seven Vials will become clear as we proceed.

As already noted, a period of multiple impacts is consistent with the belief of astronomers such as Clube and Napier who say that bombardments of the Earth by comets and asteroids can be "concentrated into brief catastrophic periods in which multiple impacts occur." [5] In fact, Clube and Napier in their book *The Cosmic Serpent* quoted passages from the Seven Trumpets of the *Book of Revelation* and said that it "is rich in illusion to what we can only see as ***astronomical events*** of the sort we have described (i.e. cometary)."[6] They made it clear that they were not championing the Bible in considering these extracts concerning the Seven Trumpets of the *Book of Revelation*; however they are at the very least "remarking on the clarity of the **astronomical associations.**" [7] Indeed, in their book *The Cosmic Winter* written eight years later, they presented a special table listing the "possible comet/impact elements" for the Seven Trumpets and some of the Vials of the *Book of Revelation*. [8]

Astronomers Clube and Napier are not the only professional astronomers to recognize and admit to the astronomical associations in the *Book of Revelation*. Astronomer Gerrit Verschuur in his book *Impact: The Threat of Comets and Asteroids* referred to the passage from *Revelation 6:13* which speaks of the stars of heaven falling to Earth as a fig tree drops its fruit as a "tantalizing snippet." He also wrote that *Revelation 8* in the Bible "is clearly a description of an impact catastrophe. [9] Astronomer John S. Lewis, in his book *Rain of Iron and Ice: The Very Real Threat of Comet and Asteroid Bombardment*, considers the passage from *Revelation 6:13* (which speaks of the stars falling to Earth as a fig tree drops its fruit) along with extracts from the first five Trumpets of the *Book of Revelation* (*Revelation 8:6* through *9:12*). He says,

> The central theme is clear and unambiguous: the events described in *Revelation* are of **astronomical origin**; and describe real physical events, not mere portents or symbols. Did John (the author of the *Book of Revelation*) somehow know more about impact phenomena than any scientist before the present decade? [10]

Before analyzing the Seven Trumpets and the Seven Vials of the *Book of Revelation* to discover exactly which verses give detailed

THE FIRST FOUR TRUMPETS AND VIALS

scientific information about comets, we should understand *why* the terms "Trumpets" and "Vials" are used to refer to these seven rounds of comet bombardment. What do the terms "Trumpets" and "Vials" have to do with comets?

In the Old Testament the Hebrew people are told to blow trumpets, that is, ram's horns (*shofar #7782* in *Strong's Corcordance*) at times of battle and war. Numbers 10:9 says, "And if ye go to *war* in your land against the enemy that oppresseth you, then ye shall blow an alarm with the ***trumpets . . . .*** * So, the Old Testament says trumpets or ram's horns are to be blown before battle and war, and since the Old Testament specifically associates comet bombardment with times of battle and war, trumpets should be blown before comet bombardments. The term "trumpet" is thus an appropriate herald for comet bombardment. Recall the direct connection between comet bombardment and times of battles and war is made in *Job 38:22-23 (NIV)* which asks, "Have you entered the storehouses of the snow (comets) or seen the storehouses of the hail (ice-comets), which I reserve for times of trouble for ***days of war and battle?***" *Job 38:22-23* asks about the storehouses of comets, which we now call the Kuiper Belt and the Oort Cloud. The association in *Job 38:22-23* between comets and "battle" and "war" is consistent with *Isaiah 13:3-6 (KJV* and *NIV)* which refers to comets as the "Lord's mighty ones." They are the "warriors" and "the host of the battle" which come "from the end of heaven" as the "weapons of his indignation to destroy the whole land." This passage was discussed earlier in Chapter 3. So, according to Biblical protocol, a trumpet should be blown when a comet comes forth to strike the Earth.

This connection between comet activity and a trumpet being

---

*   Also see *Zephaniah 1:14-16* and *Ezekiel 7:14.*

blown is also shown in *Joel 2:1-10* and *Zechariah 9:14-15*. *Joel 2:1* says,

> Blow ye the **trumpet** in Zion and sound an alarm . . . let all the inhabitants of the land tremble: for the day of the Lord cometh (judgment day), for it is nigh at hand.
>
> *Joel 2:1*

Then, verses 3-10 describe cometary activity.[*] *Zechariah 9:14-15* says,

> And the Lord shall be seen over them, and his **arrow** (**cometary material**) shall go forth as the lightning: and the Lord God shall blow the **trumpet**, and shall go with whirlwinds of the south (**comet activity**) . . . they shall devour (conquer), and subdue with **sling stones** (**cometary material**) . . . .
>
> *Zechariah 9:14-15*

Both *Joel 2:1-10* and *Zechariah 9:14-15* tell of comet activity and of trumpets being blown just as the Seven Trumpets of the *Book of Revelation* tell of trumpets being blown and then cometary activity taking place. Based on 1) these specific Old Testament instructions

---

[*] For example, verse 3 of *Joel 2* refers to comet activity by saying "a fire devoureth before them and behind them a flame burneth;" and verse 10 of *Joel 2* tells how "the Earth shall quake before them," "the heavens [solar system] shall tremble," and the sun, Moon and stars darkened.

calling for trumpet blowing before times of battle and war, 2) Old Testament reference to comets being reserved for days of battle and war, and 3) Old Testament verses specifically referring to cometary activity and trumpet blowing; then the prophesied seven rounds of comet bombardment preceded by seven rounds of trumpet blowing in the New Testament's *Book of Revelation* is consistent and appropriately referred to as the "Seven Trumpets."

As the term "trumpet" or ram's horn is related to the instrument that is to be blown before each round of battle via comet bombardment, the term "vial" (King James) or "bowl" (NAS and NIV) is related to the actual comets that will come in with each round of comet bombardment. Based on the actual usage in the Bible, the term "vial" or "bowl" or "basin" is a sort of euphemism or special expression for a comet, an object that brought terror to the ancient world. Since comets basically consist of frozen water enclosed in a rocky crust, they can be perceived of as water bowls or basins. In Chapter 11 we will see that the Bible refers to comets as the "bottles of heaven" or the "water jars of the heavens" (*Job 38:37 KJV, NAS, NIV*). This is also seen in *Job 38:29-30* and *Psalm 147:16-17* where the Hebrew word (#3713) translated as "hoarfrost" is more accurately translated as "bowls" or "basins" based on context. With this translation, *Job 38:29* would now read, "Out of whose womb comes the ice? The bowls or basins of heaven . . ." in reference to comets. Further, the word that is translated as "vial" in the King James Version throughout the 16th Chapter of *Revelation* is the Greek word *"phiale"* (#5357 in *Strong's Concordance*), which means "a broad shallow cup." The NAS and NIV translations of *Revelation 16* translate the word *phiale* as "bowl" instead of "vial." This is consistent with the Old Testament references to comets as the "bottles of heaven," the "water jars of the heavens," and the "bowls or basins of heaven" as noted above. The term "vial" is most often used in this work instead of "bowl," because the King James Version is by far the most often used of the English language Bible translations.

ROUND ONE

The First Trumpet (Shofar)*

> And the seven (cometary) messengers (*aggelos*) which had the seven trumpets prepared themselves to sound. The first (cometary) messenger (*aggelos*) sounded, and there followed hail and fire mingled with blood (death), and they were cast upon the Earth: and the third part of trees was burnt up, and all green grass was burnt up. [11]
>
> *Revelation 8:6-7*

The First Trumpet involves a cometary messenger casting "hail and fire… upon the Earth," so that a third of the trees and all the grass of the area hit are burned up. Since hail and fire are involved, this indicates a small comet or a large comet fragment hitting the land and this impact (or impacts) igniting regional forest fires. **NOTE TWO** and Chapter 5 explained how the word translated "angel" in Hebrew and in Greek (#4397 *malawk* in Hebrew and #32 *aggelos* in Greek in *Strong's Concordance*) are often better translated

---

*   The Hebrew term behind "trumpet" would be *shofar*, meaning a ram's horn, which is blown before a battle.

as "messenger" in Hebrew and Greek, and that in the ancient world "comets" were often referred to as "messengers." *

Remember that the expression "hail and fire" was first used in the Bible in the telling of the Seventh Plague of the Exodus (*Exodus 9:24*). We also saw in Chapter 5 (Joshua's Great Victory) that this "hail and fire," this "very grievous hail" or rather, "very heavy hail" which broke trees and killed man and beast (*Exodus 9:18* and *24-25*), involved cometary activity. It was also shown that the expression "hail and fire" is used not only in the Bible as a reference to cometary activity but also in the Sumerian and the Akkadian literature of the time referring to cometary activity. Sumerian and Akkadian literature contain a number of passages about cometary activity bringing fire that sets the forests and the land ablaze. For example, we saw in the Akkadian "*Epic of Gilgamesh*" that the cometary god, Ninurta, the personification of "*Imdugud,*" a cloud of fiery sling stones (a cometary cloud), appeared and then torches set the land ablaze, and the land shattered like a pot. In the Sumerian "*Lamentation Over the Destruction of Sumer and Ur*" there were "**hailstones and flames**" and large trees were destroyed, and "large stones, one after another, fell with great thuds." The cometary god, Ishkur the Bull of Heaven, opened pits (craters) into which hundreds of men fell, and both Ishkur, and the cometary goddess Inanna, the Queen of Heaven,

---

*   Recall, *Psalm 104:4 NAS* is a Hebrew **doublet** that speaks of comets and says, "He make the winds (#7307 tempests - cometary) His messengers (#4397), flaming fire (cometary) His ministers." And *Psalm 104:4* is quoted in *Hebrews 1:7 (NAS)* as "who makes his angels (#32 *aggelos*, **messengers**) winds, And His ministers a flame of fire." Also, recall that *Psalm 78:49 (NAS)* says, "He sent upon them His burning anger, fury, and indignation, and trouble. A band of destroying **angels** (#4397)"; but in the Hebrew *Tanakh* the last verse of *Psalm 78:49* reads "a band of deadly **messengers**," which makes the verse clearly cometary.

"rain(ed) flaming fire over the land." [12]

The First Trumpet's (*Revelation 8:7*) prophecy that "hail and fire" from cometary activity will cause a 'third part' of the trees and all the grass in the area hit to be burned is scientifically valid. It is now a recognized scientific fact that cometary activity can cause regional forest fires. Indeed, scientists expect that cometary activity such as this would cause widespread forest fires because it has done so in the past. At a number of sites, geologists have found "a huge amount of soot" and traces of a hydrocarbon residue called retene in the iridium rich clay layer of deposits laid down by the cosmic impacts that contributed to the demise of the dinosaurs 65 million years ago. [13] Retene can occur when coniferous trees are burnt when forest fires occur after a comet impact. On the basis of the huge amount of soot found in this boundary deposit, some scientists believe "that there was a global firestorm that burned all of the world's forests." [14] After the asteroid or cometary fragment hit Siberia in 1908, Russian scientist Leonid Kulik who led the Soviet Academy of Science expedition to the site of the explosion in 1927 wrote, "From our observation point no sign of forest can be seen, for everything has been devastated or burned...." [15] It has been estimated that the "Tunguska Event" devastated 2,000 square kilometers of forest "with a central region about half that size set ablaze by the heat." [16]

Most recently, a multi-institutional twenty-six member research team presented a new theory that a comet nearly 2.5 miles across exploded above North America about 12,900 years ago. Their theory was presented at the May 23, 2007 Joint Assembly of the American Geophysical Union in Acapulco, Mexico, and in the May 25, 2007 issue of the British scientific journal *Nature*. These researchers believe that this explosive fiery impact took place just north of the Great Lakes, and that it triggered fires that burned up forests and other vegetation from one end of the continent to the other. The wildfires left behind a distinctive black carbon rich layer of soil that contained charcoal, soot and materials that are characteristic of cosmic impacts. [17] A University of Oregon archeologist said, "This was a massive continental scale, if not global event." [18] The researchers also believe that the explosion set off a 1,000 year long cold spell that ultimately wiped out the prehistoric "Clovis" paleo indian culture and a variety of animals across North America. Woolly mammoths, ground sloths and the ancestors of the modern buffalo were among the animals that became extinct. Michael Abrams of *Discover* magazine wrote:

> The key to the new hypothesis is a thin black layer of black soil found at more than fifty North American sites. In it are grains containing iridium, an element thought to indicate extraterrestrial origins. The sediments also contain metallic and carbon spherules as well as **melted charcoal**, likely the result of forest fires that swept the continent after the impact. [19]

The evidence from the demise of the dinosaurs, Tunguska and North America show that *Revelation 8:7* is scientifically sound when it associates "hail and fire" brought by comet activity with regional forest fires and the burning of the grasses in the area where the impact is to occur.

The Bible provides other scriptures connecting cometary hail with fire equivalent to the hail and fire of the First Trumpet. *Psalm 18:12 and 13*, talk about "hail stones and coals of fire" in a cometary context, and *Psalm 148:8* talks about comets as "Fire and hail . . . . fulfilling his (God's) word." *Psalm 83:9* and *14-15 (NAS)* describe the equivalent of *forest fires* caused by impact of the First Trumpet when the Psalmist says, "Deal with them as . . . with Sisera ('the stars in their courses fought against Sisera' - *Judges 5:20*) . . . . Like the *fire that burns the forest*, And like a flame (flash) that sets the mountains on fire, So pursue them with Thy tempest (#5591 - whirlwind ancient reference to a comet), And terrify them with thy storm (#5492 - whirlwind - cometary)."

The First Vial (Bowl)*

> And the first (cometary) messenger (*aggelos*) went and poured out his vial (cometary bowl) upon the Earth; and there fell ***a noisome and grievous sore upon. . . men . . . .***
>
> *Revelation 16:2*

While the First Trumpet prophesies the First of Seven Rounds or episodes of comet bombardment from the perspective of what happens to the "Earth," this First Vial (Cometary Bowl) prophesies some of the after effects that the first round of comet bombardment will have on mankind. This First Vial tells of a cometary messenger pouring out something upon the Earth that causes "a noisome and

---

* The word translated "vial," in the King James translation, the Greek word *"phiale"* (#5357 in *Strong's Concordance*), means "a broad shallow cup." In the NAS and NIV translations of *Revelation 16*, the word *phiale* is translated as "bowl," as a reference to comets being the "water bowls of heaven."

grievous sore upon men" (KJV) ("ugly and painful sores" - NIV), which relates to the First Trumpet's (*Revelation 8:6-7*) hail, fire, and blood (death) being cast upon the Earth. [20] As the "hail and fire" or cometary material spoken of in the First Trumpet passes through the atmosphere, both the incoming comet and the explosive impact of the comet will release noxious substances into the atmosphere that will fall to Earth and upon mankind (people).

The ammonia, sulfur, nitrous oxide, cyanide, and metal radicals associated with cometary comas and tails and their explosive impacts would indeed, grievously injure man's sensitive skin. Astronomers Chapman and Morrison have analyzed the impacts that "did in" the dinosaurs. They wrote, "Calculations at the Massachusetts Institute of Technology suggest that the blast wave would have initiated chemical reactions in the atmosphere, producing large quantities of nitric acid. . . ." [21]

It is not difficult to imagine that people who have contact with the noxious and poisonous substances that can result from comet bombardment could develop "noisome and grievous sores." This concept is supported by the Sixth and Seventh Plagues of the Exodus. The Sixth Plague (*Exodus 9:8-10*) tells of "small dust" that caused boils with sores to break forth on people and animals, and the very next day, the Seventh Plague (*Exodus 9:18* and *25*) brought large hail with fire (*Exodus 9:18* and *25*). This speaks of a comet breaking up and disintegrating in the atmosphere, with noxious dust raining down upon the Earth one day and big pieces of dust laden ice (hail) coming in the next day.

We can see that not only are the First Trumpet and the First Vial scientifically sound concepts, but they are also directly connected to each other in a cause and effect relationship. *Revelation*

16:2 is scientifically sound when it associates something "poured out" that causes "sores" on humans with the "hail and fire," brought by cometary activity that causes the massive forest fires spoken of in *Revelation 8:7*.

ROUND TWO

The Second Trumpet (*Shofar*)

> And the second (cometary) messenger (*aggelos*) sounded, and as it were ***a great mountain burning with fire was cast into the sea***: and the third part of the sea became blood (death); And the third part of the creatures which were in the sea, and had life, died, and the third part of the ships were destroyed.
>
> *Revelation 8:8-9*[*]

The Second Trumpet involves a cometary messenger, a "great mountain burning with fire" being "cast into the sea" so that a third part of the creatures in the sea die, and a third part of the ships on the sea are destroyed. Here the "great mountain burning with fire" cast into the sea indicates a very large comet fragment or small comet or small asteroid hitting the ocean. This impact will bring death to a third of the creatures in the area of the ocean hit and will destroy a third of the ships in the area of the ocean affected by the impact. The Second Trumpet's (*Revelation 8:8-9*) prophecy of a fiery cosmic

---

[*]   *Revelation 8:8 NIV* ". . . and something like a huge mountain, all ablaze, was thrown into the sea."

THE FIRST FOUR TRUMPETS AND VIALS 213

body hitting the ocean is at least statistically logical since the Earth's oceans make up about three-fourth's of the Earth's surface. The ocean is the most likely place for a cosmic body to hit because the Earth's oceans are a far bigger target than the Earth's land mass. The Second Trumpet's prophesy that a third *part of the creatures* in the sea die, and that a third part of the *ships* will be destroyed is scientifically logical, considering the explosive effects and tsunamis, that a comet or asteroid hitting the ocean would produce. An example of such an ocean impact would be the cosmic body that produced the 112 mile wide Chicxulub Crater in the Gulf of Mexico off of Mexico's Yucatan Peninsula, which helped bring about the demise of the dinosaurs and other forms of life sixty-five million years ago.

    A *Time-Life* reconstruction of a six-mile wide asteroid slamming into the ocean predicted that the impact would produce energy equivalent to the explosion of five billion atom bombs.[22] Though this impact involves a much larger cosmic body than that indicated by the Second Trumpet of the *Book of Revelation*, this *Time-Life* reconstruction still gives us insight into what would happen if a "mountain sized" cosmic body struck the ocean. Basically the same type of events would happen, only on a smaller scale. The effects would be more regional than global. The *Time-Life* reconstruction describes the atmosphere being heated to over 3,000 degrees Fahrenheit and the sea water being heated to thousands of degrees Fahrenheit. It tells of 130 billion trillion gallons of seawater being thrown up into the air, a series of tsunami waves being set in motion, and "an expanding fireball of steam and molten ejecta (that) would level anything in its path within a distance of 1,200 miles." [23]

    If this impact happened in the Gulf of Mexico, as in the reconstruction, it would also trigger a huge earthquake that would, in turn, set in motion a series of tsunamis or tidal waves "nearly as high as the Rocky Mountains and three to four miles wide." [24] Finally, the scientists quoted in *Time-Life* say untold numbers of animals would drown, fish would suffocate, and delicate marine organisms would die from lack of light. [25] If a six mile wide asteroid impacting the ocean would bring about the type of effects described in the *Time-Life* reconstruction, it is not difficult to see how the much smaller ocean impact described by the Second Trumpet of the *Book of Revelation* (*Revelation 8:8-9*) could bring death to a third of the creatures in the sea and destruction to a third of the ships in the area hit. The catastrophe prophesied by the Second Trumpet of the *Book of Revelation* is at least scientifically logical in terms of cause and effect.

The Second Vial (Bowl)

> And the second (cometary) messenger (*aggelos*) poured out his vial (cometary bowl) upon the sea; and it became as the blood of a dead man: and every living soul died in the sea.
>
> *Revelation 16:3*

The Second Vial (Cometary Bowl) prophesies some of the aftereffects of the second round of comet bombardment upon the Earth's population. The Second Vial (*Revelation 16:3*) tells of the sea being hit and becoming "as blood," just as the Second Trumpet (*Revelation 8:8*) tells of the sea becoming blood (death). The Second Vial tells of a cometary messenger pouring something out upon

THE FIRST FOUR TRUMPETS AND VIALS 215

the sea that brings death to every living soul in the area of the sea where the impact occurs. The First Trumpet and the First Vial say the first round of comet bombardment will strike the land. The Second Trumpet and the Second Vial say the second round of comet bombardment will strike the sea.

In *Impact: The Threat of Comets and Asteroids*, astronomer Gerrit Verschuur reconstructs the large Caribbean ocean impact that helped doom the dinosaurs. He tells how this impact blasted "enormous masses of water outward . . . and the resulting tsunamis literally empty the basin, that is the primal Caribbean, while the region down below the Earth's crust shudders and rebounds and sends a splash of molten rock upward."[26] This is another scientific reconstruction that tells how a cometary or asteroid ocean impact could expose the sea floor, just as called for in *Psalm 18:15 (NIV)*. Verschurr's reconstruction even refers to activity in "the region down below the Earth's crust" sending some "molten rock upward," not unlike an event described in *Psalm 18:15 (NIV)* where "the foundations of the Earth (are) laid bare." When the Bible in *Psalm 18:15,* and *Revelation 8:8-9,* and *Revelation 16:3* rcfcr to a cometary ocean impact, the details that the Bible gives about such an impact are scientifically credible. The prophecy of an ocean impact as given in the Second Trumpet and the resulting affect told of in the Second Vial of the *Book of Revelation* is also scientifically possible.

The interpretation of the Second Trumpet and Second Vial meets the criteria for proper Biblical interpretation noted at the beginning of this chapter, just as the case made for the First Trumpet and First Vial does. The sole difference between the First Trumpet and the Second Trumpet and their subsequently related Vials is that the First Trumpet is a land impact and the Second Trumpet is an ocean impact. The Biblical criteria in part calls for that which is prophesied to occur to have already, in essence, occurred. The scientific reconstructions of the ocean impact that produced the Chicxulub Crater in the Gulf of Mexico that killed the dinosaurs shows that such an ocean impact has already occurred.

The criteria set forth in *Ecclesiastes 1:9* says, "that which is done is that which shall be done: and there is no new thing under the sun," and *Ecclesiastes 3:15* says, "that which is to be (prophecy) hath already been." A purely Biblical precedent for the ocean impact called for in the Second Trumpet and Second Vial comes from Noah's Flood and the recently discovered Burckle Crater in the Indian Ocean. As discussed in Chapter 4, the sources of the deep, that is, comets were broken up and those that lie in wait in heaven (comets) were turned loose. This resulted in an ocean impact that produced the eighteen mile wide Burckle Crater and produced a series of towering tsunamis that left chevron shaped deposits around the Indian Ocean. In particular one series or group of tsunamis swept across the Valleys of the Tigris and Euphrates Rivers until they slammed into the mountains of Ararat in Turkey and then fell back and flooding the land.

In addition to the Second Trumpet and Second Vial there are several other Biblical passages that relate to a cometary ocean impact. These verses include the previously mentioned *Psalm 18:15* (*II Samuel 22:16*) which tells of an impact exposing the sea bed, and *Amos 7:4* which says, "the Lord God called to contend by fire (cometary), and it devoured the great deep, and did eat up a part" (an ocean impact that leaves a crater in the seabed). Three other verses are *Nahum 1:4* which says, "the (the Lord) rebuketh the sea, and maketh it dry" (an ocean impact); *Luke 21:25 (NIV)* which tells

THE FIRST FOUR TRUMPETS AND VIALS

how during the end times there will be "perplexity at the roaring and tossing of the sea" (an ocean impact); and *Revelation 18:21* which says, "a mighty angel (cometary - messenger) took up a stone like a great millstone (asteroid or cometary), and cast it into the sea" (an ocean impact).[*]

ROUND THREE

The Third Trumpet (*Shofar*)

And the third (cometary) messenger (*aggelos*) sounded, and there fell a **great star** (***luminous body***) from heaven (the solar system) burning as it were a lamp, and it fell upon the third part of the rivers, and upon the fountains (***sources***) of waters. (11) And the name of the *star* (luminous body - *comet*) is called Wormwood; and the third part of the waters became wormwood, and many men died of the waters, because they were made bitter.

Revelation 8:10 and 11[†]

---

[*] Luke 21:25 NIV says, "There will be signs in the sun, moon, and stars (comets – in the ancient world comets were often referred to as "stars," a term that could be applied to any luminous body in the heavens). On the earth, nations will be in anguish and perplexity at the ***roaring and tossing of the sea***," presumably after an ocean impact.

[†] "Fountains" - #4077 - *pay-gay* in *Strong's Concordance* "source or supply".

In the Third Trumpet (*Revelation 8:10* and *11*), a cometary messenger, a "great star burning as it were a lamp," is to hit one of the Earth's major "sources" of fresh water and thereby poison a third of the Earth's rivers and kill the men who drink these waters. This description of a "great star (luminous body) from heaven burning as it were a lamp" falling upon the "sources of waters," that is, upon lakes, and falling upon the "third part of the rivers" indicates a comet burning brightly as it passes through the Earth's atmosphere to strike some large freshwater lakes and the rivers that flow out of these lakes, so that a third of the Earth's rivers are also affected. Since this great star is named "wormwood" (a reference to a species of plant that grows in the Holy Land, that has a strong bitter taste) and makes "the third part of the waters" bitter, and since the men who drink these waters die, this cometary material is clearly poisonous.

The Third Trumpet's (*Revelation 8:10* and *11*) prophecy of a comet hitting lakes and rivers affecting a third of the Earth's freshwater is geographically and topographically possible. There are individual lakes and associated rivers that contain upwards of twenty percent of the fresh water on the surface of the Earth. If two large watersheds were affected by the comet breaking up into different pieces as it sped toward the Earth, then one-third of the planet's freshwater supply could indeed be poisoned. One comet breaking up to hit several different locations on the Earth would be like Comet Shoemaker-Levy 9 breaking up into 21 fragments as it approached Jupiter in 1994 and then producing the long line of impact sites on Jupiter as the world watched.

Here on the Earth, new dating has shown that the Rochechouart Crater in France was created at the same time as the Manicouagan Crater in Eastern Canada. A third crater in Manitoba,

THE FIRST FOUR TRUMPETS AND VIALS

Canada was formed contemporaneously. All three ancient impact craters (and possibly a fourth and fifth) line up along the same latitude in a chain almost 2,800 miles long. [27] The Third Trumpet's prophecy (*Revelation 8:10* and *11*) that the people who drink these affected waters will die is ecologically possible. Scientists now know that cometary material can contain poisonous chemical radicals such as cyanide (CN) and hydrogen cyanide (HCN). With the heat of impact serving as a catalyst, sodium cyanide (NaCN) and other poisonous chemicals could also form.

In 1908 a known poison, cyanogen gas was detected in the tail of Comet Morehouse. It's interesting that during the 1910 return of Comet Halley it was predicted that the Earth would pass through the comet's tail and "the fear of possible poisonous gases in the tail caused more than a little concern." [28] In 1974, spectral analysis of the light emitted by Comet Kohoutek detected hydrogen cyanide, methyl cyanide, and carbon monoxide, all poisonous to humans. [29] Frozen cyanide, carbon dioxide, and carbon monoxide constitute a measurable percentage of a comet's mass. [30] Comet experts Clube and Napier note that it is now widely recognized that poisons introduced by cosmic dust have had an important role in the extinction of animals. They cite studies that identify nickel and arsenic as toxic agents introduced by impacts. [31] Since cometary material also contains sulfur, great quantities of sulfuric acid could form in the atmosphere and fall in the form of acid rain making the surface waters bitter. Also recall that the nitrogen compounds (nitrous acid and nitric acid) produced in the atmosphere by an impact are also acidic, very toxic, and very bitter; and they produce even greater quantities of acid rain. [32]

Lake Baikal in southern Siberian Russia is a surprising

example of a lake with a river flowing out of it that alone contains twenty percent of the fresh water on the surface of the planet. While Lake Baikal's surface area is only 12,150 square miles, it is the world's deepest lake (maximum depth 5,371 feet) and alone contains twenty percent of all the fresh water on the Earth's surface. Lake Baikal flows into the Yenisey River. [33] Interestingly, the cometary fragment that exploded over Tunguska, Siberia on June 30, 1908 hit in an area less than 500 miles to the north of Lake Baikal.

The Great Lakes, a group of five freshwater lakes (Lake Superior, Lake Michigan, Lake Huron, Lake Erie, and Lake Ontario) in the U.S. and Canada, taken together also hold about twenty percent of the world's fresh surface water. [34] Lake Superior, the largest of the Great Lakes is the world's largest freshwater lake in surface area, at 31,700 square miles. [35] The waters of the Great Lakes ultimately connect to the Saint Lawrence River, the Mississippi River and the Hudson River. Four of the twenty largest cities in the North American lie along the shores of the Great Lakes: Chicago, Detroit, Toronto, and Cleveland. If a comet impact hits the Great Lakes, and if the waters are poisoned, many people could perish.

Among the world's 16 largest lakes are the Aral, Baikal, and Ladoga in Russia, the Great Bear, Great Slave, and Winnipeg in Canada, and the Victoria, Tanganyika, Nyassa, and Chad in Africa. A quick examination of a globe will show that pieces of a single comet speeding toward the Earth could hit more than one of these lakes, especially those in the Northern latitudes. The Great Lakes, Lake Aral, and Lake Baikal all lie close to 45 degrees North latitude. When thinking of the Great Lakes as one possible target for the prophesied comet impacts of the Third Trumpet, it is interesting to note that some geologists believe that the Great Lakes were actually formed by multiple comet or asteroid impacts in an event not unlike the 21 fragments of Comet Shoemaker-Levy 9 hitting Jupiter in 1994. Based on geomorphology and the patterns of ejected boulders in the region, some geologists, such as Jack Szpytman of Grosse Point Woods, Michigan, believe that a very large swarm of meteorites formed the Great Lakes, with the largest of the meteorites helping to form Lake Superior. [36] The idea of the Great Lakes region as the target of previous cosmic impacts is also of interest relative to the Biblical criteria of *Ecclesiastes 1:9* and *3:15*, where "there is no new thing under the sun," and "that which is (prophesied) to be hath already been."

The Third Vial (Bowl)

> "And the third (cometary) messenger (*aggelos*) poured out his vial (cometary bowl) upon the rivers and fountains (sources) of waters; and they became blood (death) . . . thou (God) hast given them blood (death) to drink . . . ."
>
> *Revelation 16:4-6*

The Third Vial (Cometary Bowl) prophesies some of the after effects of the third round of comet bombardment upon mankind. The Third Vial tells of a cometary messenger and of something that is poured out upon the Earth's "sources" of fresh water, the lakes and rivers so that they become "blood" (death), and men are given "blood to drink." This obviously dovetails with the sources of fresh water, in our lakes and rivers being made bitter and poisonous, and those who drink these waters dying, as previously mentioned in the Third Trumpet. Biblical precedent (*Ecclesiastes 1:9* and *3:15*) for the deadly waters prophesied by the Third Trumpet and Third Vial comes from the First Plague of the Exodus (*Exodus 7:17-19*) when cometary activity caused the waters of the Nile River and throughout all the land of Egypt to turn to blood (death) as discussed in Chapter Five.

ROUND FOUR

The Fourth Trumpet (*Shofar*)

And the Fourth (cometary) messenger (*aggelos*) sounded, and the third part of the Sun was smitten, and the third part of the Moon, and the third part of the stars; so that the third part of them was ***darkened***, and the day shone not for a third part of it, and the night likewise.

Revelation 8:12

The Fourth Trumpet (*Revelation 8:12*) involves a cometary messenger darkening a third of the Sun, the Moon, and the stars. The darkening that will obliterate the view of sky indicates a comet or cometary fragment exploding and disintegrating upon entering the Earth's atmosphere (the gaseous envelope that covers the Earth). A very friable or loosely consolidated comet would tend to explode on hitting the Earth's atmosphere because this gaseous envelope is a much denser medium than space. An atmospheric cometary explosion could easily put enough dust particles into the sky to quickly blot out the Sun, the Moon and the stars for a large number of people on the planet. The effects seen would, of course, depend on where you are on the planet in relation to the location of the atmospheric explosion and on the prevailing atmospheric winds.

A number of researchers believe that the huge explosion that took place at Tunguska, Siberia in 1908 was caused by the atmospheric impact of a cometary fragment about 100 meters in diameter. This belief is based on the lack of a definitive crater, the pattern of fallen trees out from the center of the explosion, and the "scorching" of trees on their inward facing sides.[37] Laboratory experimentation and computer modeling by Russian comet experts has led British comet experts Clube and Napier to believe that this break up took place at about five miles above ground level. This is at the upper margin of the troposphere (that is the lower portion of the atmosphere where we live), where clouds form, and weather as we know it takes place.[38] Also recall the new theory that a comet nearly 2.5 miles across exploded in the atmosphere above North America about 12,900 years ago, presented earlier in the discussion of the First Trumpet.

Scientists believe, based on military satellite photography, that airburst explosions of cometary and asteroid fragments are not unusual. Up until 1993, data collected by military satellites "show that the planet is continually bombarded by big meteoroids that explode in blasts the size of atomic detonations" has been kept secret from the public.[39] A report based on this declassified government data published by the University of Arizona in the book, *Hazards Due to Comets and Asteroids*, said that, "From 1974 to 1992, the satellites detected 136 explosions high in the atmosphere, an average of eight a year. The blasts are calculated to have intensities roughly equal to 500 to 15,000 tons of high explosives, or the power of small atomic bombs." [40] The experts who wrote the report also noted "that the detection rate is probability low and the actual bombardment rate might be ten times higher, with eighty or so blasts occurring each year." [41]

The lead author of the report, physicist Dr. Edward Tagliaferri said, "There's many more of these objects impacting the Earth than we previously thought." [42] For example, the report tells of a Defense Department satellite detecting a one kiloton explosion produced by a 100 ton stony asteroid high in the atmosphere above the Pacific Ocean in the fall of 1990. [43] The explosion of this asteroid over the Pacific Ocean was first thought to be a potential nuclear event, until time consuming sophisticated analysis revealed that the exploding object was a small asteroid. The authors of the report felt, "This suggests that developing nations and potential combatants worldwide, with considerably less sophisticated equipment, might potentially misidentify one of these detonations as a nuclear attack and 'retaliate' against the country's most likely aggressor." [44] What would happen if a small comet or asteroid impacted the upper atmosphere and disintegrated over a Middle Eastern country? Would it be mistaken from a nuclear attack?

Eyewitness accounts of the Tunguska object as it crossed the sky say it left a very thick dust trail, and then a huge *fireball* covered a large part of the sky. Afterward, some people said they saw a huge cloud of black smoke appear. If the Tunguska fragment, an object of only 100 meters in diameter (slightly larger than a football field), produced these dramatic effects in the sky, it is not hard to imagine that an object of one to five miles in diameter breaking up as it hits the Earth's atmosphere would put enough dust particles into the upper atmosphere to quickly blot out the Sun, the Moon, and the stars for a third of the globe for a third part of a day and a night, as the Fourth Trumpet prophesies. Biblical precedent (*Ecclesiastes 1:9* and *3:15*) for the darkness caused by cometary activity prophesized in the Fourth Trumpet comes from the Ninth Plague of the Exodus (*Exodus 10:21-23*), when cometary activity caused "thick darkness," "darkness which may be felt" ("a darkness that can be touched" – *Tanakh*, Jewish Translation of the Old Testament) over the land of Egypt, as noted in Chapter 4. The thick darkness that could be felt or touched indicates that cometary dust in the air caused the darkness.

The Fourth Vial (Bowl)

> And the fourth (cometary) messenger (*aggelos*) poured out his vial (cometary bowl) upon the sun, and power was given unto him to **scorch men with fire.** (9) And men were **scorched with great heat,** and blasphemed the name of God, which hath power over these *plagues*, and they repented not to give him glory.
>
> *Revelation 16:8-9*

The Fourth Vial (Cometary Bowl) prophesies some of the effects of the fourth round of comet bombardment on mankind (the people). The Fourth Vial tells of a cometary messenger and of something that is poured out upon the sun so that people are scorched with fire and with great heat. The Fourth Vial obviously connects and relates to the atmospheric impact and explosion of the comet told of in the Fourth Trumpet. The Fourth Vial tells how men (people) will be scorched with fire and great heat as an obvious reference to the fireball and great heat (thermal pulse) that an atmospheric explosion of a comet would produce. The 100 meter Tunguska object, which exploded in the atmosphere over Siberia in 1908, produced a huge fireball and set tens of thousands of acres of forest "ablaze by the heat" and "scorched" trees out to a distance of twenty miles from the epicenter. An object of over one mile in diameter exploding would truly "scorch" many people with "fire" and "heat." [45] Recall in Chapter 2 how one man forty miles away from the Tunguska impact site told how his ears were scorched and how another man at the same locale told how his shirt was nearly fused to his chest by the fierce heat. In addition, entire herds of reindeer were burnt to ashes closer to the focus of the impact.

Earlier in this chapter, when considering the Second Trumpet, a *Time-Life* reconstruction of a six mile wide asteroid hitting the ocean was presented. The *Time-Life* reconstruction spoke of this impact producing energy equivalent to the explosion of five billion atom bombs, the atmosphere being heated to over 3,000 degrees Fahrenheit, and an expanding fireball that would reach out to over 1,200 miles. [46] The atmospheric impact of a small comet, as prophesied by the Fourth Trumpet and Vial would, of course, produce only a small fraction of the energy described by the *Time-Life* reconstruction. This reconstruction shows us the dramatic effects of a comet impacting and exploding in the atmosphere, where the Sun, Moon, and stars are blotted out by cometary dust, and large numbers of people are "scorched" by "fire" from the fireball arising from the explosive impact and "scorched" by the "heat" arising from the explosive impact. Using this model, we can see that the events called for by the Fourth Trumpet and Fourth Vial are scientifically sound.

There is yet another way cometary activity can cause men to be scorched with fire and great heat. A comet impacting and exploding in the atmosphere could blow a temporary hole in the ozone layer of the atmosphere. As noted in Chapter 2, the ozone (O_3) layer of the atmosphere filters out ultraviolet radiation from the sun. Unfiltered radiation can rapidly cause extreme sunburn that leads to cell damage, cataracts, and cancer. With the ozone gone for a period of time, there would be an epidemic of people with painfully scorched skin.

Referring the Biblical criteria of *Ecclesiastes 1:9* and *3:15*, where "there is no new thing under the sun," and "that which is (prophesied) to be hath already been," the atmospheric impact of a comet or asteroid fragment over Tunguska in 1908 clearly serves as a precedent for the atmospheric impact of a comet prophesied in the Fourth Trumpet. The Bible itself seems to provide another precedent for the atmospheric impact of a comet prophesied in the Fourth Trumpet. As discussed in Chapter 4, several eminent astronomers believe that the destruction of Sodom and Gomorrah told in the *Book of Genesis* involved comet activity. *Genesis 19:24* says, "Then the Lord rained upon Sodom and upon Gomorrah brimstone and fire from the Lord out of heaven." Brimstone (burning sulfur) and fire *raining* out of heaven describe the impact and disintegration of a sulfur rich comet in the Earth's atmosphere above these ancient cities. The cometary "blast" sent upon Sennacherib (*Isaiah 37:7* and *II Kings 19:7*), as discussed in Chapter 4, may in part have also involved men being scorched by fire and heat and thus, may present another Biblical precedent for the Fourth Vial.

The First Four Trumpets and Vials prophesied in the *Book of Revelation* describe four different types of cometary impact events. Each of these prophesied cometary events is, at least in principle, scientifically sound in terms of the destructive effects of each different type of impact. In these Four Trumpets and Vials we have:

1. cometary activity, characterized by "hail and fire," that involves a **land impact** which will cause regional forest fires and sores upon men.
2. cometary activity, characterized by a "great mountain burning with fire," that involves an **ocean impact** which will cause tsunamis, destroy many ships, and kill many creatures and people in and upon the sea
3. cometary activity, characterized by a "great star burning as a lamp," that involves a **lake and river impact** which will poison waters and kill people,
4. cometary activity, characterized by the "Sun, Moon, and stars" being partially blotted out by dust particles raised high into the atmosphere, that involves an **atmospheric impact** which will also cause people to be "scorched by fire and by heat."

* * * *

So, in the first four Trumpet and Vial prophecies we have descriptions of a land cometary impact, an ocean cometary impact, a lake-river cometary impact, and an atmospheric cometary impact, along with the scientifically correct destructive after effects that each type of impact will bring. Recall, how astronomer John S. Lewis, after considering the first five Trumpets of the *Book of Revelation* asked, ". . . Did John (the author of the *Book of Revelation*) somehow know more about impact phenomena than any scientists before the present decade?" [47]

Notice the **pattern** behind the catastrophic end times events conveyed by the Bible. The Bible gives information about four different types of impacts: land, ocean, lake river, and atmospheric impacts. In the first four Trumpet and Vials prophecies there seems to be a comprehensive and scientifically sound tutorial on four different types of cometary impact. This tutorial is consistent with the God of the Bible, the Lord of Hosts, the Lord of Comets saying that He knows comets. This scientifically sound tutorial on cometary impacts is consistent with the scientifically sound information the Bible also gives on the origin, composition, and working of comets, which will be the focus of Chapter 11. There is a clear sense of comprehensiveness and completeness to the information in the Bible on the subject of comets. As already noted, *Micah 7:15* tells how during the end times, God will show miracles or wonders like those that occurred when Israel came out of the land of Egypt during the Exodus. And as we have seen three of the plagues of the Exodus have provided precedents for three of the first four Trumpets and Vials. [48] We can only wonder how the other Exodus plagues relate to the last three Trumpets and Vials.

When comprehending the magnitude of what the Bible foretells for the future of Earth, it is difficult to imagine that the situation could get any worse after four rounds of cosmic bombardment, and their massive devastation of land, sea, fresh water and mankind. However, according to the Bible, these four bombardments are just the beginning. Then will come the three woes of the Fifth, Sixth and Seventh Trumpets and the corresponding Vials.

SEVEN

The Three Woes – The Fifth, Sixth and Seventh Trumpets and Vials

. . . impacts may be strongly bunched in time . . . in reality they may be concentrated into brief catastrophic periods in which multiple impacts occur.

Victor Clube and Bill Napier
The Cosmic Winter[1]

Yet once more I will shake not only the Earth, but also the Heavens (Solar System).

Hebrews 12:26 NAS

> And I will shew wonders in the Heavens and in the Earth, *blood (death)*, and fire, and pillars of smoke.
>
> *Joel 2:30*

> For in those days shall be affliction, such as was not from the beginning of the creation which God created unto this time, neither shall be (again). And except that the Lord had shortened those days, no flesh should be saved....
>
> *Mark 13:19-20*

The final three Trumpets and Vials refer to three more rounds of comet impact. The severity of the last three prophesied impacts are distinctly greater in magnitude and destruction than the first four rounds of impact. In the First Trumpet and Vial only a ***third part*** of the trees and the grass are to be burnt up. In the Second Trumpet and Vial, a ***third part*** of the sea, the creatures of the sea, and the ships are to be affected. In the Third Trumpet and Vial only a ***third part*** of the waters are to be poisoned. In the Fourth Trumpet and Vial a ***third part*** of the stars, the Sun, the Moon, and a day are to be blotted out. However, the Fifth, Sixth, and Seventh Trumpet and Vial rounds indicate much larger impact events.

These last three Trumpets have come to be called the "three woes" based on the references to these events in *Revelation 8:13, 9:12, and 11:14*. Not only are the last three rounds of impact prophesied

THE THREE WOES

to be larger than the first four rounds of impact, the sixth round of impact is to be greater than the Fifth round, and the Seventh round is to be greater than the Sixth event. The Seventh round of impact will be so large that it will be felt over the entire planet, so that every city of every nation is shaken (*Revelation 16:19*). We will also see that the Seventh round of impact is to conclude with an event that can be likened to the Earth being subjected to a "***Baptism of Fire***" from which a "new Earth" will emerge (*II Peter 3:10-13*).

ROUND FIVE
The Fifth Trumpet (Shofar)

And the fifth (cometary) messenger (*aggelos*) sounded, and I saw a star (comet) fall from heaven unto the earth: and to him was given the key to the bottomless pit (deep impact crater). (2) And he opened the bottomless pit (deep impact crater): and there arose a smoke out of the pit (crater), as the smoke of a great furnace; and the sun and the air were darkened by reason of the smoke of the pit (crater). (3) And there came out the smoke *locusts* (bacteria) upon the earth: and unto them was given power, as the scorpions of the earth have power. (4) And it was commanded them that they should not hurt the grass of the earth, neither any green thing, neither any tree; but only those men which have not the seal of God in their foreheads. (5) And to them it was given that they should not kill them, but that they should be tormented five months: and their torment was as the torment of a scorpion, when he striketh a man. (6) And in those days shall men seek death, and shall not find it; and shall desire to die, and death shall flee from them. (7) And the shapes of the locusts were like unto horses prepared unto battle; and on their heads were as it were crowns like gold, and their faces were as the faces of men. (8) And they had hair as the hair of women, and their teeth were as the teeth of lions. (9) And they had breastplates, as it were breastplates of iron: and the sound of their wings was as the sound of chariots of many horses running to battle. (10) And they had tails like unto scorpions, and there were stings in their tails: and their power was to hurt men five months. (11) And they had a king over them, which is the angel (messenger) of the bottomless pit (deep impact crater), whose name in the Hebrew tongue is *Abaddon* (destroyer), but in the Greek tongue hath his name *Apollyon* (destroyer). (12) One woe is past; and, behold, there come two woes more hereafter.

Revelation 9:1-12

The Fifth Vial (Bowl)

And the fifth (cometary) messenger (*aggelos*) poured out his vial (cometary bowl) upon the seat of the beast; and his kingdom was full of **darkness**; and they **gnawed their tongues for pain.** And blasphemed the

God of heaven because of their *pains* and their *sores,* and repented not of their deeds.

Revelation 16:10-11

The Fifth Trumpet (*Revelation 9:1-12*) describes a cometary messenger, a "star," a "hairy star," (an ancient and classical description of comets as "stars with a mane like a horse [or lion],") hitting the Earth and creating a bottomless pit. This would create a very deep impact crater, which is called a "bottomless pit," and this messenger (comet) would have the key or rather be the key to opening the bottomless pit. A great column of smoke and dust is to arise from the impact crater and the smoke and dust is to darken the Sun, which is consistent with the Fifth Vial saying that the beast's "kingdom was full of or plunged (NIV) into darkness."* The Fifth Trumpet uses the term "locusts " to refer to what we today call "bacteria," and says that the locusts (bacteria) in this smoke and dust are going to "torment" men for five months. The Fifth Vial (Cometary Bowl) prophesizes some of the aftereffects of the fifth round of comet bombardment upon mankind. For example, the Fifth Vial (*Revelation 16:10-11*) ties to the Fifth Trumpet when it explains that the "torment" of the Fifth Trumpet will be the result of very painful sores on those who came in direct contact with this bacteria laden dust (which is likened to locusts). The Fifth Trumpet goes on to give a biologically feasible description of this bacteria. In this prophecy both the Fifth Trumpet and Vial tell us that the dust of comets can contain bacteria. As

---

*   The Greek word *kap-nos* (#2586 in *Strong's Concordance*) translated as "smoke" that is in *Revelation 9:2* and *3* is of uncertain affinity; but the Hebrew word for "smoke" *awshawn* (#6227 in *Strong's*), which properly stands behind the Greek, can mean smoke, vapor, or dust.

noted in Chapter 2, the possibility of comets carrying bacteria is scientifically testable and under study. Meanwhile, we already know that long before the invention of the microscope, the Fifth Trumpet and Vial made the connection between bacteria and disease.

The Fifth Trumpet's call for a comet impact that opens up a deep crater and raises a column of dust and smoke that darkens the Sun and the air is scientifically valid. However, the prophecy (*Revelation 9:2-11* and *Revelation 16:10-11*) that bacteria laden dust that will cause painful sores on those who came in direct contact with it is, at present, scientifically controversial. Astronomer John S. Lewis in his book *Rain of Iron and Ice*, asks if John, the author of the *Book of Revelation* somehow knew more about impact phenomena than any scientist before the present decade. [2] The questions that now come to mind are: did John also know more about comet composition than any scientist before the present decade, and did John know that some comets carry pathogens such as bacteria and viruses?

In the 1970's, astrophysicist Sir Fred Hoyle, a Nobel Laureate, and a group of other world renown scientists came to believe that comets, asteroids, and meteors can carry bacteria and viruses because of the organic compounds they contain.[*] Hoyle, in a book called *Diseases From Space* (Harper & Row, 1979), speculated that virus laden comet dust has caused the widespread influenza epidemics that have struck Europe at various times since the Middle Ages. Hoyle describes a number of studies done in the 1970's that support the concept of **germs from space**. This relates to more recent studies

---

[*]   This group of other scientists included Francis Crick (Nobel Laureate for the discovery of DNA), Leslie E. Orgel (Salk Institute for Biological Studies), and Sri Lankan astronomer Chandra Wickramasinghe of Cardiff University.

THE THREE WOES

and to NASA scientists who on August 7, 1996 announced that they might have found bacteria in a meteorite from Mars.* More importantly, some scientists are now actively searching for evidence regarding microorganisms in space, and even more are worried about one of our space probes encountering some pathogen and then, upon returning to Earth, accidentally infecting life here. For more detailed information about how comets, asteroids and meteors can carry bacteria see the last chapter of this book, "What the Bible Knew First," item number 11 – How comets can carry bacteria that can cause disease.

All considered, it is significant that the Bible makes a connection in a number of places between comet activity and bacteria/virus caused pestilence. In *Ezekiel 38:22*, "pestilence" is associated with "great hailstones, fire, and brimstone" as brought by comet activity. In *Habakkuk 3:5*, "pestilence" is associated with "burning coals" or meteorites as brought by comets. *I Chronicles 21:12* associates "pestilence" with both the "messenger of the Lord" and the "sword of the Lord," where both phrases are references to comets. In the Bible this pestilence caused the death of 70,000 people

---

*   One study supported the concept of germs from space by noting the similarity between the spectra of a warm dust cloud in space with those of certain bacteria, and with Hoyle's earlier comparative measurements ("Germs in Space," April 1991, *Sky and Telescope*, p. 357). Another study done by J. Mazio Greenburg, an astrophysicist at the University of Leiden in the Netherlands, concluded "a naked cell could survive in space for as many as ten million years if it is protected from radiation by a thin shell of ice" (John Horgan, "In the Beginning," *Scientific American*, February 1991, p. 124). A more recent group of studies show some types of bacteria can survive in Antarctica, and thus, could also survive the cold of outer space. In Antarctica some microbes survive the extremely cold winters by actively freezing and going into a suspended state called "cryobiosis" (Adam Rogers, "Eking Out Life in the Ice," *Newsweek*, July 6, 1998, p. 62.).

of Israel (*I Chronicles 21:14*). Then *I Chronicles 21:16* referring to the "pestilence" the Lord sent upon Israel, tells how King David saw the "messenger of the Lord stand between the Earth and the heaven, having a drawn sword in his hand, stretched out over Jerusalem," which seems to describe a comet with a tail (sword) reaching from heaven to Earth standing in the sky above Jerusalem. Astronomers Clube and Napier in their book *The Cosmic Serpent* refer to this same verse, *I Chronicles 21:16*, as giving a description of a comet.[3]

Finally, recall from Chapter 5, the comet activity where "the messenger of the Lord" produced a "blast" that killed 185,000 invading Assyrians outside the walls of Jerusalem. Chapter 5 considered how this "blast" could have brought poisonous gases, but this blast" may have just as well carried a deadly fast acting plague like disease. In *The Antiquity of the Jews*, the Jewish historian Josephus based on testimony from the Greek historian, Herodotus (c. 490-424 BC) wrote, "God had sent a pestilential distemper" upon these Assyrians.[4] If Josephus and Herodotus are right, then just as cometary activity referred to as "the messenger of the Lord" and "the sword of the Lord" brought deadly pestilence laden dust during the time of David (c. 1000-961), it may be that the "the messenger of the Lord" and the "blast" sent by the Lord brought deadly pestilence laden dust during the time of Hezekiah (c. 715-687 BC). The Fifth Trumpet and Vial which call for a comet bringing pain inducing bacteria or virus find precedent in the cometary activity that brought these agents during the time of David, and possibly the deadly blast during the time of Hezekiah; and probably during the Sixth plague of the Exodus, where dust over the land of Egypt caused boils on man and beast (*Exodus 9:8-12*). The Fifth Trumpet, just like the first four Trumpets, meets the Biblical criteria of *Ecclesiastes 1:9* and *3:15*, where "that which is to be hath already been," and "there is nothing new under the sun."

Precedent for the Fifth Trumpet and Vial not only involves cometary activity having brought pestilence in the past, but also the precedent of a star (comet) falling form heaven to Earth, opening up a bottomless pit (deep impact crater), smoke arising out of the pit (crater), and the Sun and air being darkened by reason of the smoke that arose out of the pit (crater). Direct evidence for such an impact in the Middle East comes from the recently discovered two mile wide Amarah Impact Crater in Southern Iraq that was discussed in Chapter 1. The massive impact would have been equivalent in power to the detonation of dozens of modern nuclear bombs. Even though the impact occurred in a then shallow sea, the crater produced by the impact would have initially been over two miles wide and more than 400 feet deep. A towering mushroom cloud of steam, molten material and dust would have risen up darkening the Sun and sky of the region for days or more. This had to be one of the greatest catastrophic events ever witnessed by mankind. Reference to this huge Amarah impact is found in the literature of the survivors.

The ancient Near Eastern story that best records the account of the Amarah impact is the historiographic poem known as *The Curse of Akkad*. A number of scientists have also come to recognize this connection, and geologist Sharad Master cites the connection between the dust deposits produced by the impact and the fall of the Akkadian Empire. This poem tells about an important historical event that precipitated the sudden collapse of the short lived ancient Akkadian Empire (ca 2360 BC – 2180 BC) situated in modern day Iraq, as result of a great catastrophe. The Akkadian Empire brought together city states from two different cultures who spoke two very different languages, Akkadian and Sumerian. *The Curse of Akkad* was written in Sumerian, the sacred language of the Akkadian Empire. The city of Akkad was the Akkadian Empire's capital city.

The Curse of Akkad explains that as a result of a defiant act on the part of man, divine wrath fell upon the new city. It tells how some of their gods then struck the entire land and caused the fall of the new empire. Most of the gods the Sumerians and Akkadians believed in were personifications of natural phenomena and were cometary in nature. Many of the terms and expressions in this ancient work make it clear that the destruction sent by the gods was caused by cometary activity. For example, *The Curse of Akkad* tells how the city of Akkad was to be "prostrated like a city ravaged by Ishkur." Ishkur, the "fiery wild Bull of Heaven" was a cometary god who could destroy the land by raining down torrents of small and large stones and snort to open pits into which hundreds of men fall, and kill with the tip of his tail. *The Curse of Akkad* and related works tell of "flashing potsherds raining from the sky," "raining dust that rose sky high" and how "many stars were falling from the sky." [5]

Yale University archeologist Harvey Weiss conducted excavations at a number of Iraqi sites from around 2200 BC and confirmed the sudden collapse of the Akkadian Empire, the world's first empire. Weiss' team found large amounts of dust in the layers of their excavations that correspond to the time period when the newly built capital city of Akkad and Akkadian towns of all sizes were being abandoned. The dust deposits Weiss and his team found have come to be recognized by some as having a cometary signature. Weiss told a reporter for the journal *Science*, "For the first time we have identified abrupt climate change (cooling and drought) directly linked to the collapse of a thriving civilization." [6] In a report in

Science about the collapse of civilizations in third millennium Mesopotamia, Weiss and his team detailed how the abrupt climatic change led to "imperial collapse, regional desertion, and large scale population dislocation." [7]   Note that the *Curse of Akkad* tells how after the gods struck "the large field and acres produced no grain," and "the heavy clouds brought not rain. [8]   Dr. Mike Baillie, a professor of paleoecology at Queens University in Belfast says that just a single comet impact large enough to create the Amarah Crater in Iraq, "would have caused a mini nuclear winter (colder temperatures and less rainfall) with failed harvests and famine, bringing down any agriculture based populations . . . ." [9]   After citing how *The Curse of Akkad* and two other cuneiform works attribute the collapse of Akkad to many stars falling from the sky and flaming potsheds raining from the sky, Professor Weiss wrote, "The abrupt climatic change at 2,200 BC, regardless of an improbable impact explanation, situates hemispheric social collapse in a global, but ultimately cosmic context." [10]

The Amarah Impact Crater (see Chapter1) provides irrefutable physical evidence that a cosmic impact of enormous size took place in the Middle East during Biblical times. Like *The Curse of Akkad* and other Sumerian/Akkadian works, the Bible should also make note of this impact, which would have been felt across the entire Middle East. If the Bible makes references to either the physical or historical events associated with the Amarah Crater, and also indicates that God caused the impact to take place, then we would have a Biblical precedent (*Ecclesiastes 1:9* and *3:15*) for several end time prophesies which involve land impacts, especially impacts occurring in the Middle East. (For example, in the Sixth Trumpet and Sixth Vial we will read about four messengers or comets bound for a river in the Middle East – the Euphrates River.)

In the next chapter we will see that the Amarah Impact Crater is the "smoking gun" for an important Bible story from the *Book of Genesis*. This Bible story clearly ties to the sudden collapse of the Akkadian Empire, and as explained earlier, some scientists now attribute this sudden collapse to the Amarah impact. So often overlooked and dismissed, this Bible story contains enough credible information that places the account on solid ground in terms of physical evidence (crater and dust deposits) and in terms of archeological/historical evidence. This solid foundation of evidence is especially important because this Bible story also provides a reason why the Bible prophesies that wrath and judgment will be brought against the powerful empire that will arise during the end times.

As noted, although the Bible in a number of different passages calls for comets bringing bacteria and pestilence in the form of infectious boils, painful sores, and related skin diseases, at present this concept in scientifically unresolved. Taking a closer look at exactly what is said in the Fifth Trumpet is most revealing. The Fifth Trumpet has additional information about comet related pathogens that argues for the Bible being right about comets bringing pestilence. After the Fifth Trumpet (*Revelation 9:1-12*) tells of a comet impact that will occur and the dust that will be raised by this impact, it may in fact be providing us with the description of a specific type of organism that will be found in the dust resulting from the disintegration of this comet.

Amazingly, the Fifth Trumpet gives this biologically sound description of bacteria or viruses **long before** the invention of the microscope in the 17th century permitted scientists to discover these invisible agents. The Fifth Vial adds to the Bible's discourse on micro organisms given by the Fifth Trumpet saying this pathogen causes pestilence in the form of tongue gnawing painful *sores*, which the Fifth Trumpet explains will be long lasting (five months) and produce pain like the bite of a scorpion. It is astonishing that the Fifth Trumpet and Fifth Vial make this connection between bacteria and disease **long before** Louis Pasteur and others in the 19th century discovered diseases are caused by microorganisms. Using the Fifth Trumpet and the Fifth Vial's physical description of bacteria and painful sores that it causes, it may be possible to identify the specific bacteria involved and even the specific disease. Of course, entirely new types of bacteria and diseases may be indicated. Interestingly, the possible bacteria and the disease indicated by the Fifth Trumpet and the Fifth Vial establish that bacteria is already known to cause certain types of diseases, diseases that the Bible connects to comet activity.

A biologically sound description of a specific type of bacterium is found in the Fifth Trumpet in the description given for the "locusts" that will come out of the smoke and dust arising from the impact crater. In other words, the bacteria in the cometary dust are referred to or likened to "locusts" because, like locusts, the bacteria laden dust will go wherever the wind carries it. It is clear that the term "locusts" is used in a symbolic manner, and not literally, because real locusts eat grass, green things, and trees. The Fifth Trumpet (*Revelation 9:4*) specifically says that these "locusts" ***do not*** harm the grass, green things (plants - NIV), and trees. Further, real locusts obviously do not have a king over them. *Proverbs 30:27* correctly states that "the locusts have no king" when referring to the insects, yet the Fifth Trumpet (*Revelation 9:11*) specifically says that these "locust" ***do have*** a king over them. These smoke and dust borne critters can not be literal "locusts" because, after the eighth plague of the Exodus when locusts came and ate every green thing on trees and plants (*Exodus 10:15 NIV*), *Exodus 10:14 NIV* says "Never before had there been such a plague of locusts, nor would there *ever* be again." Also, interpreting "locusts" literally in the Fifth Trumpet would then place the Bible in conflict with itself because it says a great plague of locusts will never come again. The bacteria that rise up out of the smoke and dust of the impact crater are symbolically likened to "locusts" only because, like "locusts," this mix of smoke, dust, and bacteria will go wherever the wind carries it.

It becomes clearer that the "locust" referred to in the Fifth Trumpet are likely bacteria when the physical description given for the locust in the Fifth Trumpet is compared to the physical description of some known types of bacteria. See **NOTE SIX** for a more detailed analysis of the Fifth Trumpet's description of bacteria.

The Fifth Trumpet and Vial prophecies are not the first time the Bible has given a prophecy about pestilence causing skin disease. *Isaiah 3:17* and *24* tell how pestilence brought by comets will bring skin disease that will "scab the crown of the head of the daughters of Zion" and bring "baldness" and "stink" ("putrefaction" - NAS) and "burning." God in *Deuteronomy 28:27* and *35* promised to smite Israel "with the scab (festering sores - NIV) and with the itch . . . that can't be healed, from the sole of thy foot unto the top of thy head," if they became disobedient to his commandments. The King James translation renders, *Isaiah 3:17* and *24* as:

> Therefore the Lord will smite with a *scab* the crown of the head of the daughters of Zion, and the Lord will discover their secret parts . . . (24) And it shall

come to pass, that instead of sweet smell there shall be *stink*; and instead of a girdle a rent; and instead of well set hair *baldness*; and instead of a stomacher a girding of sackcloth; and *burning* instead of beauty.

Isaiah 3:17 and 24

The NIV translation renders *Isaiah 3:17* as, "Therefore the Lord will bring sores on the heads of the women of Zion; the Lord will make their scalps bald."

The Fifth Trumpet concludes in *Revelation 9:11* by saying that these locust (bacteria) had a "king" over them which is the angel (messenger) of the bottomless pit, whose "*name*" in the Hebrew tongue is *Abaddon* (destroyer) but in the Greek tongue his name is *Apollyon* (destroyer). Remember that the Greek word that is often translated into English as "angel" (*aggelos*, #32 in *Strong's Concordance*) can also be translated or rendered as "messenger." Once again in this case it should be "messenger" since the verses describe a "star," that is, a comet falling from Heaven to Earth, which produces the "bottomless pit" or a deep impact crater. The Greek word that is often translated into English "king" (*basileus*, #935 in *Strong's Concordance*) can also be translated as "*controller*," since it comes from a root word relating to the base of power.

It should also be noted that the root word that is often translated in English as "over" (*epi*, #1909 in *Strong's Concordance*) as in "And they had a king *over* them, which is the angel (messenger) of the bottomless pit" is a preposition that can also pertain to "*direction*." With this in mind, *Revelation 9:11* is more accurately translated to convey that these bacteria depicted as "locusts" had a "controller" "directing" them. This controller is the same "messenger" of the bottomless pit or deep crater with the key or means to open the bottomless pit or deep crater referred to in *Revelation 9:1*. The scripture tells us that the name of this messenger (and what it will do) means destroyer in both Hebrew and Greek. In summation, we are told that the direction and movement of the comet brought bacteria is controlled by a cometary messenger (a comet with the appropriate "*name*" of "destroyer") that will produce a deep impact crater.

Isaiah 40:26 says the God of the Bible "calleth them (the "host," "starry host" - NIV, that is, the objects of heaven, and in particular comets) all by *names*," and *Psalm 147:4* says, "He telleth the number of the stars (the luminous bodies of heaven); he calleth them all by their *names* ('he gives names to all of them' - NIV)." In this clearer translation of *Revelation 9:11*, we find a statement that is consistent with the Fifth Trumpet's (*Revelation 9:1-12*) telling of a "named" comet falling from Heaven to the Earth creating a deep impact crater out of which smoke, dust, and bacteria arise. Then, the bacteria contained in the comet are carried along by the smoke and dust as it is blown by the wind, just as migrating locusts are carried along by the wind.

ROUND SIX

The Sixth Trumpet (*Shofar*)

> And the sixth (cometary) messenger (*aggelos*) sounded and I heard a voice . . . (14) Saying to the sixth (cometary) messenger *(aggelos)* which had the trumpet (shofar-ram's horn), **Loose** the *four*

THE THREE WOES

(cometary) messengers (*aggelos*) which are **bound** [for] the great *river Euphrates*. (15) And the *four* (cometary) messengers (*aggelos*) were **loosed**, which were **prepared** for an hour, and a day, and a month, and a year, for to slay the third part of men. (16) And the number of the army of the horsemen were **two hundred thousand thousand** (200 million): and I heard the number of them. (17) And thus I saw the *horses* in the vision, and them that sat on them having *breastplates of fire*, and of *jacinth (blue)*, and *brimstone (burning sulfur)*: and the **heads of the horses** were as the heads of lions: and out of their **mouths issued fire and smoke** and *brimstone* (burning sulfur). (18) By these three was the third part of men killed, by the *fire*, and by the *smoke*, and by the *brimstone*, which issued out of their *mouths*. (19) For their power is in their *mouth*, and in their *tails*: for their *tails* were like unto serpents, and had heads, and with them they do hurt. (20) And the rest of the men which were not killed by these plagues yet repented not of the works of their hands, that they should not worship devils, and idols of gold, and silver and brass, and stone, and of wood: which neither can see, nor hear, nor walk: (21) Neither repented they of their murders, nor of their sorceries, nor of their fornication, nor of their thefts.

Revelation 9:13-21

The Sixth Vial (Bowl)

And the sixth (cometary) messenger (*aggelos*) poured out his vial (cometary bowl) upon the great *river Euphrates*; and the water thereof was dried up, that the way of the kings of the east might be prepared . . . " (14) to gather them to the battle of that great day of God Almighty . . . (16) And he gathered them together into a place called in the Hebrew tongue *Har-Mageddon* (mountain rendezvous of troops).

Revelation 16:12, 14, and 16

The Sixth Trumpet (*Revelation 9:13-21*) and the Sixth Vial (*Revelation 16:12, 14, and 16*) involve four cometary messengers (four comets) probably from the same progenitor comet, and 200 million boulder size cometary fragments traveling in a steam along with these comets, striking the Euphrates River area. (Remember that the Greek word often translated into English as "angel" can also be translated as "messenger" – see Chapter 2, **NOTE TWO**, and Chapter 6 – The First Trumpet.) The impact of these comets in the Euphrates River area is to cause this great river to dry up. Note how the Sixth Trumpet and the Sixth Vial clearly tie together by naming the same specific site, the Euphrates River. The Sixth Trumpet tells how the "great river Euphrates" will be the impact target for the incoming comets, while the Sixth Vial tells how the "great river Euphrates" will dry up as a result of these impacts. The Sixth Vial (Cometary Bowl) prophesizes some of the effects the sixth round of comet bombardment is to have on men and their activities. Also note that the Sixth Trumpet tells how on a certain day, an extraordinary number of men, "the third part of men" will be slain by the comets **bound for** the Euphrates River, while the Sixth Vial tells how on a certain day, men "of the whole world" will be "gathered to the battle." The Sixth Trumpet says that the fire (burning coals), smoke, and brimstone (burning sulfur) that comes out of the "mouths" or vents of these comets along with the break up of the "heads" of these comets and the material in the "tails" of these comets will kill "the third part of men."[*] This Sixth Trumpet prophecy gives a scientifically accurate description for the physical appearance of active or 'outgassing' comets, coming in to strike the Earth, which the Sixth Trumpet likens to "horses" with heads "as the heads of lions," and "mouths" that issue "fire and smoke and brimstone." It also describes the millions of cometary fragments (mini comets) as "horsemen" with breastplates of fire and of jacinth, and brimstone; and comets as "serpents" with "heads" and "tails." *(See Illustration A)*

The Sixth Trumpet's (*Revelation 9:13-21*) prophecy of four comets and 200 million boulder sized cometary fragments hitting the Euphrates River area in one day is at least scientifically possible. While falls of over 100,000 cometary fragments or meteorites have rarely been recorded, and the fall of millions of stones is somewhat conceivable; only recently has the idea of four comets hitting the Earth on the same day become imaginable. A little more than ten years ago, astronomers believed that large cosmic impacts were widely separated in time, and the idea of four comets hitting in one day would have been considered ridiculous.

[*] These four comets most likely would be the products of a larger parent comet that broke up into these four smaller comets, and the 200 million boulder-size cometary fragments travelling with these comets.

Today astronomers know that cosmic impacts can actually be closely bunched in time. In late June of 1975 seismographs left by the Apollo astronauts recorded a swarm of one ton car sized boulders that hit the Moon over a five day period. [11] Multiple impacts also produced the chain of thirteen craters on Ganymede, the largest of Jupiter's Moons. [12] Some scientists believe multiple impacts formed the Great Lakes, [13] and that three or more impacts wiped out the dinosaurs 65 million years ago (as discussed in the previous chapter). There were 21 separate one-half kilometer in size comet fragments from the breakup of Comet Shoemaker-Levy 9 that struck Jupiter during July of 1994, and "many fragments (of the 21 separate fragments) exhibited *tails* as if each were a small comet." [14]

A few weeks after the 1994 Jupiter impacts, Comet Macholz 2 was discovered and identified as a comet comprised of six major fragments or components that all were traveling along the same orbital path. [15] Just as the Sixth Trumpet calls for four comets coming at the same time, Comet Macholz 2 can be recognized as six comets coming at the same time. Since the common orbital path of the six components of Comet Macholz 2 crosses the orbital path of the Earth, we must acknowledge that it is possible for the Earth to be hit by multiple comets in a short period of time by related comets derived from the breakup of a larger comet. Four multiple impacts could occur within a day, as called for in the Sixth Trumpet of the Book of Revelation.

To understand the Sixth Trumpet, it is important to emphasize that these four cometary messengers are four comets that are "**bound for** the great river Euphrates." The comets are **not** "**bound in** or **at** the great river Euphrates," which would make a senseless sentence that is inconsistent with the pattern of discourse about comets presented throughout the Seven Trumpets of the *Book of Revelation*.* That the messengers or comets of the Sixth Trumpet are bound "for" the river Euphrates is further indicated when the Sixth Vial (*Revelation 16:12*) tells how the sixth "messenger poured out his vial (cometary bowl) upon the great river Euphrates and the water thereof was *dried up*, that the *way of* the kings of the east *might be prepared*." Comets hitting the Euphrates River would instantly break up the riverbed and cause the river to dry up; a dry riverbed would indeed prepare the way for invaders to enter Israel from the east.

The Sixth Vial is like the prior five vials in which the "messengers" pour out their vials, their cometary contents, upon the particular areas that are hit in each of the corresponding five trumpets.

1. In the First Vial, the "messenger" (cometary) pours out his vial upon the **Earth,** and the First Trumpet tells of a comet hitting the **Earth.**
2. In the Second Vial, the "messenger" (cometary) pours out his vial upon the *sea*, and the Second Trumpet tells of a comet hitting the *sea*.
3. In the Third Vial, the "messenger" (cometary) pours out his vial upon the **rivers and lakes**, and the Third Trumpet tells of a comet hitting the **rivers and lakes**.
4. In the Fourth Vial, the "messenger" (cometary) pours out his vial upon the ***atmosphere***, and the Fourth Trumpet tells of a comet exploding and disintegrating in the upper ***atmosphere***.
5. In the Fifth Vial, the "messenger" (cometary) pours out

---

* The Greek word *epi* (#1909 in *Strong's Concordance*) which some have translated as "in" or "at" in *Revelation 9:14* is a preposition which can also be translated as "toward" or "for."

his vial upon the **land** and the Fifth Trumpet tells of a comet hitting the **land** and creating a deep crater.
6. In the Sixth Vial, the "messenger" (cometary) pours out his vial upon the **river Euphrates**, and the Sixth Trumpet tells of comets hitting the **river Euphrates**.

Finally, we see that the (cometary) "messenger" in the Seventh Vial pours out his vial into the air which brings *great hail*, *a great earthquake*, and the cities of the *nations* being hit; and the Seventh Trumpet tells of a cometary activity which brings *great hail*, an *earthquake*, and the *nations* of the world being hit. There is an obvious connection or pattern behind the Seven Trumpets and the Seven Vials, the processes of comet impact are at work!*

The Bible is also internally consistent when in the Sixth Trumpet (*Revelation 9:15*) it says, "And the four (cometary messengers) were loosed, which were prepared for an hour, and a day, and a month, and a year . . . ." This description of the appointed time is consistent with a number of earlier passages in scripture about **when** comets come. *Job 38:22-23 (NIV)* says the "storehouses of the snow . . . and the storehouses of the hail," that is the comet storehouses we call the Kuiper Belt, the Hills Cloud, and the Oort Cloud, are reserved "for *times* of trouble, for *days* of war and battle." *Isaiah 40:25-26 (NIV)* says the "Holy One . . . brings out the starry

---

*   There is also conceptual/linguistic consistency behind the various expressions for "comets" used in the Trumpets and Vials. "Messengers" is a reference to "comets" who pour out their "vials" or "bowls" or "basins" which represent a second type of reference to "comets." Remember that the Bible also refers to "comets" as the "bottles of heaven" and the "water jars of heaven" (*Job 38:37 NAS* and *NIV*) and that comets are in fact giant ice balls, or snowballs or water balls whose frozen contents could be "poured out" after they enter the earth's atmosphere and melt.

host (comets) one by one, and calls them (out) each by name." *Isaiah 45:11-12 (NIV)* tells how the "Holy One . . . stretched out the heavens; . . . (and) marshaled their starry hosts (including comets)." *Psalm 147:4* says, "He telleth the number of the stars (the luminous bodies in heaven including comets); he calleth them all by their names."

Impressively, the Sixth Trumpet gives a scientifically accurate description of the physical features of outgassing comets. Some of this information has only recently been discovered, and some of this has only been observed with the close approach of a space ship. The Sixth Trumpet in *Revelation 9:17* says the "horses" had heads "as the heads of lions." This brings to mind the jagged mane that frames a lion's head and is a reasonable description of the silhouette of a comet after it begins to give off gas. Recall that *kometes astares* the Greek term for "comet" means "**hairy** stars." Astronomer John S. Lewis in his book *Rain of Iron and Ice* makes note of "many classical descriptions of comets as 'stars with a mane like a **horse**.'" [16]   *Revelation 9:16-17* also says an "army of horsemen" sat on the "horses." An "army of horsemen" is reasonable imagery for a great multitude of incoming comets and cometary fragments. Comets where "a fire devoureth before them, and behind them a flame burneth" are also described as "horses" and "horsemen" in *Joel 2:3-4*, which adds "The appearance of them is as the appearance of horses, and as horsemen."

When *Revelation 9:17* says these "horsemen had breastplates of fire, and of jacinth, and brimstone," the scripture is providing additional scientific information about the physical appearance of this incoming cometary material. In the "breastplates of fire, and of jacinth, and brimstone (the ancient term for sulfur)," we have a perfect description of an active comet's coma, which a comet develops after it enters the inner solar system from deep space and begins to outgas. This "breastplate" or "coma" represents the enormous round to elliptical shaped *cloud* of gas and dust that forms around the comet's "body," which is then called a "nucleus." This "breastplate" or "coma" or "cloud" is many, many times larger than the comets' body (nucleus). *(See Illustration A)*

If a cometary tail also develops from the outgassing, the cometary "coma" or "cloud" can be referred to as the comet's "head." Spectral analysis of a comet's "breastplate" or "coma" or "cloud" or "head" shows that it contains "brimstone" or burning sulfur, just as *Revelation 9:17* refers to "breastplates ... of brimstone." The color of a comet's cloud is "jacinth" or deep blue, which is the color of sulfur when it burns in oxygen, and *Revelation 9:17* refers to "breastplates of ... jacinth." Fluorescing carbon monoxide ions also color a comet's coma blue. A comet's cloud can have a small inner feature that astronomers call a "central condensation," or "false nucleus." This "false nucleus" has a "fiery" appearance similar to that of a white star seen against the blue background of the cloud, which *Revelation 9:17* refers to as "breastplates of fire." When *Revelation 9:17* says these "horsemen" or comets had "breastplates of fire, and of jacinth and brimstone," it is referring to an active comet's cloud, the sulfur contained in this cloud; the blue appearance of this cloud, which is the color of burning sulfur in oxygen and the fiery or white false nucleus (central condensation) also appearing within this cloud.

The above passage from *Revelation 9:17* gives four separate facts about active comets. *Revelation 9:17* gives yet another scientific fact about active comets, a fact astronomers have only recently discovered. This **quintessential** fact about active comets is that active comets have "mouths" or "vents" in their crust through which they outgas the elements, ions and dust contained in their frozen waters as the comet is warmed by the Sun. In other words, active comets have mouths or vents or holes in their crust out of which issue gas and dust. Astronomers thought that an active comet's entire surface gave off gas and dust until in 1986 the European spacecraft Giotto passed close by Comet Halley and photographed the comet. The photographs showed gas and dust coming out of only a few small holes or vents in Comet Halley's crust. [17]   In the case of Comet Encke, one astronomer writes about it "issuing jets of gas and dust from two active source areas whenever these icy **vents** were exposed to sunlight." [18] The Sixth Trumpet in *Revelation 9:17* gives a specific and accurate scientific fact about active comets when it says, ". . . out of their **mouths** issued fire and smoke and brimstone."

In *Revelation 9:19*, the Sixth Trumpet confirms this scientific fact about active comets having vents or mouths when it says, "For their power is in their **mouths**." Gas rapidly coming out of a small hole constitutes a powerful gas jet, much like the gas jet that powers a rocket and gives the rocket its thrust. A *Sky and Telescope* article about the discoveries that resulted from Comet Halley's visit in 1985-1986 stated "*gas jets* were *unknown* until April 1986." [19] Yet, the Bible correctly referred to active comets' "mouths" and gas jets in the *Book of Revelation* long before astronomers made these discoveries. The Sixth Trumpet of the *Book of Revelation* also accurately recognizes the physics behind the gas and dust jetting out of a comet's vents or mouths, when in *Revelation 9:19* it says "their power is in their mouths."

In addition to recognizing the power in a comet's mouth or vent, the Sixth Trumpet also accurately recognizes the physics behind an active comet's tail, when in *Revelation 9:19* it says that "their power is in . . . their tails" ("their power is in their **mouths** and in their *tails*"). The material in a comet's "tail" is the extension of the material being jetted out of a comet's mouth. Active comets can have both a dust tail and a much longer gas tail. A gas tail can extend from six million miles to sixty million miles behind a comet's head (coma or nucleus). A comet's tail (or tails) whose direction is affected by the solar wind, can act like the rudder of a boat and affect the speed at which a comet travels through space. Astronomers have learned that Comet Encke often didn't arrive according to their orbital calculations because of the atypical movement generated by Comet Encke's tail. [20] The powerful gas jets of material from the comet's tail and the material jetting out of a comet's vents, constitute two sources of power for an active comet. So, when *Revelation 9:19* says, "their power is in their mouths and in their tails," it is an astonishing accurate description of active comets.

Recall that the Sixth Trumpet in *Revelation 9:17* likens an active comet's overall appearance to "horses" with heads "as the heads of lions," *Revelation 9:18* and *19* adds:

> By these three was the third part of men killed, by the fire, and by the smoke, and by the brimstone, which issued out of their ***mouths***. For their power is in their mouth and in their tails: for their ***tails*** were like unto serpents, and had heads, and with them they do hurt.
>
> *Revelation 9:18-19*

This is a familiar image of an active comet with both a "head" and a "tail." The comparison to a serpent is fair because, like a serpent, a comet's "tail" can be curved. An active comet can "do hurt" ("do harm" - NAS) with its "head" and its "tail" by virtue of it impacting the Earth, and by virtue of the fire, smoke, and brimstone that issues from the "mouths" or vents in its head and becoming part of the tail. When *Revelation 9:19* says these comets can "do hurt" with their "heads" and "tails," it is internally consistent with *Revelation 9:18* which says, "By these three was the third part of men killed, by the fire, and by the smoke, and by the brimstone, which issued out of their mouths." "Heads" that "do hurt," and "fire" that kills indicate the deaths that an explosive impact and fireball would cause. "Tails" that do hurt speaks of the deaths that the fiery cometary fragments in the tail would cause. Add to this toll the deaths resulting from the firestorms, earthquakes, tsunamis, and volcanic eruptions set in motion by the four fiery impacts called for in the Sixth Trumpet.

"Smoke" that kills indicates the deaths that the cometary smoke or gas would cause because comets can contain poisonous gases such as nitrogen dioxide (NO_2) and hydrogen cyanide (HCN). "Brimstone" that kills indicates the deaths that would be caused by the sulfur dioxide (SO_2) and toxic acid rain which forms when SO_2 reacts with moisture in the air. It is not difficult to imagine how the cometary bombardment prophesied in the Sixth Trumpet would kill "the third part of men" who encountered it. It is more difficult to imagine how anyone could survive it. In his book *Rain of Iron and Ice - The Very Real Threat of Comet and Asteroid Bombardment*, astronomer John S. Lewis presents ten different cosmic bombardment scenarios. These cometary bombardments were generated by computer simulations of impacts of different types and sizes over a time span of one century, and the staggering death toll that would result is astounding. [21]

Biblical precedent (*Ecclesiastes 1:9* and *3:15*) for the fire, smoke, and brimstone (caused by cometary activity) prophesied in the Sixth Trumpet comes from the destruction of Sodom and Gomorrah (*Genesis 19:24-29*) when cometary activity ("from the Lord") caused fire and brimstone to rain out of heaven and "the smoke of the country" to go "up as the smoke of a furnace" (impact produced mushroom cloud). A Biblical precedent for the massive comet impacts to hit the Near East and the deadly smoke (dust) also prophesied in the Sixth Trumpet comes from the aforementioned Biblical story that ties to the sudden fall of the Akkadian Empire and the two mile wide Amarah Impact Crater that hit the ancient Near East. This Biblical story will be taken up in detail in the next chapter.

As a result of the impacts of the comets described in the Sixth Trumpet, the Sixth Vial says the waters of the Euphrates River will dry up and make it possible for the nations from the East to be able to cross over and gather for "the battle of the great day of God Almighty" at a place called *har-mageddon* in Hebrew. We shall see that in Hebrew *har-mageddon* literally means a "mountain rendezvous of troops." The nations (from the East) that will gather to battle in Israel could follow the traditional route Alexander the Great used and cross the Euphrates riverbed around ancient Thapsacus, Syria (and then enter Northern Israel's mountains from either the south or north of Thapsacus). [22]

In the Sixth Vial we have an army from the east gathering together with other armies for the day of a great battle to take place in Israel. One question that arises is what role will the Seventh Trumpet and Seventh Vial play in this great battle popularly and imprecisely, called Armageddon? (We will see that the Bible tells of even more horrific cometary activity that is to cause an earthquake of global proportions and trigger a volcanic eruption on Mount Hermon in northern Israel, close to where most of the world's armies are gathered together.)

    The author supposes there are some people who will balk at his explanation of the events described in the Trumpets and Vials as being solely caused by cosmic forces. Dozens upon dozens of books have been written giving a variety of explanations for the imagery used for the end times tribulation in the *Book of Revelation*. However, as covered in Chapter 2 neither a theory of Intercontinental Ballistic Missiles nor a theory of an attack of never before seen nightmarish demons provide a Biblically accurate and scientifically sound explanation for the locusts that are shaped like horses wearing crowns in the Fifth Trumpet nor the Sixth Trumpet's 200 million horsemen riding horses with tails like serpents and mouths that issue fire, smoke, and brimstone. Clearly, these descriptions are metaphorical or symbolic, not literal. **It is the information that these metaphorical or symbolic descriptions convey that is literal.** These Biblical descriptions use symbology to help convey scientifically sound information about natural catastrophic phenomena.* Once again a scientific understanding of comets helps one to understand Biblical end times prophecy.

---

*   The messages of the Bibles in these verses are literal, but the types of expressions used in the verses to deliver these literal messages varies. For example, these expressions can be symbolic, literal, or idiomatic. Ancient Hebrew and the languages used by the ancient Sumerians, Akkadians, Babylonians, Assyrians and Egyptians often used symbology and idioms to convey important information. The only 100% literal language is mathematics. The Bible makes comment on this issue in *Psalm 49:4, 78:2, Proverbs 1:6, Hosea 12:10,* and *Matthew 13:34-35.*

ROUND SEVEN

The Seventh Trumpet (Shofar)

And the seventh (cometary) messenger (*aggelos*); and there were great voices in heaven, saying, 'The kingdoms of this world are become the kingdoms of our Lord, and . . . (19) there were lightnings, and voices, and thunderings, and an earthquake, and great hail.

Revelation 11:15-19

The Seventh Vial (Bowl)

And the seventh (cometary) messenger (*aggelos*) poured out his vial (cometary bowl) into the air . . . (18) And there were voices, and thunders, and lightnings; and there was a great earthquake, such as was not since men were upon the earth, so mighty an earthquake (from an impact), and so great (19) And the great city was divided into three parts, and the cities of the nations fell: and great Babylon came into remembrance before God, to give unto her the cup of the wine of the fierceness of his wrath. (20) And every island fled away, and the mountains were not found. (21) And there fell upon men a great hail out of heaven, every stone about the weight of a talent ("about one hundred pounds each" NAS, NIV), and men blasphemed God because of the plague of the hail; for the plague thereof was exceeding great.

Revelation 16:17-21

The Seventh Trumpet (*Revelation 11:15-19*) and the Seventh Vial (*Revelation 16:17-21*) entail a very large comet approaching the Earth and raining down 100 pound hailstones, before it finally slams into the Earth. They tell of a massive impact that will produce the largest earthquake that has occurred since people have been on the Earth. The shaking and ground movement from the earthquake are to be so great that the cities of the nations are to fall, and somehow the islands and mountains of the world are to disappear. Hundreds of faults would occur after an earthquake of this size, and the prophesy specifically tells how the "great city" (Rome?) is to be divided into three parts.

The Seventh Vial prophesizes some of the after effects the seventh round of comet bombardment will have on the planet's surface and on mankind. Note that the Seventh Trumpet and Seventh Vial both call for many of the same events to occur to the Earth and on the human race. They both call for lightning, voices (sounds or blasts), thundering, a great earthquake, and great hail.* However, as it was explained in Chapter 2, no fault exists in the Earth that can generate an earthquake big enough to produce the amount of

---

*   The Greek word *phone*, word #5456 in *Strong's Concordance* which is translated as "noise" can also be translated as "sounds" or "blasts" – NAS "sounds" and NIV "ramblings." Cosmic bodies moving at thousands of miles per hour through the atmosphere typically produce a series of powerful sonic booms or blasts. Also note the editors of the *Time-Life* book on comets and asteroids describe the impact of a six-mile wide asteroid as producing "a stupendous crack of thunder."

shaking and movement called for in the Seventh Vial. Geologically speaking, a fault generated earthquake cannot cause all the cities of the nations to fall. Only one thing can cause this level of global destruction. Only an impact from a large cosmic body *can* produce an earthquake big enough to cause all the cities of the world to fall. A magnitude 14 or greater impact generated earthquake is the only possible explanation.

In 1811, a deep fault caused a magnitude eight earthquake in New Madrid, Missouri. One resident of the then sparsely populated area wrote, "The whole land was moved and waved like the waves of the sea. With the explosion and bursting of the ground, large fissures were formed . . . ."[23] We probably cannot imagine what a magnitude 14 earthquake caused by a massive comet impact would do! Recall the quote cited in Chapter 2 from astronomer Duncan Steel's book *Rogue Asteroids and Doomsday Comets*. Duncan Steel of the Anglo-Australian Observatory wrote:

> It is difficult to overstate the almost unimaginable energy that is released when a massive asteroid or comet hits the Earth. Merely stating that the explosive power is far greater than all the world's nuclear arsenals combined does not properly convey matters. The reader may think that such combined power might simply result in a larger area being flattened than that which a nuclear bomb devastates . . . In fact, the impact of a large asteroid or comet is quite different from that. Were one to land in Southern California, for example, all of Los Angeles along with several kilometers of the rock from the Earth's crust

beneath it would be picked up and largely vaporized, lumps raining down on Hawaii and New York an hour or so later. Not that Honolulu or New York City would be left standing by then. Phenomenal seismic shocks following the impact would have already shaken them flat.

> *Rogue Asteroids and Doomsday Comets: The Search for the Million Megaton Menace That Threatens Life on Earth* by Duncan Steel, Ph.D. [24]

This is a definitive statement by an astronomer that a massive cosmic impact could produce an earthquake with seismic shock waves large enough to shake cities flat, just as the Seventh Vial describes (*Revelation 16:19*). We will see that the impact indicated by the Seventh Trumpet and Vial will be vastly greater than the one Dr. Steel described above. From a planetary perspective, the massive impact indicated would literally rock the Earth on its axis (*Isaiah 13:13,* and *24: 18-20*), and everything on the Earth that could be shaken would be shaken (*Haggai 2:6-7,* and *Hebrews 12:26-27*).*

---

*   Isaiah 13:13 "Therefore I will shake the heavens, and the earth shall remove out of her place, in the wrath of the Lord of hosts, and in the day of his fierce anger." Isaiah 24:18-20 "Behold, the Lord maketh the earth empty, and maketh it waste, and turneth it upside down, and scattereth abroad the inhabitants thereof." (18-20) And it shall come to pass, that he who fleeth from the noise of the fear shall fall into the pit; and he that cometh up out of the midst of the pit shall be taken in the snare: for the windows (those that lie in wait) from on high are open (loosed), and the foundations of the earth do shake. The earth is utterly broken down, the earth is clean dissolved, the earth is moved exceedingly. The earth shall

Now as other scientific explanations for different aspects of the prophecies of the Seventh Trumpet and the Seventh Vial are considered, an obvious question is: what could possibly make the islands and mountains of the world disappear? Is there any known geological mechanisms that could *literally* make all of the islands and mountains of the Earth disappear? The answer is yes! As explained in Chapter 2, geologists and planetary scientists have recently learned that *only* a cosmic impact can cause a "mighty earthquake," such as the one described (or foretold) in the Seventh Trumpet and Vial, an earthquake equivalent of 12 to 15 on the Richter Scale. Only a horrendously large impact can produce the energy needed to make islands and mountains disappear. But, again what is the geophysical or astrophysical mechanism?

If a "giant comet," such as 120 mile wide Chiron, were to hit the Earth, it could *punch* or *penetrate through*, **YES THROUGH**, the Earth's crust and release most of its energy, deep inside the Earth. The shock waves produced would of course, cause a mighty earthquake and would cause the Earth's relatively brittle crust to crack and break up in places far from the impact site. However, the massive amount of heat, that is, the thermal energy released by such an impact deep inside the Earth, would have far more damaging consequences. A 120

---

reel to and fro like a drunkard, and shall be removed like a cottage; and the transgression thereof shall be heavy upon it; and it shall fall, and not rise again." *Haggai 2:6-7* "For thus saith the Lord of hosts; Yet once, it is a little while, and I will shake the heavens, and the earth, and the sea, and the dry land. And I will shake all nations, and the desire of all nations shall come; and I will fill this house with glory, saith the Lord of hosts." *Hebrews 12:26-27* "Whose voice then shook the earth: but now he hath promised, saying, Yet once more I shake not the earth only, but also heaven. And this word, Yet once more, signified the removing of those things that are shaken, as of things that are made, that those things which cannot be shaken may remain."

mile wide projectile coming in at 40,000 miles per hour exploding deep inside of the Earth would suddenly *superheat* the Earth's already hot molten interior and cause the crust above it to literally melt.* The Earth's relatively brittle and thin crust (the thickness averages 18 miles) in one sense, acts as a barrier and serves to insulate us from the Earth's hot molten interior. The crust consists of 12 main pieces or plates, and it is upon these plates that we find the Earth's land mass, and its islands, mountains and valleys. All of these crustal plates ride upon the Earth's 1,800 mile thick hot mantle.

The sudden superheating of the Earth's already molten interior would cause the Earth's thin crust to be heated to the point of it melting or dissolving.† The Earth's crust would also be heated by hot magma from the mantle flooding out lava onto its surface through cracks created by the impact. Picture lava surging out of thousands of volcanoes, and surface cracks. These innumerable magma outflows of hot lava would heat the Earth's crust from above, while the deep impact would heat the Earth's crust from below. The Earth's surface would rapidly melt. As the crust dissolved, all of the islands and mountains of the world (which lie upon the crust) would disappear or become level with the gravity tending geoid. The islands and mountains of the world would melt and disappear or level out just as the Seventh Vial prophesizes. The Earth's surface would become a sea of molten burning rock. Everything that was upon the Earth would be burned up or dissolve. After the Earth's

---

* The three main layers of the Earth are the crust, mantle and core. The Earth's crust of solidified rock is light, and brittle and can be broken. The Earth's crust formed after it's molten interior stopped giving off great amounts of heat and the young atmosphere helped cool the Earth's surface, and a balance of sorts was established.

† The Earth's crust is just 1% of the thickness of the Earth's mantle.

surface cooled down, a new crust would form.* In essence there would now be a "new Earth," an Earth with a brand new surface, that is, a resurfaced Earth. *(See Illustration I)*

As far fetched and incredible as this "mountain melting mega impact" scenario might seem, planetary scientists have recently recognized that it could actually happen. This recognition is based on new discoveries made as a result of data received from our satellites in space. Since a growing number of scientists believe that "giant impacts" have already taken place on the Earth, it is safe to say that an event of this sort could conceivably happen again. Most astonishing is that scripture actually talks about the different types of geological phenomena that would occur in a "giant impact mountain melting" event.

The breakthrough that led scientists to believe a mountain melting mega impact type of event is possible came when geoscientists Ralph Von Frese and Laramie Potts from Ohio State used gravity measurements taken by NASA's Clementine and Lunar Prospector satellites to map the Moon's interior. The Moon, like the Earth has a crust, mantle, and core. To their great surprise they found that a giant cosmic impact on the far side of the Moon, created a depression so deep that it **penetrated** all the way through the Moon's crust and went into the Moon's interior or mantle. They also found that this large impact also drove a portion of the Moon's mantle into the Moon's metallic core, which is about 700 miles below the surface. Dr. Von Frese said, "People don't think of impacts as things that reach all the way to the planet's core." [25] The Moon's interior shows how it absorbed and transferred the blow from this powerful impact.

---

* Many scientists believe that the thin crust of the hot planet Venus is relatively new. They estimate that the planet underwent a rapid global resurfacing event about 300-500 million years ago.

One news headline read, "The Impact Which Went Through the Moon." [26] Von Frese said, "The core bulges, as if core material was pushed in on the far side and pulled out into the mantle on the near side. Above that, an outward facing budge in the mantle, and above that on the Earth facing (near) side of the Moon sits a bulge on the surface." [27]

Considering the way all these features lined up, Von Frese and Potts believe that a large cosmic object hit the far side of the Moon and sent a huge shock wave ***through*** the Moon's core, that traveled all the way to the opposite side of the Moon. On the opposite side of the Moon, this shock wave produced a bulge and caused magma from the Moon's interior to flow out as lava and flood the Moon's surface creating oceans of frozen magma which are called the "mare" or "lunar seas." How this vast quantity of magma got to the Moon's surface had long been a mystery. These lunar seas (obviously a misnomer) of frozen surface magmas or lava and the bulge make up the lunar feature popularly referred to as the "Man on the Moon."* Von Frese and Potts also believe there was an earlier impact this time on the near side of the Moon, that also caused magma to flow out from the Moon's interior onto the surface as lava, opposite to the point of impact: that is, "antipodal" to the point of impact. Von Frese and Potts reported their findings in the journal "Physics of the Earth and Planetary Interiors." [28] Chris Matzner, a professor of astronomy and astrophysics at the University of Toronto, said that big impacts like the one that penetrated the Moon "were common

---

* "The 'man in the Moon' is a collection of dark plains on the Earth-facing side of the Moon, where magma or lava from the Moon's mantle once flowed out onto the surface . . . the dark plains (lunar seas) are a remnant of that early active time – 'a frozen magna ocean.'" *Science Daily*, February 10, 2006, as above.

in the early solar system and represent the end of the planet building process."29

While Von Frese's and Potts' surprising discovery has not yet won over all planetary geologists, the newly discovered features found on the Moon do cast new light on the Earth's history and future. Von Frese and Potts believe that it is likely that Earth experienced similar impacts. It is conceivable that an extremely large comet or asteroid could penetrate the Earth's crust, superheat the Earth's interior, and also force vast quantities of hot magma from the Earth's interior to flow lava out onto the surface on the side of the Earth opposite to the point of impact (antipodal). The net effect of these two different processes (interior heating and surfacing magma flow) would ultimately cause the Earth's crust to melt and dissolve.*

It seems that the Earth has already experienced penetrating impact events that caused the heating and upwelling of massive amounts of magma from its interior. Many scientists believe the volcanoes of the Hawaiian Island chain came from a "hot spot" that was created by an impact similar to the way the big impact that penetrated deep inside the Moon created a hot spot with magma flowing to the surface on the opposite side of the Moon.†

In addition to the Hawaiian "hot spot," a recently discovered crater that lies almost a mile below the ice of Antarctica provides

---

* Rocks at the bottom of the Earth's Crust can range from 1000 to 1500° F. The temperature of the magma in the Earth's mantle increases with depth from 1000° to 7200° F. In general rocks melt at around 2200° F.

† Dr. Von Frese said, "Evidence of impacts here (Earth) is obscured, but there are hot spots like Hawaii. Some hot spots have corresponding hot spots on the opposite side of the Earth. That could be a consequence of the effect." "Kaboom! Ancient Impacts Scarred Moon to Its Core, May Have Created 'Man in the Moon,'" *Science Daily*, February 10, 2006. www.sciencedaily.com/releases/2006/02/060210091105.htm.

further evidence that the Earth experienced a penetrating impact that caused massive amounts of magma from the Earth's interior to rise up into the crust. While smaller in size than the massive lunar impact mapped by Von Frese and Potts, this impact, nevertheless, had a similar one two punch of crust breaking and upward magma movement.

The same geoscientists who discovered the penetrating impact on the Moon, Ralph Von Frese and Laramie Potts, working in collaboration with NASA scientists and scientists from Russia and Korea, recently discovered a crater in the Wilkes Land region of East Antarctica, south of Australia. Gravity fluctuations measured by NASA satellites that looked below Antarctica's ice detected a 200 mile wide "plug" of once molten mantle material that had risen up into the Earth's crust. Such a plug of dense mantle material is called a "mascon," short for "mass concentration." On the Moon, most mascons are believed to be the result of magma from the Moon's interior (mantle) that rose up to fill the basins created by large impacts as lava mare. Radar images of the ground beneath the ice showed that this Antarctic mascon was centered inside a 312 mile wide crater, a crater large enough to hold the state of Ohio! Satellite gravity data and airborne radar data were in agreement. It is clear that the mascon and the crater together provide evidence that another very big impact took place on Earth. Dr. Von Frese said, "If I saw this mascon signal on the Moon, I'd expect to see a crater around it . . . and when we looked at the ice probing airborne radar, there it was. There are at least twenty impact craters this size or larger on the Moon, so it is not surprising to find one here. The active geology of the Earth likely scrubbed its surface clean of many more" via processes of erosion and deposition as well as crustal subsidence at plate boundaries. [30]

The main point of the Wilkes Land Crater in Antarctica is that a large impact caused molten magma to rise up from the Earth's interior to its surface. Now, while lava didn't flow out beyond the mascon and the crater, Von Frese's research team believes that the Wilkes Land Impact might have also caused the flood vulcanization or lava flows on the opposite side of the Earth that produced the Siberian Traps. [31] The Siberian Traps (or trappen, German for steps) involved a gigantic outflow of molten material from the Earth's interior that flooded an area about the size of the lower 48 United States with dark basaltic lava that is thousands of feet in thickness. [32] This was over a million times the amount of material released by the May 18, 1980 eruption of Mount St. Helen, "enough lava to cover the entire Earth to a depth of ten feet." [33] Thus, the Earth's Antarctic impact is not unlike the deep penetrating lunar impact discussed earlier that caused the massive "Man on the Moon" magma flows on the opposite side of the Moon, with surfacing lavas solidifying in the intense cold of a very thin lunar atmosphere.

Dr. Thomas Wagner, the director of the Antarctic Geology and Geophysics program of the National Science Foundation said, "The Wilkes Land discovery is crucial in understanding Earth's history." [34] The thing that is staggering to grasp is that the Wilkes Land impact occurred much earlier and was much larger than the impact that killed the dinosaurs. Dr. Von Frese and Potts believe that it probably caused the "Great Dying," the biggest extinction in the Earth's history. The "Great Dying" occurred during the Permian-Triassic period, when almost ninety percent of life on land and in the ocean was wiped out. The Wilkes Land Crater in Antarctica is more than twice the size of the Chicxulub Crater near Yucatan, Mexico, the crater associated with the impact that killed the dinosaurs. The "Chicxulab" impactor was about six miles wide, while the Wilkes Land impactor is believed to have been about thirty mile wide and thousands of times more massive. Dr. Von Frese and Dr. Potts plan to go to Antarctica to confirm their findings.

It seems that a major part of the energy produced by the Wilkes Land Impact went into breaking up the Earth's crust. Prior to the impact, it is believed that all of the Earth's continents converges in a super continent, that is, a single great land mass, called Gondwanaland, that was surrounded by a single great ocean. A number of scientists suspect that it was the Wilkes Land impact that initiated the breakup of this super continent. They hypothesize that the impact created a rift and began to drive the continents apart. This rift cut through the Wilkes Land Crater, and as it expanded into a rift valley, it pushed Australia northward from Antarctica.

A preliminary study by Dr. Von Frese shows that the biggest volcanoes on Mars (which are like the Earth's Hawaiian volcanoes) are diametrically opposite or antipodal to big impact craters on the opposite side of Mars.[*] Likewise, the largest feature on the surface of highly battered Mercury, the 800 mile wide impact crater called the Caloris Basin is diametrically opposite or antipodal to a large region of broken up terrain (the so-called 'Weird terrain') on the other side of the planet. The impactor or projectile that produced the Caloris Basin was probably over sixty miles wide in size, and it is clear that the shock waves generated traveled around and through the planet and caused the terrain on the opposite side of Mercury to break apart and melt. This impact also caused volcanic activity that

---

[*]   Personal communication from Dr. Ralph Von Frese on September 14, 2006.

resulted in Mercury's "smooth plains."[35] Like the Moon, Mercury essentially has no atmosphere and thus no erosion and deposition to conceal evidence of big impacts. For the Earth on the other hand, plate tectonics and erosion have destroyed most of the evidence for big impacts and the subsequent impact caused volcanic activity.

We have covered what geoscientists Von Frese and Potts found regarding the big impact on the Moon that caused vast quantities of hot magma from the interior to flow out onto the Moon's surface as lava opposite or antipodal to the site of impact. We also considered the discovery of the thirty mile wide Wilkes Land impactor that produced a crater 300 miles in diameter, and conceivably caused the massive Siberian lava flows on the opposite side of the world. What would a 100 mile wide projectile or impactor do to planet Earth? What would a 300 mile diameter impactor do? Would a huge impact penetrating deep into the Earth's interior produce a blow big enough to break up the Earth's crust in a number of places and cause the release of enough magma to turn the entire surface of the Earth into a molten sea of lava? Could it temporarily raise the temperature of the Earth's interior high enough so that the Earth's relatively thin crust (eighteen mile average thickness with a range of three miles to thirty miles) would literally dissolve or melt? Or, could these two possibilities work together to produce the same ultimate effect – the **resurfacing** of planet Earth, just as Venus was resurfaced many millions of years ago? It should also be noted that the surface of the Earth was temporarily made molten after the cosmic impact that led to the formation of the moon from the material ejected by the impact.

In the internet Cambridge Scientific Conference (www.Cambridge Conference.com), a University of Colorado Planetary

THE THREE WOES

scientist asked what would happen if a massive Oort Cloud comet were to strike the Earth and penetrate deep into the planet's interior? After doing the math and physics, he concluded that the Earth would become like Venus and be "***reborn.***" (Venus has a relatively new surface or crust as a result of a very big impact that struck just 300 to 500 million years ago.) He said that the surface of the Earth would become a molten sea of lava that would eventually cool down and form a new crust. [36]

The Bible's Seventh Vial calls for all of the mountains and islands of the world to disappear. Now, we have seen how an exceptionally large comet or asteroid penetrating through the Earth's relatively thin crust and exploding deep in the Earth's interior could produce this event.[*] For a short while, the entire planet would be on fire and engulfed in flames as a result of the impact. The heat produced would be so intense that the Earth would become a flaming crucible, and *everything* on its surface would be burned up. The experts retained at *Time-Life* estimate that an impactor of just six miles in diameter would cause the temperature of the atmosphere to reach about 50,000 degrees Fahrenheit as it was consumed, while the temperature of the oceans would reach about 100,000 degrees before they were vaporized. [37]

In addition to the "mountain melting mega impact" mechanism presented here, there is another way to make all of the islands and mountains of the Earth disappear as called for in the Seventh Vial (*Revelation 1:20*). A "crust blasting impact" could immediately blast away the planet's crust. This new type of geological mechanism has recently been proposed by three teams

---

[*]   Not unlike the "bottomless pit" of *Revelation 9:1-2*, or "the foundations of the world were discovered" of *Psalm 18:15*.

of scientists; one from the Massachusetts Institute of Technology and NASA's Jet Propulsion Laboratory, one from the California Institute of Technology, and one from the University of California - Santa Cruz. [38] These three teams presented separate papers in the June 26, 2008 issue of the British Journal "*Nature*." All three papers explained why the northern hemisphere of the planet Mars looks so dramatically different from the southern hemisphere. The northern hemisphere of Mars is smoother and lower lying than the southern hemisphere, which is pockmarked and filled with ragged highlands. The crust in Mars' lower-lying northern hemisphere is approximately thirty kilometers or sixteen miles less thick than the crust on Mars' southern hemisphere. Based on gravity changes, magnetic differences and surface measurements made by two NASA spacecraft, these scientists all believe that a huge comet or asteroid smashed into Mars about four billion years ago and blasted away most of Mars' northern crust. This collision left a giant elliptical shaped crater or impact basin, similar to what would be expected if a huge comet or asteroid had hit Mars at an angle. [39] This impact basin called the Borealis Basin, measured 5,300 miles across and 6,600 miles long, and it is "the largest such structure thus far identified in the Solar System." [40] The Borealis Basin covers about forty percent of the planet's surface.

"All three teams believe that a single mega impact took place. The California Institute of Technology team refers to such mega impacts as *"planetary scale impacts".* Based on the giant impact basins produced, these scientists now recognize that in addition to Mars' Borealis Basin, "planetary scale impacts" are also responsible for the South Pole-Aitken Basin on the Moon, the Caloris Basin on Mercury, and the Hellas Basin on Mars. [41] According to the calculations of the California Institute of Technology team, "a 1,000 mile wide object travelling at more than 13,000 miles per hour – or twenty-four times faster than a jet liner – would hit Mars at an angle between thirty and sixty degrees. The collision would be equal to seventy-five to 150 trillion megatons of TNT." [42]

It is important to note that this "crust blasting impact" model and the "mountain-melting mega impact" model for making islands and mountains disappear are not mutually exclusive or incompatible. Clearly, an impact big enough to blow off a large part of the Earth's crust, would also suddenly raise the heat in the Earth's interior and cause molten rock to well up. Depending on the size of the impactor and the nature of the impact these two natural geological mechanisms could work together to make all of the islands and mountains of the Earth disappear overnight. Consider that the California of Technology team's paper said:

> Planetary scale impacts penetrate into the mantle. The resulting rarefaction wave completely removes the surrounding crust, which reimpacts elsewhere on the planet or is ejected to space . . . the ejected material is distributed globally. Melt production and distribution are also (like crustal removal) strongly dependent on impact energy, velocity and angle. [43]

In their papers the California Institute of Technology team also said that their simulations "provide new insights pertinent to global scale impact processes thought to prevail in the early Solar System."[44] These simulations dealt with crustal surface melting, crustal removal, and planetary disruption as a function of impact energy and impact angle.[45] Note that crustal melting (*II Peter 3:10-13*), crustal removal (*Revelation 16:18-20*) and planetary disruption (the Earth reeling "to and fro" on its axis – *Isaiah 24:17-21* and *Isaiah 13:13*) are all associated with the huge cometary impact of the Seventh Trumpet and Vial.

Obviously, the impact of a 100 mile wide or more projectile would cause an earthquake that would make planet Earth "reel to and fro like a drunkard" (*Isaiah 24:20*) in space. Before telling of the islands and mountains disappearing, the Seventh Vial says, "There was a great earthquake such as was not since men were upon the Earth, so mighty an earthquake and so great" (*Revelation 16:18*). As previously explained, the Seventh Vial also tells how the shock waves and shaking produced by this impact generated earthquake will cause all of the cities of the nations to fall.

The Seventh Trumpet like the Seventh Vial refers to this same earthquake and indicates that it is to take place on the "**Day of the Lord.**" *Isaiah 13* and *24* also make reference to an earthquake on the day of the Lord, and they tell how the entire planet will be shaken and rocked from the blow it receives on that day. *Isaiah 13:9* and *13* say, "Behold the **day of the Lord** cometh, cruel both with wrath and fierce anger to lay the land desolate . . . (13) Therefore I will shake the heavens, and the **earth shall remove out of her place**, in the wrath of the Lord of hosts, and in the day of his fierce anger." Referring to this same day *Isaiah 24:18-21* says, that comets are loosed, "and the foundations of the Earth do shake. (19) The Earth is broken asunder. The Earth is **split through** (NAS), the Earth is **moved exceedingly**. (20) The Earth shall *reel to and fro* like a drunkard, and **shall be removed like a cottage** . . . (21) . . . in that day that the Lord shall punish the host (armies) of the high ones that are on high (the invading armies of antichrist's empire at Armageddon), and the kings of the earth upon the earth."

The Seventh Vial concludes by reconfirming the cometary nature or cause of the catastrophic events it prophesizes. The Seventh Vial describes 100 pound hailstones falling from heaven, just as great hailstones fell during Joshua's victory over the Amorites at Gibeon (*Joshua 10:5-14*). The deadly fall of cometary hailstones called for by the Seventh Vial would occur before the massive comet impact and impact generated great earthquake. Sonic booms from these 100 pound hailstones coming in to strike the Earth could be one of the sources of the "voices" or "sounds" or "blasts" that both the Seventh Trumpet and Vial say will be heard. The Seventh Vial's reference to these cometary hailstones is also consistent with the clearer translation of *Isaiah 24:18-20* just noted, which describes an impact that causes the Earth to "reel to and fro like a drunkard" in space, and causes the Earth to be "broken asunder" or "broken up" and "split through" as a result of comets being loosed. The term "split through" may be a reference to an impact that penetrates the Earth's crust and explodes in the Earth's interior (mantle).

The "great hail" referred to in both the Seventh Trumpet and the Seventh Vial could be produced from cometary debris or large fragments that break off and disintegrate from the huge incoming comet. This initial bombardment event would create great fear in the armies of the nations that the Bible says will be gathered for battle against Israel at a place called Armageddon (*Revelation 16:14-16*).

Isaiah 24:18 describes cometary bombardment as the cause of this massive blow when it says, "the windows from on high are open and the foundations of the Earth do shake." The phrase "the windows on high are open" is better translated as "those that lie in wait (comets) in heaven are loosed." Recall, that the clearer translation of this phrase which appears in both *Genesis 7:11* and *Isaiah 24:18* was explained in the discussion of the Flood in Chapter 4 and in the discussion of Joshua's Great Victory in Chapter 5. The bottom line here is that large cosmic impacts can cause "the foundations of the world (to) shake" and the earth to "move exceedingly" (*Isaiah 24:18-20* above); while rainfall does not cause planetary shaking and moving.

Few Bible scholars have felt confident enough to offer definite explanations for the following scriptures. However, in light of the growing information that astronomers and planetary geologists have learned about cometary impacts over the last twenty years, few can question their meaning now or offer other explanations that fit so well Biblically. *Deuteronomy 32:22* refers to God's anger, and tells how He "shall consume the Earth . . . and set on fire the foundations of the mountains." *Judges 5:5* says, "the mountains melted before the Lord." *Psalm 97:5* says, "The hills (mountains) melted like wax at the presence of the Lord."* *Isaiah 64:1* says, "Oh that thou wouldest rend the heavens, that thou wouldest come down, that the mountains might flow down at thy presence."† *Zephaniah 3:8* relates the conflagration of an impact to the judgment upon the nations of the Earth and says, "Therefore wait ye upon me, saith the Lord, until the day that I raise up to the prey for my determination is to gather the nations, that I may assemble the kingdoms, to pour upon them mine indignation,

---

*   The word for "mountains" and "hills" involve the same Hebrew word.

†   As explained in chapters 3, 4 and 5, God is sometimes personified as a comet in the Bible. This is the case in *Isaiah 64:1*, so in this verse we have reference to the "mountain-melting mega-impact" phenomena.

THE THREE WOES 279

even all my fierce anger: for all the Earth shall be devoured with the fire of my jealousy." It is important to recognize how *Zephaniah 3:8* quoted above connects all the Earth being devoured by fire with God pouring out His wrath upon all of the nations and kingdoms that have gathered together and assembled on this terrible day. *Psalm 46:6* says, "the heathen (nations – NAS) raged, the kingdoms were moved: he uttered His voice, the earth melted." Recall that *Revelation 16:14-16* also tells how the kings of the nations of the world will be gathered together for battle on the day of the Lord, "the great day of God Almighty."

We have seen how the details about cosmic impact provide the key to understanding that the prophecies of the Seventh Trumpet and Seventh Vial are talking about the impact of a "giant comet." The details of these prophecies call for a comet that is over 100 miles in diameter and possibly as big as 300-600 miles in diameter.* The impact of a comet of this size would be many times bigger than all of the impacts of the first six Trumpets and Vials combined. The impact

---

* Recall the 120-mile wide comet Chiron that is in unstable orbit between Saturn and Uranus, mentioned in Chapter 2. A number of other comets that are large enough to cause the enormous impact prophesied in the Seventh Trumpet and Vial have been found in the Kuiper Belt. The Kuiper Belt consists of a large population of comets at the edge of the Solar System beyond the orbit of Neptune which lead up to the Oort Cloud of comets. The Hubble Space Telescope has directly measured the size of more than 500 very large comets in the Kuiper reservoir of comets. Many had diameters of 186 miles or more. One icy body named "Quaor" had a diameter of about 800 miles – "A Cold New World – The Hubble Space Telescope has measured the diameter of a distant world more than half the size of Plato," NASA, Marshall Space Flight Center, *Government Headlines*, October 7, 2002. And "Hubble Spots An Icy World Far Beyond Pluto," *Science Daily Releases*, October 8, 2002. And "Hubble Detects Long-Sought Comet Population Beyond Neptune," *Space Science Telescope Institute*, Washington, D. C., June 14, 1995.

of the Seventh Trumpet and Vial seems to be a continuation of the comet bombardment called for in the Sixth Trumpet and Vial. The Seventh Trumpet and Vial seem to talk of the giant progenitor comet from which the earlier four smaller comets of the Sixth Trumpet and Vial break off and come in to strike the Earth. The prophesied pattern of disaster caused by a series of comet impacts during the end times is consistent with the pattern of disaster caused by comet impacts that occurred during Old Testament times.

While there is no precedent (*Ecclesiastes 1:9* and *3:15*) recorded during Biblical times for an impact of the great size described in the Seventh Trumpet and Vial, there is indeed a precedent for an impact of this size in the Earth's distant past. Although the recently found Wilkes Land Impact that produced the 300 mile wide crater under the ice of Antarctica could provide a precedent of sorts for the impact called for by the Seventh Trumpet and Vial. But in truth, it is not quite big enough. However, there was an impact that was big enough.

Most scientists now believe that our Moon was born of a truly "giant impact" that the Earth experienced shortly after the Earth was formed. Scientists have come to believe that the Earth was slammed into and rocked by a gigantic comet or planetismal like body about 600 miles in diameter that caused the Earth to eject a large volume of molten and vaporized debris. This debris went into orbit around the Earth, and then accumulated to form the Moon. This theory received a major boost in 2001 by computer simulations that demonstrated how a single impact could render the current Earth-Moon system. In these new computer simulations, this enormous impact would have also established the Earth's equator and its rate of spin. Previous computer simulations have already made it clear that an impact of this size would have produced so much heat that the early Earth would have been left with a world wide sea of molten rock hundreds of miles deep. Some astronomers refer to the theory of such a collision as the "Big Splash." [46]

The simulations about the Moon's formation were performed by Dr. Robin Canup of the Southwest Research Institute and Dr. Erik Asphaug of the University of California at Santa Cruz. Their work was reported in the journal *Nature* and in *Time* magazine. These sophisticated simulations took "factors such as gravity, impact shock, melting, and vaporization" into account. [47] Beyond insight into how our Moon was formed, Dr. Asphaug explains, "It's now known that giant collisions are a common aspect of planet formations." [48] Echoing this thought, University of Arizona Professor Michael Drake, the principal investigator for NASA's mission to asteroid RQ36 says that, "Toward this end of the formation of the planet, as with all the rocky planets, you basically have Earth being hit with objects the size of Mars . . . That's pretty energetic. It's enough to melt and vaporize the outer part, and possibly the whole planet, depending on how the impacts occurred." [49]

While the exact details on how the Moon was formed are still being worked out, Dr. Thorsten Kleine of the University of Muenster in Germany notes that his data indicates that "collisions caused almost complete melting of Earth resulting in a scenario called magma (lava) ocean, in which the Earth was covered by a layer of magma (lava)." [50] If giant collisions or impacts once caused the surface of the Earth to become molten, another giant impact could do so again fulfilling the prophecies of the Seventh Trumpet and Vial and *II Peter 3:10-13*.

Beyond the scriptures relating to the Seventh Trumpet and

Seventh Vial discussed so far, there are still other scriptures that tell of the frightening events to take place leading up to the time of the Seventh Trumpet and Vial, and the day the Bible refers to as the "Day of the Lord" (and variations thereof). In particular, these scriptures relate the catastrophic events that are prophesied to occur to the political events that are prophesied to take place on the "Day of the Lord"; just as *Zephaniah 3:8* connects the frightening events of this day to the nations and kingdoms of the world being gathered together and God pouring out His wrath upon them. These scriptural connections will be discussed in the ninth chapter.*

* * * *

While the first Four Trumpets and Vials involved comet impacts that seem quite catastrophic, they are to be considered moderate in magnitude compared to the magnitude and destructive effects of the final three prophesied Trumpet and Vial impacts. The Sixth round of bombardment is to be greater than the Fifth round, and the Seventh round greater than the Sixth. The Seventh round is to be so destructive that no place on Earth will remain untouched.

We saw that the Fifth Trumpet and Vial called for comets bringing in bacteria or a virus that torments men with sores and pain. This is reminiscent of the Biblical accounts of the plague

---

*   *Zephaniah 3:8* "Therefore wait ye upon me, saith the Lord, until the day that I raise up to the prey for my determination is to gather the nations, that I may assemble the kingdoms, to pour upon them mine indignation, even all my fierce anger: for all the Earth shall be devoured with the fire of my jealousy."

during King David's reign, the deadly blast during the time of King Hezekiah, and the Sixth plague of the Exodus, where the dust over the land of Egypt caused boils on man and beast.

The Sixth Trumpet and the Sixth Vial involve four comets, and 200 million boulder size fragments traveling in a stream along with these four comets, striking the Euphrates River area. The Sixth Trumpet describes fire, smoke, and brimstone coming from the "mouths" or vents of these comets which is a scientifically accurate description of active or 'outgassing' comets. These four incoming comets and the millions of cometary fragments (mini comets) travelling with them are described as "horsemen" with breastplates of fire and of jacinth, and brimstone; and as "serpents" with "heads" and "tails."

The Seventh Trumpet and the Seventh Vial entail at least one truly great impact, from a very large comet, that first rains down 100 pound hailstones from the fragments that precede it, before it finally slams into the Earth. Scripture describes a horrific event that will produce the largest earthquake in mankind's history, with islands and mountains disappearing, and the Earth being rocked on its axis. A "mega impact" of an extremely large comet slamming into the Earth would cause earthquakes of unprecedented proportions. This massive body could conceivably penetrate the Earth's crust, superheat the Earth's interior, and also force vast quantities of hot magma from the Earth's interior to flow out onto the surface as lava, which taken together could cause the Earth's crust to melt and dissolve. As the crust melted, all of the islands and mountains of the world (which lie upon the crust) would disappear. Also based on the recent discovery that a huge comet or asteroid smashed into Mars and blasted away most of the crust on Mars' northern hemisphere;

a "mega impact" could blast away part of the Earth's crust. Scientist have come to call such mega impacts "planetary scale impacts." They note how "planetary scale impacts" penetrate into the mantle, and can cause crustal melting, crustal removal, and planetary disruption. Based on the giant impact basins produced, they now recognize that "planetary-scale impacts" have occurred on Mercury, Mars and the Moon as well as possibly Earth. The "crust blasting impact" model and the "mountain melting mega impact" model are not incompatible and they could work together in making the Earth's islands and mountains suddenly disappear. In having the earth's islands and mountains disappear, the Bible gives a scientifically reasonable accounting of geological mechanisms unknown to geoscientists until the last decade.

After this giant impact, eventually the Earth's surface would cool, and a new crust and new atmosphere would form. In essence there would now be a new atmosphere and a new Earth, just as the Bible speaks of a new heaven (or atmosphere) and a new Earth.

II Peter 3:10-13 says:

> But the **Day of the Lord** will come as a thief in the night, in which the heavens shall pass away with a great noise, and the elements shall melt with fervent heat, the Earth also and the works that are therein shall be burned up. (11) Seeing then that all these things shall be dissolved (destroyed)... (12) Looking for and hasting unto the coming of the Day of God, wherein the heavens (air, atmosphere or sky) being on fire shall be dissolved (destroyed) and the elements

shall melt with fervent heat? (13) Nevertheless we, according to his promise, look for new heavens, and a new Earth . . . ."⁵¹

EIGHT

The Tale of the Tower –
The Truth about What Really Happened at the Tower of Babylon

> Because tales of enormous floods and fire from the sky seem incredible to us, such stories have generally been interpreted as the amusing ravings of primitive people. We tend to interpret ancient legends, myth, and folklore in terms of the way the world appears to us today. It is now only beginning to dawn on us that the skies observed by the ancients may have been different from what we see today. . . People experienced impacts that shattered lives and even civilizations.
>
> <div align="right">Dr. Gerrit L. Verschuur
<i>Impact: The Threat of
Comets and Asteroids</i> [1]</div>

> Evidence for the cause of the decline of ancient civilizations is difficult to come by, but when contemporaneous pieces of the puzzle are compared, the *scenario of impact disaster* looks intriguing if not convincing.
>
> Dr. Gerrit Verschuur,
*Impact: The Threat of
Comets andAsteroids* [2]

> Its central theme ["*The Curse of Agade*"] concerns national catastrophe as a direct consequence of *divine wrath kindled by a defiant act* on the part of man.
>
> SamuelNoah Kramer, who
translated the Sumerian work
"*The Curse of Agade*" (Akkad) [3]

Let us begin with a story of an incident that is supposed to have actually happened in a children's Sunday school class. The teacher had taught a lesson about the fall of Adam and Eve in the Garden of Eden. Then she passed out paper and instructed the kids to draw a picture illustrating part of the Bible story. After ten minutes, she quietly walked around the tables looking at the various pictures and complimenting the children on their art work. When she looked down at one five year old boy's picture, she was surprised. He had drawn a bright red convertible car. The robed driver had long hair and a white flowing beard, while a near naked man and woman sat

THE TALE OF THE TOWER

in the back seat. The teacher leaned down and quietly reminded the boy that the picture was to illustrate the Bible story. He answered that it did! Perplexed, the teacher asked the boy what part of the story he was illustrating. He said that the drawing was a picture of when God ***drove*** Adam and Eve out of the Garden of Eden.

What this little story shows is that the meaning of just one word can easily be misunderstood and that interpretations based on this misunderstanding can dramatically alter the original message of a story. This confusion has happened with another Bible story. In this story a few words and phrases have been assigned incorrect meanings, and interpretations have been made based on these misunderstandings that have changed the original message of the story. This misinterpretation has run into the brick wall of historical facts, revealing the misinterpretation and yet the wheels of this convertible keep spinning.

* * * *

Based on the information discussed in Chapters 1 and 7 about the Amarah Crater and the associated dust deposits, the author believes few scholars would have a problem connecting the Amarah Crater to the sudden destruction of the Akkadian empire as told in the Sumerian story *The Curse of Akkad* (Akkad is the same as Agade and the Bible's Accad). But, the discovery here regarding which Biblical story is related to the Amarah Impact Crater and relates to the evidence archeologists have found for the Akkadian empire and "imperial collapse, regional desertion, and large scale population dislocation," will come as a surprise to many and may be difficult for some to accept. [4] *(See Illustration B)*

It isn't the story of Noah's Flood. Noah's Flood did not happen during the same time period as the Amarah Crater. Although the story of Noah's Flood involves a catastrophe of a large enough magnitude to have created the massive impact that produced the Amarah Crater, the dates aren't right. According to Biblical chronology the Flood occurred around 3,000 BC, while the Amarah Crater and the collapse of the Akkadian Empire occurred nearly 1,000 years later around 2200 BC.

Could it be the story of Sodom and Gomorrah? While the time line for the destruction of Sodom/Gomorrah (1800 BC) and Akkad are close, it is not the story of Sodom and Gomorrah, because of the difference in the magnitude of the impact. The destruction of the cities of Sodom and Gomorrah is too small a catastrophe to compare to the devastation the two mile wide Amarah Crater represents. The Amarah impact with energy equivalent to that of hundreds of Hiroshima sized atomic bombs would have vaporized Lot's wife, as well as Lot and his retreating family.*

The Biblical story that *does* relate to the catastrophic collapse of the Akkadian Empire and the Amarah Crater is the story of "The Destruction of the Tower of Babel" (*Genesis 10:8-10* and *11:1-9*). As traditionally told, this is a story about a time when the whole world spoke the same language, and people came together to make a "name" or "authority" for themselves by building a new city and a new temple tower at Babel in the "*land of Shinar*." However, because

---

*   Sharad Master, "Ummal Binni Lake, a possible Holocene impact structure in the marshes of Southern Iraq: Geological evidence for its age, and the implications for Bronze-age Mesopotamia," for *Environmental Catastrophes and Recoveries in the Holocene* – Conference, Professor Suzanne Leroy and Dr. Iain Stewart organizers, Dept. of Geography and Earth Sciences, Brunel University, Uxbridge, UK, Aug 29 – Sept 2, 2002. On WEB at http://atlas-conferences.com/egi-bin/abstract/caiq-is.

the people became one and had one language, nothing they imagined they could do would be restrained from them. The God of the Bible then went down and confused or mixed their language so that they could not understand one another's speech. People's speech now sounded like babble to one another. This traditional mistranslation and misinterpretation says that, as a result of the destruction of this universal language, an event popularly called "the confusion of tongues," people stopped building the new city and the new temple tower, and the Lord scattered the people over the face of the whole world.

There are several problems with this version of the story. First, this traditional translation and interpretation of the Tower story derived hundreds of years ago does not fit with what we know from other scriptures in the Bible (*Genesis 10:10* and *Zechariah 5:11*) that also talk about the "land of Shinar." The second problem is that the traditional interpretation of a universal language at the time of the Tower does not fit archeological and historical records. In addition, and most importantly, we get a very different account about what happened from writings of the people from "the land of Shinar, who were building the new tower."

However, the more accurate translation of the Bible's Tower story (presented here for the first time) does fit with what we know from other scriptures that speak of the new city and the new temple tower that were being built in the "land of Shinar." Equally as important, this clearer translation of what happened is consistent with the historical records as well as the cuneiform writings of the people referred to in the story. The more accurate translation of the Tower story reveals and establishes the truth behind this story and places this important Biblical event in the fabric of history.

For those familiar with the Bible, the idea that the Tower of Babel was destroyed by cometary impact will have many wondering how could this be possible? The Bible doesn't say the Tower was destroyed by cometary activity, or does it? Previous chapters of this book have detailed how the God of the Bible has used comets as "the weapons of His wrath" (*Isaiah 13:5 NIV*). Chapter 3 showed there are various scriptures throughout the Bible that say God controls comets and uses comets to bring catastrophe. In Chapters 4 and 5 we looked at six Old Testament catastrophes that involved cometary activity. Chapters 6 and 7 detailed explanations for the cosmic caused catastrophes described in the *Book of Revelation's* Trumpet and Vial prophecies. Surely, we must recognize and acknowledge that in the Bible **there is a pattern** for the God of the Bible using comets as his weapons of wrath. In this chapter we will see that the clearer interpretation of the Tower of Babel story is connected to the other catastrophic events recorded in the Bible, and that the theory the Towel of Babel involved cometary activity can be substantiated.

The historical record clearly shows that the people referred to in the Tower story were Sumerian and Akkadian. Fortunately for us, these people left written records behind. These cuneiform records are helpful in determining a more accurate translation of the Tower story. In addition, physical evidence has been found that confirms their account of what caused them to stop building the new city and the new tower and abandon the area. Taken together these events resulted in the sudden collapse of their new empire.

Before the author herein began research of the Bible, the story of "The Destruction of the Tower of Babel" had been one of the stories that kept him from taking the Bible too seriously. The story does not fit with what archeologists know about the ancient cultures of the Near East and about the origins of the languages of the world. There could have been no universal language at the time of the story because different populations of people speaking different languages in Europe, Asia, Africa, the Pacific, and the Americas had existed long **before** the first temple towers (*ziggurat* from *zigura* "to raise up") were built in Mesopotamia.

Specifically, two totally different languages, Sumerian and Akkadian (Babylonian and Assyrian are *later* dialects of Akkadian) were spoken in the "land of Shinar" (*Genesis 10:10, 11:2* and *Zechariah 5:11*) long **before** the Tower of Babel (Babylon) was built there. Evidence for this has come from archeological sites that have yielded tens of thousands of cuneiform tablets written in both Sumerian and Akkadian, some dating to over 1,000 years **before** the new Tower and the new city referred to in the Bible story were built. Further, the phrase "land of Shinar" is actually a direct reference to these two cultures where two very different languages were spoken. The word "Shinar" is merely the spelling out in English (transliteration), of a Sumerian word that means "Sumer-Akkad" just as ancient literature repeatedly makes reference to "the land of Sumer and Akkad." * In fact, in the land of Shinar, that is, in the land of Sumer and Akkad (Akkad was to the north of Sumer), different languages were not a point of confusion, since the Akkadians, the Semitic conquerors of the Sumerians, for the most part adopted the Sumerian culture and made Sumerian their literary and religious language. The schools for scribes at the new city of Akkad (*Genesis 10:10* and *11:4*) made

---

*   The unquestioned dean of Sumeriology, Samuel Noah Kramer wrote that scholars usually identify the word "Shinar" with Sumer, but it "actually stands for the Sumerian equivalent of the compound word 'Sumer-Akkad.'" Samuel Noah Kramer, *The Sumerians: Their History, Culture and Character*, The University of Chicago Press, Chicago, 1963, p. 297.

the study of Sumerian their basic discipline, and there were bilingual dictionaries and even manuscripts where each line in a Sumerian composition was followed by the Akkadian translation. (Recall that the Sumerians, their religion and their relationship to the Akkadians and the Babylonians was discussed in detail in the first part of Chapter 3 – "The Lord of Hosts.")

How could the Bible itself be so wrong about the origins of the different languages of the world? The Bible isn't wrong, don't the problems lie with the mistranslation and misinterpretation of certain lines from the original text? Reevaluating the translation and interpretation of the original Hebrew text in light of newer more accurate knowledge is *not* an attack or disrespect to these who have done the difficult work of translation. Sadly, mistakes were made in the early translations and interpretations of the Tower story, and they have been repeated under the cloak of tradition. That is the more tragic story! Yet, there is always something more to learn. Consider a passage from a novel by Chaim Potok called *In the Beginning*. In it the main character David, thinking to himself that how what he had studied was not enough, says:

> The medieval commentators used the most advanced knowledge of their day to understand the *Torah* [the first five books of the Bible]. But they did not have the tools we have today. They did not have anthropology, archeology, comparative religion, linguistics, a true grasp of the texture of history. I do not know what kind of commentaries Rashi, Ibn Ezra, the Ramban, and the others might be writing were they alive today; they might be helping me penetrate the precious

rectangles of *Torah* I hold in my hands. I wanted to say all that, but I remained silent. For my father was right; what I had studied with Mr. Bader and in the yeshiva was not enough. It could not be the entire truth. 5

This reevaluation of the Tower story is not a case of forcing or manipulating the Biblical story to conform to the historical account. Rather, we are allowing a clearer understanding of history and science to either support or contradict the Scriptures. We find that "Anthropology, archeology, comparative religion, linguistics (and) a true grasp of the texture of history" enhances our understanding of the Bible's story of the Tower, and substantiates the Tower story by connecting it to historical records. Over the years, archeological excavation and the translation of ancient texts have consistently confirmed information in the Bible. This is true as well for the Biblical text regarding the existence of the Sumerians and the Akkadians and the collapse of the Akkadian Empire.[*] The conflict between science and the Bible regarding the origin of the different languages of the world is due to errors in the translation of the story of the Tower of Babel.

These errors in translation and interpretation become

---

[*] For example, the Babylonian King Belshazzar and the Assyrian King Shalmaneser were only known from the Biblical text until excavations in the late nineteenth century turned up cuneiform tablets and building inscriptions that spoke of these Kings. And the Bible tells how the little known ancient nations of Meshech, Tubal, and Togarmah in Asia Minor traded slaves, vessels of brass, horses, and mules (*Ezekiel 27:13-14*). Yet it wasn't until the early twentieth century, when all the records of the ancient Assyrian Court were translated, that historians read how these nations existed and traded the very same things that the Bible said they traded.

apparent once research begins concerning the literal meaning of the words used in the original Hebrew text for this story. In order to properly translate and interpret the Bible's Tower of Babel story ("Babel" and "the land of Shinar" from *Genesis 11:1-9* are also referred to in *Genesis 10:10*), it is important to understand several expressions or idioms that were used in the ancient Near East. Careful study reveals a very different story from the orthodox translation and interpretation of the Tower story. Given a corrected or clearer translation of these passages, we can now understand what really happened at the Tower of Babel. As stated, the corrected translation of these events is supported by cuneiform writings, the historical record, and by physical evidence. These corrections show that this is not a story about the destruction of a universal "language" to stop the building of a temple tower, but rather a story about the destruction of what we call today an "empire" to keep this empire from becoming too powerful.

The new city (Akkad) that was being built was the capital of the new Akkadian empire. The new temple tower that was being built nearby at the ceremonial center of Babylon was the home of the new empire's patron deity, the Sumerian goddess, Inanna, the "Queen of Heaven." The temple tower was to help **unify** the two different cultures of the empire (Akkadian and Sumerian) under a single state religion. Temple towers were part of the Sumerian religion, they were dwelling places for the gods and places where the priests could attend to the needs of the gods. The politically astute Akkadians who conquered the Sumerians adopted the Sumerian religion.

To further demonstrate this religious conversion (besides building a new tower), the Akkadian emperor Sargon ceremoniously married the Sumerian goddess, Inanna in a *hieros-gamos*, a "holy marriage," where the goddess bestowed all power to rule on earth to him, just as all power had been given to this single deity when she was elevated to the position of "Queen of Heaven," the leader of the Sumerian pantheon of gods.* Sargon also made his daughter (Enheduanna) both the high priestess of the Sumerian goddess Inanna, and the human spouse of the Sumerian moon god, Nanna (Sin), the father of Inanna and the patron god of the Sumerian city of Ur. In short, the Sumerian religious order was reflected in the political order of the new Empire. This marriage of religion and politics helped Sargon consolidate and focus the power of the new Empire.

When the Akkadian Empire soon collapsed, the state religion that was based on the worship of a woman, the "Queen of Heaven," survived and became the religion of the Babylonians and the Assyrians, the descendants of the Akkadians. In other words, the religion practiced by the later Babylonians and Assyrians was based on the worship of the Sumerian goddess Inanna, who the Babylonians and Assyrians called Ishtar.† They worshiped the goddess whose house was the temple tower at Babylon as the "Queen of Heaven" (*Jeremiah 44:17-19* and *25*) who led "the host of heaven" (*Jeremiah 8:2* and *19:13*). Sometimes she was referred to as "the harlot of heaven," which brings to mind the woman associated

---

* As noted in Chapter 3, Professor Thorkild Jacobsen wrote that the delegation of power to this single deity is described in the bilingual (Sumerian and Akkadian) *"Elevation of Innana"* – *The Treasures of Darkness – A History of Mesopotamian Religion*, Thorkild Jacobsen, Yale University Press, New Haven, NY, 1976, p. 234.

† The Neo-Babylonian Empire (612-536 B.C.) officially worshipped a pantheon of cometary gods, where Ishtar was the Queen of Heaven, and her consort Marduk, was the King of Heaven. By the time of the Neo-Babylonian Empire, Ishtar had a planetary expression as the planet Venus in addition to her original and divine expression as a comet.

with "***Mystery Babylon***" who was called the "***great whore***" or "***great harlot***" (NAS), and the "***mother of harlots***" in *Revelation* (*Revelation 17:1-7*). Over time, the religious order of the woman came to be reflected in the political order and political history of a number of succeeding (different) empires. For more on the Sumerian religion and the Queen of Heaven, see Chapter 3 – "The Lord of Hosts," and **APPENDIX A – Sumerian/Babylonian Cometary Gods.**

In the decades that followed, the temple or *babylon* that was built for the woman gave its name to the city that grew up around it and to the entire region – Babylonia. Basically, the land of Shinar and Babylonia represent the same geographical area, and the people who lived in this area were called Babylonians. It is important to know that the so-called "***Whore of Babylon***" is non-other than the Sumerian goddess Inanna, the supreme deity of the world's first empire, whose name in Sumerian literally means "***Queen of Heaven.***"

So, how did the mistranslation and misinterpretation of the Tower story come about? Since the ancient Hebrew language did not use vowels, the traditional translation of the Tower story has taken the Hebrew consonants *b-b-l* to denote the word "babel" and then assumed this word meant "confusion caused by language differences." This is nothing but a play on words that is solely based on the word "babel" being similar to the actual Hebrew word (*balal*) for "confusion" or "mixing."* But, despite the popularity of this ***contrived*** etymology, "babel" is ***not*** a Hebrew word, nor should it even appear in the story of the Tower in the first place.

The Hebrew consonants *b-b-l* actually represent the spelling out in Hebrew, the transliteration of the Akkadian word "*babylon*" (a

---

*   In modern Hebrew the actual word for the phenomena of speech known as "babble" is "*siyach*" not "*babel*."

THE TALE OF THE TOWER

Greek variant of the Akkadian *babilum*). 6 The Hebrew consonants *b-b-l* appear in the Old Testament 282 times, and **only twice** are they rendered as "Babel," both instances (suspiciously) being in regard to the Tower. The other 280 times these consonants are rendered as "Babylon." More recent translations than the King James Version of the Bible, such as the New American Standard (NAS) and the New International Version (NIV), at least recognize that the consonants translated as "Babel" should be translated as "Babylon."

Archeologists have long known the name "Babylon" is an Akkadian word that means "gate of god," or "house of god," and that every temple tower was called a ***babylon,*** a gate or house of god.* Thus, the Hebrew consonants *b-b-l* in this story actually refer to a place that came to be called "Babylon" because of the ***babylon,*** the temple tower that was built there. The Moody Bible Institute in *The Wycliffe Bible Commentary* says that, "The best Hebrew lexicographers claim that it [*Babel* translated Babylon] could not have come from the Hebrew *balal* to 'confuse' or 'mix,' but that it meant 'gate of God.'"7 *The Baker Encyclopedia of the Bible* says "a popular etymology [for Babel or Babylon] replaced the original meaning of the name." 8 The Akkadian word *babilum* itself is a translation of the Sumerian word for "gate of god," the Sumerian word *Ka-dingir-ra*. In some ancient Babylonian texts, the city of Babylon is in fact sometimes referred to as *Ka-dingir-ra*.†

---

*   To the Sumerians and the Akkadians a "gate of god" also represented a "door of god," a "gate of heaven," and a "house of god"; places where man and god met. *Proverbs 8:34* talks about watching daily at God's "gates" and at his "doors." In *Genesis 28:12-17* Jacob after dreaming of a ladder or stairs reaching to heaven and seeing God says, "How awesome is the place, this is none other but the **house of God,** and this is the **gate of heaven.**"

†   For example see "Historical Documents – *Cyrus*" in *Ancient Near*

The key to determining that this story is actually about the destruction of an empire comes from learning the meaning of certain types of political idioms or expressions peculiar to the ancient Near East. For example, in *Genesis 11:1* the original Hebrew text of this story literally says "the whole land was of **one lip**" (e.g. see *Septuagint*). This has been traditionally translated as "the whole earth was of **one language**." However, in the ancient Near East "one lip" is an idiom that means "one government." (This would be similar in our culture today if it were said, "In 1776 the 13 colonies spoke with **one lip or voice or mouth** about independence from England.") Confirmation of this interpretation comes from other terms that are used a number of times in the original Hebrew text of the story that refer to "one government," "one command" and the people being "one" or "united." The concepts of "one government," "one command," and being "united" convey the concept of "empire."

Thus, while the traditional translation of *Genesis 11:1* says, "And the whole earth was of one language and of one speech"; a historically correct translation of *Genesis 11:1* says, "And the whole land was of one government and one commander." Obviously, the concept of "one government" and "one commander" go together in the description of an "empire." The new translation aptly describes the Akkadian Empire that built the new tower at Babylon according to the written records these people left behind. It was the political level of an empire that let the people become "one" and make a "name" or "authority" for themselves (*Genesis 11:6* and *11:4 NIV*), not the building of another temple tower. With empire comes great power.

Eastern Texts – Relating to the Old Testament, edited by James Pritchard, Princeton University Press, 1969, pp. 315-316.

Genesis 11: 1 Translation

Literal	Traditional	Historically correct
"And the whole land or earth was of **one lip** and of **one word, cause** or **command**."	"And the whole earth was of **one language** and of **one speech**."	"And the whole land was of **one government** and **one commander**.

Coincidentally, additional confirmation for this clearer translation of *Genesis 11:1* comes from a Babylonian document that used the same type of idiom that was used in the Bible's story of the Tower of Babel. This document tells how Sargon, the founder of the Akkadian Empire, conquered a number of countries. Then it literally says, "He made its (the land's) **mouth be one**," which has been translated by a leading expert in the language as "*He established there a central government.*" [9] In this case, the idiom "**mouth be one**" was used to refer to "central government," which is another way to convey the concept of "empire."*

Even if one doesn't know that "*one lip*" is an ancient idiom meaning "*one government*," there is another way to determine that this is a story about the destruction of an empire and not a universal language. The Hebrew word that literally means "*lip*" (*saw-faw* #8193 in *Strong's Concordance*) that is traditionally translated as "language" or "speech" in the Tower story (*Genesis 11:1, 6, 7* and *9*) can also be taken to be a Hebrew word that means "gathering" (*saw-fakh* #5596 in *Strong's Concordance*), where "one gathering" is consistent with

---

* Interestingly, the NIV in two instances translates the Hebrew word translated as "*lip*" (#8193 in *Strong's Concordance*) as "*mouth*."

"one government" under one commander as in empire.* This is possible because the three Hebrew consonants used for the Hebrew word meaning "lip" (#8193 in *Strong's Concordance*) are the same Hebrew consonants used for the Hebrew word meaning "gathering" (#5596 *Strong's Concordance*), the only difference being in the vowel signs.

It is important to understand that the writing of ancient Hebrew only used consonants. Consonants were written down, vowels were not written down. It was not until about 600 AD that a complete system of vowel signs was added to the text of the Old Testament by the scribes of the Massoretes ("transmitters"). So aside from context, it is impossible to distinguish between certain ancient Hebrew words that contain the same set of consonants as in the case with words #8193 and #5596 in *Strong's Concordance*.

Thus, in *Genesis 11:1* we can have "one gathering" or "one government" ("one lip") under "one commander," where either translation fits the historical context of the Akkadian Empire as begun and commanded by Sargon. This empire brought people of different cultures and languages together into a single politico-religious entity in the land of Shinar. *Genesis 10:10* tells how the land of Shinar, that is, the land of Sumer and Akkad, included the Sumerian city of Erech (Uruk) and the Akkadian cities of Akkad and Babylon. The lesson here is that translations of the Biblical manuscript should neither be done in a historical vacuum, nor manipulated to fit one's theological preconceptions. We can't have Biblical translations that ignore historical context, nor translations based on fabricated events that are inconsistent with the course of real historical events. When the

---

* For example, the Hebrew word *saw-fakh* (#5596 in *Strong's Concordance*) is translated as "gathered together" in *Job 30:7 KJV* and *NAS* and "unite" in *Isaiah 14:1 NIV*.

Tower story is correctly translated and interpreted, the Bible should be credited with accurately telling the story of the collapse of the historical Akkadian Empire.

In defense of the traditional translation of the story of the Tower of Babel, it should be realized that it was made long before scholars knew very much about the ancient Near East.[*] The traditional translation was made before scholars learned that there was a Sumerian culture in ancient Mesopotamia and before it was known that the Bible locations of Ur and Erech were really Sumerian cities, not Babylonian cities. The acclaimed Sumerian expert Samuel Noah Kramer wrote, "The very name Sumer had been erased from the mind and memory of man for more than two thousand years. The discovery of the Sumerians and their language was quite unlooked for and came quite unexpectedly . . . ."[10] More importantly, the traditional translation of this story was made before archeologists discovered that the Sumerians and Akkadians had lived side by side in the land of Shinar and that at least two different languages, not one, were spoken in the land. As a result, this inaccurate traditional translation laid the ground work for what amounts to a myth about the origin of languages – a myth that in time science and the historical record have proven to be untrue. Once again, what we don't find with the correct translation is conflict between science, history and the Bible. Rather, we find the Bible providing accurate information hundreds of years before archeologists uncovered evidence that confirmed what the Bible says.

[*] It should be understood that every translation from the original text of an ancient document represents the translator's interpretation of what he or she thinks is being said. Since ancient documents such as the Bible were not written in a cultural or historical vacuum; the better the translator understands the cultural and historical contexts of what is being said, the more accurate the translation.

Beyond having certain cities in common and the sudden destruction of empire, there are other factors that connect the Bible's story of the Tower with the historical empire of Sumer and Akkad, which is also known as the Akkadian Empire. The clearer translation of *Genesis 11:1* is now consistent with the next line of the story and the archeology behind this line. *Genesis 11:2 (NAS, NIV)* tells how men journeyed eastward to the land of Shinar and settled there. Based on inscriptions and loan words (where words from one culture come to be used in the language of another culture) archeologists know that around 4000 BC Semitic nomads (Amorites) from the Syrian desert and the Arabian peninsula began to migrate into the land of Shinar and settle there.

These Semites, now called the Akkadians, joined the Sumerians who were already in the land, and adopted their sedentary life style and culture. [11] The Akkadians also adopted the Sumerians' cuneiform system of writing to produce clay tablets written in their own language. As noted in Chapter 3, the late dean of Sumeriology, Samuel Noah Kramer said that by the time the Akkadians came to be called Babylonians,

> They took over Sumerian culture and civilization lock, stock, and barrel. Except for the language, the Babylonian educational system, religion, mythology, and literature are almost identical with the Sumerian . . . And since these Babylonians, in turn, exercised no little influence on their less cultured neighbors, particularly the Assyrians, Hittites, Hurrians, and Canaanites, they, as much as the Sumerians

themselves, helped to plant the Sumerian cultural seed everywhere in the ancient Near East. [12]

It is amazing that this information first given in the Bible has been confirmed thousands of years later by the excavation of archeological sites and the translation of ancient writings.

Just as research has led to the connection between the historical Akkadian Empire's city of Akkad and the Bible's city of Akkad, it has also shown that the historical Akkadian ruler Sargon was one in the same as the ruler the Bible recalls as Nimrod. The Bible says it was Nimrod who founded a kingdom or empire in the land of Shinar, the land of Sumer and Akkad that included the ceremonial center at Babylon and the cities of Erech and Akkad (*Genesis 10:8-10* and *11:1-4*), while the historical records show that Sargon founded an empire in the land of Sumer and Akkad that included Babylon and the cities of Erech and Akkad. *(See Illustration B)*

Emeritus professor of Assyriology at the University of London, D. J. Wiseman notes that a text written by Sharkalisharri, the last king of Akkad, and Sargon's great grandson (ca. 2200 BC), supports the idea that Sargon built a ceremonial center and temple tower at Babylon. In this text *Sharkalisharri*, (which means "king of kings" in Akkadian) mentions that he repaired the temple-tower (meaning that the temple tower had already existed) at the sacred site of Babylon and that this sacred site apparently took its name from the "babylon," the "gate or house of god" that was built there.*

---

*   D. J. Wiseman, "Babel," in *New Bible Dictionary*, J. D. Douglas editor, Tyndale House, Wheaton, 1982, pp. 110-111. A Neo-Babylonian "historical text" written in cuneiform called the "*Sargon Chronicle*" tells how "he (Sargon) took away earth from the pits of Babylon (to reach clean earth) and he built upon it another *babylon* (a gate of god) beside the town

When *Zechariah 5:7-11* refers to the building of a *"house"* for a *"woman,"* that represents *"wickedness,"* a *"winged"* goddess in the land of Shinar, these verses probably refer to Sargon's building of a *"house"* or *temple towe*r to the *"woman,"* the *"winged"* (cometary) goddess Inanna at Babylon in the land of Shinar.* This would have been the same temple-tower the Bible says was being built by Nimrod at Babylon in *Genesis 10:8-10* and *11:1-4* and *9*.

From these things we see that the name "Sargon" and the name "Nimrod" are just two different names for the same person who founded the empire in the land of Sumer and Akkad that included Babylon and the cities of Erech and Akkad. When Akkadian rulers took the throne it was customary for them to take on a throne name, and "Sargon," which means "true king" in Akkadian, was Nimrod's throne name. From the historiographic document called "*The Sumerian King Lis*t" we know that Sargon's father was a gardener. It is unlikely he would have named his son "true king."

When the Bible (*Genesis 10:9*) says that Nimrod "was a mighty hunter," the Bible is clearly identifying him as an Akkadian king, because all the kings of the Akkadian culture were celebrated as "mighty hunters." Based on cuneiform inscriptions, archeologists know that the ritual hunting of lions was a traditional pasttime, a "lordly sport" for Akkadian Kings.† Writing about the person the of Agade (Akkad)." "The Sargon Chronicle," A. Leo Oppenheim, translator, Babylonian and Assyrian Historical Texts in *Ancient Near Eastern Texts - Relating to the Old Testament*, Pritchard, 1969, p. 266.

* *Zechariah 5:7* refers to a *"woman,"* and *Zechariah 5:9* refers to *"two women."* The *"woman"* would be Inanna, and the *"two women"* would be the Sumerians' Inanna, and the Akkadians's Ishtar who were basically one in the same in "the land of Shinar," the land of Sumer and Akkad.

† The inscription on the famous hunting relief of Ashurbanipal (669-633 BC) housed in the British Museum says: "I am Ashurbanipal, King of the Universe, King of Assyria ... endowed with surpassing might.

THE TALE OF THE TOWER

Bible calls Nimrod, Dr. D. J. Wiseman of the University of London, a leading authority on Assyriology, notes that "many scholars compare him (Nimrod) with Sargon of Akkad c. 2300 BC, who was a great warrior and huntsman and ruler. . . ."[13] While the author is not the first archeologist to connect Sargon and Nimrod as the same person, he may be the first archeologist to show that the Bible's story of the Tower is about the destruction of the Akkadian Empire, a destruction where the God of the Bible in accord with His pattern, used comets as the weapons of His wrath. See **Exhibit 2 – "Comparison of the Bible's story of the Tower of Babel and the Akkadian Empire."**

Remember, in *The Curse of Akkad*, the heavenly gods brought about the destruction of the Akkadian Empire by cometary activity because of a defiant act on the part of man (see Chapter 7 and the Fifth Trumpet). It is a story of divine wrath bringing about the collapse of an empire. The dean of Sumeriology, Samuel Noah Kramer, who translated *The Curse of Agade* (Akkad), wrote, "Its central theme concerns national catastrophe as a direct consequence of **divine wrath kindled by a defiant act** on the part of man."[14] The Biblical account of the tower is also a story of divine wrath bringing the collapse of an empire. The Bible story tells how the God of the Bible, because of a defiant act on the part of mankind, brought about the destruction of the new empire and scattered the people who were building the new city and the new Tower. The use of comets to destroy

---

The lions which I slew: I seized a fierce lion of the plain by his ears . . . I pierced his body with my lance . . . in my *lordly sport* I seized a lion of the plain by his tail . . . I smashed his skull with the club of my hand . . . in my *lordly sport* they let a fierce lion of the plain out of his cage and on foot with my spear . . . I stabbed him later with my iron girdle dagger and he died . . . I shattering the might of the lions . . . ." from Daniel D. Luckenbill, Ph.D. *Ancient Records of Assyria and Babylonia, Vol. II – Historical Records of Assyria*, University of Chicago Press, 1926, pp. 391-392.

the empire associated with the Tower of Babel even fits a theory recently put forth by some eminent astronomers. They suggest that there is a "pattern of world history" involving the periodic "collapse of empires" as a result of "cosmic impacts." [15]

The Curse of Akkad, written in Sumerian provides an independent and much older account of the same basic set of events spoken about in the Bible's story about the Destruction of the Tower of Babel. In *The Curse of Akkad* the cometary goddess Inanna, the Queen of Heaven, who like a warrior hastening to his weapon, went forth against Akkad in battle and combat to attack it. [16] *The Curse of Akkad* and related works tell how there were "flashing potsherds raining from the sky," "many stars falling from the sky," so that "the raining dust rose sky high." [17] The account of cometary impact as told in *The Curse of Akkad* was apparently written within a few centuries after the catastrophic event.

Archeologists have found copies of *The Curse of Akkad* inscribed on clay tablets dating back to 1800 BC. In the last 20 years tests have been performed on dust deposits excavated from abandoned Akkadian sites from around 2200 BC that have indicated the presence of trace elements associated with cometary activity. The discovery five years ago of the two mile wide Amarah Crater in Southern Iraq has provided the physical evidence to confirm suspicions that cometary activity destroyed the Tower of Babel (see Chapter 7 and the Fifth Trumpet).

The Amarah impact, which produced energy equivalent to that of hundreds of Hiroshima sized atomic bombs, took place just 125 miles or so to the south of where ancient Akkad lay. The historical corroboration with the Akkadian Empire, the direct physical evidence from the Amarah Crater, and the dust deposits now provide a powerful one-two punch in establishing the reality of one of the Bible's first stories of catastrophe. Recalling Chapters 4 and 5 which dealt with a number of Old Testament catastrophes, here is added support for the credibility of the Bible's stories of catastrophe, as well as added support for the historical accuracy of the Bible. This breakthrough in determining what really happened at the Tower of Babel also provides some answers to why the Bible says certain events will take place during the end times, and establishes precedent for some of the events the Bible says the God of the Bible has orchestrated to take place during the end times.

It is the discovery that the Bible story of the Tower of Babel is about the destruction of an empire that links it to the destruction of the Akkadian Empire and thus the Amarah Crater. Equally important, this Biblical account holds a warning to mankind, a warning about empire and the God of the Bible's opposition to empire. An empire is a type of government involving people of different lands under the rule of a single political authority, typically having an emperor as chief of state. With such absolute authority comes great power. Having people from different lands under one rule gives an empire the ability to marshal vast resources to accomplish its goals. This would enable an emperor to pursue anything he imagines or wishes. Having one language does not enable a leader to do anything he imagines he can do (*Genesis 11:6*), but having absolute power over all people and lands gives him everything he needs.

The idea that the building of a temple tower was in and of itself threatening to the God of the Bible lacks credibility. It should be noted that the remains of many temple towers have been found in Mesopotamia (approximately 28), and archeologists know that the Tower of Babel was not the first, nor the last, temple tower to be built there. So, the defiant act that kindled divine wrath was not the building of the tower, but rather the building of the world's first empire, since it is ruling an empire that would let a brutal dictator begin to do anything he imagined to do.

A related reason the Bible gives for the God of the Bible's opposition to empire, where one man could rule over all the nations of the earth, is that this position of authority is reserved for the God of the Bible. The Bible says that God is to come in the person of the Messiah, the Christ, to "establish" an empire or kingdom that will last forever. *Isaiah 9:6-7* tells how "the government shall be upon His shoulder" and He will rule from "the throne of David." *Daniel 2:44 NIV* tells how during the end times "the God of heaven will set up a kingdom (empire) that will never be destroyed nor will it be left to another people." *Daniel 7:13-14* tells how the Son of Man (the Messiah) will be given "dominion, and glory and a kingdom (**empire**), that all people, nations, and languages should serve him. . . ."*

Not only did the emperors of the Akkadian Empire expand the empire and seek to rule over many nations, they also aspired to be worshipped as living gods. Even though Sargon was an Akkadian, he made the Sumerian religion the "official" religion of the Akkadian Empire. As stated earlier, by virtue of his holy marriage (*hierosgamos*) to the patron goddess of the new empire, the goddess Inanna bestowed all power to rule on earth to him, and Sargon declared himself to be a god-king. Thus, through marriage to the "woman", a goddess, the emperors of the Akkadian empire were worshipped as

---

*   In *Daniel 7:14* delineating that the "kingdom" (#4437 *malkuw*, dominion, **empire**, kingdom, realm) of the Messiah will include all people, nations and languages," we have the equivalent of the modern working definition of "empire," as a political entity of two or more nations.

living gods (ruler-worship).

This consuming desire to rule with omnipotent power and to be acknowledged as a deity brings to mind several empires throughout history, including the Greek Empire and Alexander the Great's desire to conquer and rule over the entire civilized world as a living god, the Roman Empire and Augustus Caesar's desire to rule and be worshipped as a living god; and the German Third Reich (Empire) and Adolph Hitler's desire to rule over all the world. (More recently consider the dictator Saddam Hussein and his ability to amass expensive weapons and his desire to rule over the entire Middle East.) Just as the story of the Tower of Babel involved a dictator who led a powerful empire that would defy God, Bible prophecy says that during the end times the antichrist will arise and form a worldwide empire and defy God by seeking to rule over the whole world and be worshipped as God. Thus, the message of an all powerful ruler who claims himself to be a god in the story of the Tower of Babel relates to the empire the Bible says will be established by the antichrist during the end times.

The Bible says that the antichrist, a dictator who "will succeed in whatever he does" (*Daniel 8:23-24 NIV*), will eventually seek to be worshipped as God and exalt himself above God.* *II Thessalonians 2:3-4* says this antichrist will even sit in the rebuilt Temple of God in Jerusalem "proclaiming himself to be God." Just as *The Curse of Akkad* and the Bible's story of the Tower tell how a defiant act on the part of mankind kindled divine wrath, resulting in cosmic catastrophe destroying the empire, the Bible says that cosmic catastrophe will destroy the empire of the antichrist. The Seven Trumpets and the

---

*   This is not unlike the Bible telling how Lucifer, or the devil wanted to raise his throne above God's throne in heaven, and sit in heaven and "be like the most high" – *Isaiah 14:12-15*.

Seven Vials of the *Book of Revelation* detail the basic sequence of cosmic impacts that will accomplish this destruction.

There is another reason the Bible gives for the God of the Bible being opposed to empire. An empire with its **state religion** takes away each person's freedom to choose the god he will serve. The God of the Bible has given each person free will and the freedom to choose who or what they will serve or believe. In the empire begun by Sargon, the state religion was primarily the worship of a "woman," a goddess, and in the empire of the antichrist, the state religion will also involve the worship of a "woman," a goddess. Ultimately, the empire of the antichrist will involve worship of the antichrist who will attempt to rule as a living god, just as Sargon did.

Just as the Akkadian Empire, the first empire of mankind involved the worship of a woman, according to the *Book of Revelation* the final empire will also involve the worship of a woman until the empire disavows her (*Revelation 17:16*). Actually this woman, who is called the "The Great Whore," and described as "***Mystery, Babylon The Great, The Mother of Harlots (Idolatresses) and Abominations of the Earth***" in *Revelation 17:1-7*, is none other than Inanna, the "Queen of Heaven" (*Jeremiah 7:18, 44:17-19, and 25*), the "wicked" and "winged" woman for whom a house or tower was built in the land of Shinar at Babylon (*Zechariah 5:7-11, Genesis 10:10, and 11:4-9*). The essence of this idolatrous religious system would be carried forward in time from Nimrod/Sargon's empire to the end times empire of the antichrist. The Bible says that God would cause the destruction of all who worship this "woman," and all who have all invaded Israel, that is, the Neo-Babylonian, Persian, Greek, and Roman Empires (*Daniel 2:24-45, and 7:3-27*). The Bible also says that God will bring the destruction of the antichrist's empire which will represent the third and final expression of Roman Empire (the ancient Roman Empire, the Holy Roman Empire of the Crusades, and the "revived" Holy Roman Empire in the form of the European Community) after it invades Israel during the end times (*Daniel 2:24-45, 7:3-27, 11:36-45,* and *Revelation 13:1-3*). *(See Illustration C)*

Knowledge of certain ancient Near Eastern idioms has led to a more accurate translation of the Bible's story about the Tower of Babel. This clearer translation is supported by the cuneiform writings of the people involved, the historical context, and physical evidence. In turn, this new understanding of what really happened at the Tower of Babel provides new insight into some of the events foretold in the Bible's end times prophecies. We should now be able to see more of the connections between the catastrophes in the Old Testament and those prophesied for the end times. For example, as noted earlier we now have a new perspective of what Jesus meant when he compared the day of his return to **both** the day of Noah's Flood and the day of the destruction of Sodom and Gomorrah, a day "when the sun shall be darkened and the moon shall not give her light and the stars (comets) shall fall from heaven, and the heavenly bodies (NIV) shall be shaken" (*Matthew 24;29, Luke 17:24-30,* also see *Isaiah 34:4,* and *Revelation 6:13-17*). The scientific common denominator for these days, seems to be comet activity.

As this study of the tale of the Tower shows that the God of the Bible used comets to bring the collapse of the pretentious Akkadian Empire, what took place is consistent with the pattern of God using comets as the weapons of His wrath. If we acknowledge that there is indeed a pattern of cometary activity chronicled in the Bible, finding a different explanation other than cometary activity for the number of Bible prophecies that describe the destruction of the empire of the antichrist during the end times will be quite a feat. History matters. Further, it seems that the same type of religious and political machinations that brought on God's wrath and resulted in the destruction of the historical Akkadian Empire by comets, will again bring on God's wrath and result in the destruction of the antichrist's Empire by comets.

*   *   *   *

The Amarah Crater constitutes a two mile wide "smoking gun" for a cosmic bombardment that occurred during Biblical times in the Bible's backyard. Students of the Bible must look into the subject of cosmic bombardment in understanding our past. Bible skeptics should be slow to categorize the Bible's stories and characters as myths. The more accurate translation of the Bible's story of the Tower of Babel made it possible to recognize that it was an account of the sudden collapse of the short lived historical Akkadian Empire. Further, the Bible's many scriptures about the Old Testament comet caused catastrophes help us to recognize the descriptions of the destruction of the Akkadian Empire as recounted in *The Curse of Akkad*. *The Curse of Akkad* and the Biblical story of the Tower represent two different accounts by two different cultures of the **same** disastrous catastrophe. The discovery of the Amarah Crater and the associated dust deposits found in abandoned Akkadian sites provide **physical evidence that cannot be ignored.**

When the collapse of the historical Akkadian Empire and the Bible's Tower story were compared, we found two separate accounts of the same events occurring in the same land during the same time period. The historical record shows that instead of the whole Earth being of one language and one speech, the whole land or region was of one government and one commander, just as the Bible text says when it is more accurately translated. It was one government under one commander that made the people unified, not one language. One government under one commander speaks of empire, and with empire comes absolute authority and power. Both the Akkadian Empire and the empire of the Tower Story represent the same empire. They both included the Sumerian city of Erech, the building of a new city at Akkad, the building of a new temple tower at Babylon (*babilum*), and the same sudden collapse of the empire resulting in abandonment and scattering.

In the Akkadian work, *The Curse of Akkad* this collapse is attributed to "divine wrath" which is brought by cometary activity. In the Bible story this collapse is also attributed to "divine wrath," where cometary activity can be inferred from the overall Biblical text. The Amarah Crater and associated dust deposits in abandoned sites should allow us to acknowledge that the Bible once again has recorded a verifiable event that is both historically and scientifically correct in a story many have previously dismissed as myth.

In conclusion, this chapter is not only important because it toppled another long held misinterpretation of a Bible story, or showed that here is another incident recorded in the Bible where God brought about destruction using His cometary weapons of wrath. What is most important is how this *Genesis* story relates and connects to the *Book of Revelation*. In other words how events in the first book of the Bible relates to events prophesied to occur in the last book of the Bible. The Tower of Babel story is about an all powerful leader (Sargon) posing as divine and his empire, whose multi-national and multi-cultural people were bound together by a common religion involving the worship of a false deity. Sargon's empire was destroyed and its survivors scattered as a result of a comet impact. End times Bible prophecy indicates that this same set of circumstances will again occur with the antichrist as a leader of a vast political empire with one religion who at one point will seek to be worshipped as God. Cometary impact at that time will destroy his empire with horrific finality.

EXHIBIT 1

Translations of
the Story of the Destruction of the Tower of Babel
(*Genesis 10:9-10* and *11:1-9*)

Literal	*Traditional*	*Historically Correct*
Genesis 10:9 He was a great hunter before Jehovah; so it is said; Even as **Nimrod** the great hunter before Jehovah.	He was a mighty hunter before the Lord; wherefore it is said, Even as **Nimrod** the mighty hunter before the Lord.	He was a great hunter before the Lord; so it is said; Even as **Nimrod** (a.k.a. **Sargon**) the great hunter before the Lord.
10:10 And the beginning of his kingdom was **Babylon**, and **Erech**, and **Akkad** - all of them (*Tanakh*) in in the land of **Shinar**.	And the beginning of his kingdom was Babel (**Babylon**- NAS NIV), and **Erech**, and **Akkad** - all of them (NIV) in the land of **Shinar**.	And the beginning of his kingdom was **Babylon, Erech,** and **Akkad**- all of them in the land of **Sumer-Akkad.**
11:1 And the whole land or earth was of **one lip** (*Septuagint*) and of **one word**, cause, **command** or **commander.**	And the whole earth was of **one language** and of **one speech.**	And the whole land was of **one government** and **one commander** (**Sargon** a.k.a. **Nimrod** of *Genesis 10:9*).

11:2 And it was as they traveled eastward, they found a level valley in the land of **Shinar**; and they lived there.	And it came to pass, as they journeyed eastward (NAS, NIV), that they found a plain in the land of **Shinar**; and they dwelt there.	And it came to pass, as they journeyed eastward, they found a plain in the land of **Sumer-Akkad**; and they dwelt there.
11:3 And said each one to his neighbor 'come, let us make bricks and burn thoroughly' And they had brick for stone, and there was asphalt for them for mortar.	And they said one to another, Go to, let us make brick, and burn them thoroughly. And they had brick for stone, and slime had they for mortar.	And they said one to another, Come let us make brick and burn them thoroughly. And they used brick for stone, and they used tar for mortar.
11:4 And they said, Come, let us build for ourselves a city, and a tower with its top in the sky or heavens and make for ourselves an **authority** or **name**, lest we be scattered on the face of all the land our earth.	And they said, Go let us build us a city and a tower, whose top may reach unto heaven: and let us make us a **name**, lest we be scattered aboard upon the face of the whole earth.	And they said, Come let us build for ourselves a city and a tower with its top in the sky and make for ourselves an **authority** (**empire**), lest we be scattered upon the face of the land.

THE TALE OF THE TOWER

11:5 And Jehovah came down to see the city and the **tower** which the children of men built.	And the Lord came down to see the city and the **tower**, which the children of men builded.	And the Lord came down to see the city and the **tower** (*ziggurat*) which the children of men built.
11:6 And Jehovah said, See the people is one or united and the **one lip** to all them, and this they are beginning to do, and now not will be restrained from them all which they have imagined to do.	And the Lord said, Behold the people is one, and they all have **one language**, and this they began to do: and nothing will be restrained from them, what they have imagined to do.	And the Lord said Behold, the people are united, and they all have **one government** and this they began to do: and now nothing will be impossible for them.
11:7 Come, let us go down and there **fail** (*balah*) their **lip**, so that they may not **show forth** their **one lip**.	Go to, let us go down, and there **confound** their language, that they may not **understand** one another's **speech**.*	Come, let us go down and there **cause to fail** their **government** that they may not **regard** or **show forth** their **one government**.

---

*   In *Genesis 11:7* the word traditionally translated as "confound" is the Hebrew word **balah** (#1086 in *Strong's Concordance*). **Balah** is a prime root meaning "to fail." Examination of the Hebrew manuscript shows that the word used in this verse is **balah**, **not** *balal* as so many commentators report.

11:8 So Jehovah **dashed in pieces** them from there over the face of all the **land** or **earth**, and they stopped building the city.	So the Lord **scattered** them abroad from thence upon the face of all the **earth**: and they left off to build the city.	So the Lord **dashed in pieces** them (by **comet impact** – Babylon was less than 125 miles away from the massive impact that left behind two mile in diameter Amarah Crater) from there over the face of all the **land**: and they stopped building the city.
11:9 After this or on this account (*Septuagint*) it's name is called **Babylon**; surely Jehovah **mixed** (*balal*) the **lip** of all the **land** or **earth**: And from there did Jehovah scatter them abroad on the face of all the **land** or earth.	Therefore is the name of it called Babel (***Babylon*** -NAS, NIV); because the Lord did there **confound** the **language** of all the **earth**: and from thence did the Lord scatter them abroad upon the face of all the **earth**.*	After this is the name of it was called **Babylon**: surely the Lord did there **break up** the **government** of all the **land**: And from there did the Lord scatter them over the face of all the **land.**

Note: In *Genesis 11:1, 6, 7* and *9* the word translated as "**lip**" (#8193 in *Strong's Concordance*) could also be translated as "**gathering**" (#5596 in *Strong's*

---

* In *Genesis 11:9* the word traditionally translated as "confound" is the Hebrew word ***balal*** (#1101 in *Strong's Concordance*). ***Balal*** is a prime root meaning "to overflow," and by implication "to mix."

Concordance). As "***one lip***" is an idiom that means "***one government***," "***one gathering***" also conveys the meaning of "***one government.***" Either translation fits the historical context of the Akkadian Empire and its sudden destruction by comet catastrophe.

EXHIBIT 2

Comparison of the Bible's story of the Tower of Babel and the Akkadian Empire

Biblical Tower of Babel	*The Akkadian Empire*
New Empire (*Genesis 10:9-10 & 11:1-9*)	**New** Empire
Founded by **Nimrod** (personal name)*	Founded by **Sargon** (throne name meaning "true king")†
One government under **one commander** (*Genesis 11:1*)	**One government** under **one commander**
In the land of "**Shinar**," which means "**Sumer- Akkad**" in Sumerian, where **two** very different languages were spoken and written	In the land of **Sumer-Akkad** where **two** very different languages were spoken and written
People journeyed to the **east** *Genesis 11:2 (NAS, NIV)*	Semitic people journeyed to the **east** to become the Akkadians
New city of **Akkad**	New city of **Akkad**

---

*   Nimrod is the personal name of the historical person whose throne name was Sargon, which means "***true king***" in Akkadian.

†   Same as above.

THE TALE OF THE TOWER

Empire included Akkad (Akkadian), Erech (Sumerian), and Babylon (Akkadian)	Empire included Akkad (Akkadian), Erech (Sumerian), and Babylon (Akkadian)
New Temple Tower of Babel ('*Babylon*' NAS, NIV)	New Temple Tower of **Babylon*** (a *ziggurat*)
Defiant act brings divine wrath upon the Empire	*Defiant act brings divine wrath* upon the Empire†
Sudden destruction of the Empire (Comets are the weapons of God's wrath – *Isaiah 13:5*)	*Sudden destruction* of the Empire by comet activity – Amarah Crater
People *scattered*	People *scattered*
Bible date *ca. 2200 BC*	Archeological date *ca. 2200 BC*

Physical Evidence

- Two mile wide Amarah Impact Crater in Southern Iraq with

---

* In Akkadian "***babylon***" means "***gate of god***," or "***house of god***" (*Genesis 28:12-17*). Temple Towers or "***babylons***" were built both before and after the events of *Genesis 11* and the rise and fall of the short-lived Akkadian Empire. Over time the Akkadians came to be called Babylonians (and Assyrians) and the region came to be called **Babylonia**.

† *The Curse of Akkad*, an Akkadian historigraphic poem (ca 2000 BC) tells how the cometary gods in their wrath destroyed the Empire by cometary activity.

impact deposits that date to this time period.
- Soil layers containing dust particles with a chemical cosmic signature, resulting from the Amarah impact have been found in suddenly abandoned Akkadian sites in the region.

NINE

Armageddon, The Seven Seals, The Day of the Lord, and A Rock Cut Without Hands

War! War! According to currently popular interpretations of the *Book of Revelation*, battles and nuclear wars are to characterize the end times. But, we have seen that all of the great fiery catastrophes described in the Seven Trumpets and Vials are to be brought about by cometary impacts rather than thermonuclear explosions. We learned that nuclear explosions, even if every one of the world's nuclear arsenals were combined and exploded simultaneously, are not powerful enough to cause the global sized catastrophes that are to characterize the end times. Most importantly, there is no Biblical precedent for man's weapons serving as God's instruments of destruction.* On the other hand, we have seen how the Biblical God repeatedly refers to comets as the weapons of His wrath (Chapter 3). In addition, we have seen how scientific study of the great catastrophes of the Old Testament show that they were caused by comets (Chapters 4 and 5). Only cosmic impacts can deliver 100 pound hailstones. Only cosmic impacts are powerful enough to rock the Earth on its axis, and cause earthquake shaking violent enough to flatten all of the cities of the world. Only cosmic impacts that penetrate the Earth's crust can heat the interior of the Earth enough to make all the mountains and islands of the world disappear, and finally engulf or baptize the entire planet in fire as called for by the last Trumpet and Vial. *(See Illustration I)*

From this baseline of understanding that the Seven Trumpets and Vials are about a series of comet impacts, it is relatively easy to expand our understanding of other parts of the *Book of Revelation* and other end times events described in the Old and New Testaments of the Bible. To get the correct overview of the end times, it is essential to know the Bible says that it is God and not man, who will orchestrate the events of the end times, just as God orchestrated the events during the Exodus. God will use the comets that He says He has "prepared for an hour, and a day and a moth, and a year" as part of His original plan of creation (*Revelation 9:15* and *Isaiah 40:26*). God will deliver His wrath and judgment upon mankind through the power of nature, the natural forces of comets. (*Isaiah 40:25-26,*

---

* Prophetic fulfillment as solely God's responsibility says, "God does not need man's modern inventions to work his will" (Paul Boyer, *When Time Shall Be No More: Prophecy Belief in Modern American Culture*, Harvard University Press, Cambridge, Massachusetts, 1992, pp. 131 and 143). Also there is no Biblical precedent for man's weapons being used by God for destruction.

Isaiah 13:3-7, Psalm 103:20-21, Psalm 104:4 and *Psalm 148:8*. See Chapter 3 for corrected translations of these verses.) He will not be breaking the "ordinances" or "laws" of Heaven and Earth which He says He ordained and cannot be broken (*Job 38:33, Jeremiah 31:35* and *33:25*). These are the same immutable scientific laws that our scientists now use and place their trust in. As the size of the impacts increase, it seems that God will be giving people a chance to recognize what is happening and why. This will be an opportunity for people to recognize that "this is the finger of God," just as Pharaoh's magicians, told him after the Third plague of the Exodus (*Exodus 8:19*).

Chapter Outline

I **Armageddon**
    A. What does Armageddon mean?
    B. Armageddon at the beginning or at the end of the tribulation period?
    C. Who is NOT Magog?
    D. Who was Magog?
    E. Who will be the end times Magog and why?
    F. Rome's relationship to Magog
    G. Where does Armageddon take place?
    H. The Winepress running with blood

II *The Seven Seals Connection to the Seven Trumpets and Seven Vials*

III *The Great and Terrible Day of the Lord*

IV *The Stone Cut Without Hands*

V *The Timing of the Events*

Armageddon

So where does the so-called "mother of all battles," the battle of Armageddon, fit into the end times scenario? The last verses of the Sixth Vial tell of the kings of the whole world being gathered to battle on the "great day of God Almighty" into a place called "Armageddon." The Sixth Trumpet and Vial *(Revelation 9:13-21* and *Revelation 16:12-16)* tell of four comets "bound **for** the great River Euphrates" *(Revelation 9:14)*, that on impact will cause the "waters thereof" to dry up, so that the kings of the east can cross the dry river bed easily as they head toward Israel *(Revelation 16:12)*. The Euphrates River lies to the northeast of Israel in Syria and Iraq. This sets the stage for armies from the east to join armies coming from the north in order to invade Israel. The Sixth Vial *(Revelation 16:14-16)* tells how the kings of the Earth and their armies will be gathered for battle at this place on the great day of God, a place called in the Hebrew language, Armageddon.

What does Armageddon mean?

Where is this place called Armageddon? It has generally been thought that the Hebrew word "Armageddon" *(Har-mageddon)* means "mountain Megiddo" and that "Armageddon" designates a mountain at the ancient Israeli town of Megiddo. However, there is NO mountain **at** Megiddo! The ancient site of Megiddo is only associated with a ten acre, sixty foot high hill archeologists call a "tell" which is made up of many layers of ancient ruins. Few scholars have acknowledged that there is another translation for the word

"Armageddon" (*Har-mageddon*). In Hebrew names often give literal information about the place being named.* So, instead of taking the word "mageddon" to be the name of a town, the literal meaning of the word can be used. In Hebrew *mageddon* means "a place of troops" or "an assembly of troops" or "a rendezvous of troops" (#4023 in *Strong's Concordance*).† Therefore, instead of the commonly recognized translation of Armageddon (*Har-mageddon*) as "mountain Megiddo," in the context of *Revelation's* events, it should be translated as **"a mountain assembly of troops"** or **"a mountain rendezvous of troops."**

A "mountain rendezvous of troops" is consistent with the Sixth Vial that describes the armies of the world being gathered together in a specific place on the day of the Lord. The Hebrew word *har* of *Har-mageddon* tells us that this place of gathering is indeed in the mountains. So, on Armageddon the armies of the world will gather together and assemble in the mountains of Israel. But which mountains? Exactly where is this battle to be fought? To answer this question one needs to search the Scriptures and study those passages that call for an end times battle in the mountains of Israel. According to the details in *Revelation*, this end times battle: a) should take place in the mountains; b) occur on the great day of God Almighty (*Revelation 16:14*); c) involve the nations of the world being gathered together (*Revelation 16:14-16*); d) involve God's wrath destroying these armies (Seventh Trumpet – *Revelation 11:18*, Seventh Vial - *Revelation 16:19*); and, e) involve a big cometary impact resulting in great hail, a great earthquake, mountains not

---

*   For example, Bethel – house of God; and Bethlehem – house of bread.

†   It is interesting to note that *Zechariah 12:11* refers to Valley of Megiddon.

being found (disappearing), islands fleeing, and fire engulfing the Earth (*Revelation 11:18-19* and *16:18-21*).

The end times battle that meets all of these criteria is the battle described in *Ezekiel 38-39* of the Old Testament. Five different times *Ezekiel 38-39* makes reference to a great battle in the **mountains**: 1) *Ezekiel 38:8* says, ". . . in the latter years thou shalt come . . . **gathered** out of many people against the **mountains** of Israel . . .", 2) *Ezekiel 38:21* says, "And I will call for a sword against him throughout all my **mountains** . . .," 3) *Ezekiel 39:2* says, "I will . . . **bring thee** upon the **mountains** of Israel," 4) *Ezekiel 39:4* says, "Thou shalt fall upon the **mountains** of Israel . . . ," and 5) *Ezekiel 39:17* **Assemble yourselves,** and come; **gather** yourselves on every side to my sacrifice that I do sacrifice even a great sacrifice upon the **mountains** of Israel."* **Note the repeated theme of an enemy gathering or assembling in the mountains,** just as *har mageddon* means a "**mountain assembling of troops.**" Clearly, these verses from *Ezekiel* contain the same theme of troops gathering and assembling themselves for battle as *Revelation 16:14-16*. (See Illustration H)

<u>Armageddon at the beginning or at the end of the tribulation period?</u>

The same prophecy scholars who believe that *har mageddon* means "Mountain Megiddo" (as opposed to a "mountain rendezvous of troops") also think that the battle of *Ezekiel 38-39* occurs at the beginning of the tribulation period. They do not connect the battle of *Ezekiel 38-39* with the battle of *Har-mageddon* that is to occur at

---

*   In the *Septuagint Ezekiel 39:2 says,* "I will **assemble thee** and guide thee . . . and I will bring thee upon the **mountains** of Israel."

the end of the tribulation period. They believe that the *Ezekiel 38-39* battle involves a Russian led invasion of Israel at the beginning of the tribulation, rather than an invasion of Israel by the nations of the world at the **end** of the tribulation period.

One of the main problems with having the battle of *Ezekiel 38-39* occur at the beginning of the tribulation period lies in who the Scriptures say will be destroyed. In this battle God's "hot anger" is aroused and He pours out His "fiery wrath" (*Ezekiel 38:17-19 NIV*) on the invading multi-national force and "annihilates" them.* Note that the Seventh Trumpet in *Revelation 11:18* says, "And the nations were angry and thy **wrath** is come and the time of the dead, that they should be judged, and . . . shouldest destroy them which destroy the earth" (also see *Revelation 6:17, 16:19,* and *19:15*). Also, *Zephaniah 3:8* tells of God's "fierce anger" and how He will pour out His "wrath" (NIV) on the nations gathered and assembled for battle on "the day."† Now, if God destroys or annihilates all the armies of the invading multi-national force at the beginning of the tribulation period, then what armies will the nations of the world have to gather, assemble and invade Israel at Armageddon on "the great day of God Almighty," at the end of the tribulation period? If God were to annihilate the invading armies at the beginning of the tribulation period, then what armies would dare try another invasion? It should

---

* In the *KJV Ezekiel 39:2* says that God will "leave but the sixth part" of the invaders. Interestingly *Strong's Concordance* says that the Hebrew word *shaw-shaw* translated as "sixth part" (word #8338) actually means **"annihilate."** Strong's says the error arose because of confusion with word #8341 which means "sixth."

† *Zephaniah 3:8* says, "Therefore wait ye upon me, saith the Lord, until the day that I rise up to the prey: for my determination is to gather the nations, that I may assemble the kingdoms, to pour upon them mine indignation, even all my fierce anger: for all the earth shall be devoured with the fire of my jealousy" (zeal).

be clear that *Ezekiel 38-39* describe the battle of Armageddon, which is to take place at the end of the tribulation period!

Ezekiel 38-39 also tells of the presence and appearance of God to save the Jews and "be *known* in the eyes of many nations" (*Ezekiel 38:20, 23* and *39:7, 21,* and *29)* on this day. *Matthew 24:29-30* states that Jesus is to return at the *end* of the tribulation period. If *Ezekiel 38-39* occurs at the beginning and not at the end of the tribulation period, does this mean there will be two "second comings?"

In *Matthew 24:28* Jesus tells of a great sacrifice where there will be "carcasses" and "vultures" (NAS, NIV) gathered, and *Ezekiel 39:17-20* tells of a sacrifice in the mountains where birds of prey will gather to eat the flesh of mighty men and horses. *Revelation 19:11-16* tells of Jesus' Second Coming upon a white horse; *Revelation 19:17-19* then tells of the armies of the kings of the earth gathered to make war against him and gives an invitation to all of the fowls to gather "together unto the supper of the great God. . . That ye may eat the flesh of kings and the flesh of captains, and the flesh of mighty men, and flesh of horses . . . ." Are there to be two such sacrifices, one from the battle of *Ezekiel 38-39* at the beginning of the tribulation period, and one from Armageddon at the end of the tribulation period? Or, are these nearly identical scriptures describing the same event?

Once the nations of *Ezekiel 38-39* are correctly identified, we have another reason why the battle of *Ezekiel 38-39* will occur at the end of the tribulation period. As the Bible calls for two or three witnesses to establish a matter, *Ezekiel 38-39* constitutes the Old Testament prophecy of this great end times battle, while *Revelation's* battle of Armageddon constitutes the New Testament prophecy of this great end times battle.

Who is NOT Magog?

So who are the nations referred to in *Ezekiel 38-39* and who do they represent? Do they really represent a Russian led multi-national invasion of Israel during the end times as most Bible prophecy scholars believe? In *Ezekiel 38-39* the prophet is told to set his "face against Gog, the land of Magog, the chief prince of Meshech and Tubal and prophesy against him. And say, 'Thus saith the Lord God; Behold I am against thee O Gog, the chief prince of Meshech and Tubal'" (*Ezekiel 38:2, 3* and *39:1*). In *Ezekiel 38:5, 6*, Persia, Ethiopia, Libya, Gomer, and Togarmah are also said to be with Gog. For decades, and still today, most Bible prophecy scholars have erroneously assumed that the ancient nation of Magog in western Asia Minor (modern Turkey) is related to the ancient Scythians, a group of Iranian speaking nomadic tribes from central Asia north of Iran, that ranged across the steppes to the Ukraine and Russia. Again in error, they assume that the ancient Scythians were the progenitors of the modern Russians. Then for a third strike, it has been assumed that Meshech and Tubal (two nations of Asia Minor) associated with Magog in *Ezekiel 38-39* are related to the Russian cities of Moscow and Tobolsk. *(See Illustration E)*

The erroneous idea that Russia is the "evil empire" from the north destined to attack Israel during the end times can be traced back to a small group of 18th and 19th century theologians, who based their assertions about Russia's origins on historical references that were purposefully altered. For a detailed presentation of the nations of *Ezekiel 38-39*, see **APPENDIX B - Russia is Not Magog**. Although evidence that contradicts this conclusion has become common knowledge within the academic and other arenas, this misconception is still prevalent in end times Bible prophecy books.

Actually, among academia today, no professional archeologist or historian associates Magog with the Scythians nor the Scythians with the origins of the Russians. The May 2000 issue of *National Geographic,* reflecting modern scholarship, explains clearly and simply that a group of Scandinavian Viking traders called the **Rus** began the Russian state during the mid AD 800's. It says, "The Slavs and Finns there (near today's St. Petersburg) called them **Rus**, after the Finnic term for Swedes, ***Ruotsi*** . . . These ***Rus eventually founded the first Russian state***, centered on Kiev, in today's Ukraine - and ***gave their name to Russia***, a cultural inconvenience the Soviet historians were compelled to dispute for decades."[1]

The name "***Rus***" or "***Rhus***" appears in the writings of Bishop Troyes in 839 AD. According to the 12th century document known as the *Primary Chronicle*, the land around Kiev was named "***Rus***" and the inhabitants called "***Russes***" in 852 AD. [2] *National Geographic* goes on to tell how ***Novgorod*** (east of St. Petersburg), an early Russian capital was founded by the ***Rus***, and how in the city of Kiev, the ***Rus*** Prince Vladimir converted to Christianity in 988.[3] It is also interesting to note that the renouned Arab chronicler Ibn Fadlan wrote how in 921 AD during the course of his journey to meet with the King of the Upper Volga Bulgar he "met a people called the ***Rus***, a group of Swedish origin acting as traders in the Bulgar capital."[4] (The Upper Volga Bulgar lived on the banks of the Volga River in western Russia. The Volga is Russia's largest and most important river.) In his account (a Risala) which has great historical value Ibn wrote, "I have seen the ***Rus*** as they came on their merchant journeys and encamped by the Volga. I have never seen more perfect physical specimens, tall as date palms, blonde and ruddy . . . they are big men with white bodies." ⁵ *(See Illustration G)*

Today no contemporary history book or textbook accepts the unfounded popular prophecy book claims about Russian origins, instead they clearly show that Russian origins are Scandinavian. Even the author's daughter's high school history textbook detailed how Russian origins are Scandinavian.* Simply put, the Scythians, who were never historically identified as the people representing Magog, had ceased to be a cultural group at least 500 years before the Russians became a recognized culture. Russians trace their origins back to Scandinavia. There is neither a genetic nor a cultural connection between the Scythians and the Russians.

Misconceptions about Russia being the end times invading empire of *Ezekiel 38-39* create an unfortunate international misunderstanding and can provoke unwarranted political action. For example, this error led President Ronald Reagan and others to refer to the Soviet Union as an "evil empire" and "the focus of evil in the modern world." ⁶ Before he became president, prophecy student **Ronald Reagan** spoke at a dinner with California legislators in 1971 and concisely summed up his view of Russia's end times role:

> *Ezekiel* tells us that Gog, the nation that will lead all of the other powers of darkness against Israel, will

---

*　For example, this high school text says, "The arrival of the Vikings is recorded in the *Primary Chronicle*... According to the *Chronicle*, in about A.D. 860 the Slavic people from the northern forest village of Novgorod asked Vikings from Scandinavia for aid: 'Our land is great and rich, but there is no order in it. Come to rule and reign over us.' The Viking leader Rurik accepted the invitation. The Slavs called the Vikings and the area they controlled *Rus*. Trading with the strong, plundering the weak, they (*Rus*) moved south from Novgorod to Kiev, where they founded a political state." *World History: The Human Experience*, McGraw-Hill Company, Gerville, Ohio, 1997, pp. 260-262.

come out of the north. Biblical scholars have been saying for generations that Gog must be Russia. What other powerful nation is to the north of Israel? None. But it didn't seem to make sense before the Russian revolution, when Russia was a Christian country. Now it does, now that Russia has become communistic and atheistic, now that Russia has set itself against God. Now it fits the description of Gog perfectly. [7]

Ironically, in a book issued by a major Christian publisher, *Foes From the Northern Frontier*, Dr. Edwin Yamauchi, a history professor at Miami University, explains that modern Russia's origins are Scandinavian (the ***Rus***).[8] Like other historians, Yamauchi explains how none of the ancient nations referred to in *Ezekiel 38-39* can possibly be related to modern Russia. Further, Yamauchi says:

> Even if one were to transliterate the Hebrew ***rosh*** as a proper name (as does the NAS) rather than translate it as 'chief' (as does the KJV, NIV, and Hebrew *Tanakh*), it can have nothing to do with modern 'Russia.' This would be a ***gross anachronism*** for the modern name is based upon the name ***Rus***, which was brought into the region of Kiev, north of the Black Sea, by the ***Vikings*** only in the Middle Ages. [9]

Yet, just as few within the Christian community seem to be aware of what is now common knowledge that Russia is not Magog, even fewer are aware of the loss of credibility their misidentifying the nations of *Ezekiel 38-39* has produced in the secular academic world. Historian Paul Boyer at the University of Wisconsin writes, "So hackneyed had this scenario (Magog and Sycthia as Russian) become by the 1980's that its proponents hardly bothered with the geographic and linguistic evidence marshaled by earlier writers." [10] Similarly, Daniel I. Block, a professor of Old Testament at Wheaton College in Wheaton, Illinois in *The Book of Ezekiel* writes,

> The popular identification of Russia is ***impossibly anachronistic*** and based on faulty etymology, the assonantal similarities (similar sounds) between Russia and Rosh being purely accidental... ***The name Russia of northern Viking derivation*** was first used for the region of the Ukraine in the Middle Ages. [11]

<u>Who was Magog?</u>

The nations of the world that ultimately invade Israel are the nations of the end times empire led by the antichrist (*Daniel 11:36-45, II Thessalonians 2:3-9, I John 2:18, Revelation 17:8-14, Revelation 19:19,* and *Revelation 20:10*). In establishing his empire, the antichrist will use religion as the common glue that binds together different nations of the world; much like the ancient Akkadian ruler Sargon used the Sumerian religion and the worship of the "Queen of Heaven" (Inanna/Ishtar) to bind together the nations of his Akkadian Empire (see Chapter 8) together, or as the emperors of the Holy Roman Empire (900 AD – 1800 AD) used Roman Catholicism to bind the nations of western Europe together. The antichrist is to establish a new worldwide religion where he is to be worshiped as a living god and is to reign from God's throne in the rebuilt temple in Jerusalem (*II Thessalonians 2:4*). We know a number of Old and New Testament prophecies foretell the destruction of this multinational empire by comets; just as the Akkadian Empire that constructed the temple tower at Babylon was destroyed by a comet impact (see Chapter 8). The two mile wide Amarah Impact Crater in southern Iraq constitutes the "smoking gun" for the destruction of this empire.

Today, archeologists and historians know the exact identities of the nations of *Ezekiel 38-39* based on the dealings these ancient nations had with the Assyrian Royal Court.[12] Remarkably, records of the ancient Assyrian Royal Court describe these ancient nations in the same way that the Bible does in *Ezekiel 38-39* and in *Ezekiel 27:10-15* and *32:26*. For example, *Ezekiel 27:13 (NAS)* tells how Meshech and Tubal traded "vessels of bronze," and the Assyrian records state how the Assyrians took "bronze vessels" as booty in Meshech (Asshurnasirpal II, 885-860 BC) and "bowls with gold handles" as booty in Tubal (Sargon II, 722-705 BC).[13] Meshech's capital was near the ancient city of Gordion (south of modern Ankara in Turkey) and archeological evidence gathered from burials there confirms the ancient accounts; and incidentally attest to the excellence of Meshech's metallurgy.[14]

These Assyrian records give the locations of Magog, Meshech, Tubal Togarmah, and Gomer. *Ezekiel 38:2-5* correctly indicates that they were ancient nations of ancient Asia Minor (Anatolia in modern-day Turkey). These nations spread in a geographic arc from west to east to the north of Israel. On the other hand, Gomer (*Ezekiel 38:5* - Cimmerians) was an enemy of the Assyrians that invaded ancient Asia Minor by coming down from the north around the 8th century BC. The Assyrians called the barbarous invading Cimmerians (Gomer) "creatures of hell." [15] *(See Illustration E)*

These nations all coexisted in Asia Minor (modern Turkey) at a time when Magog (also known as Ludu or Lydia in both the Bible and the Assyrian texts) was led by a militant leader called **Gog** (685-652 BC), about 100 years before the *Book of Ezekiel* was written. *Gog* is the Hebrew spelling for the name of this militant leader from western Asia Minor, who was known as Gyges of Lydia to the Greeks.[16] This same leader was known as "Gugu, king of Ludu" and "Gugu, King of Lydia" to the Assyrians. [17] The Assyrians made great use of **eponyms** (words or names derived from the name of a person) and even published eponym lists. In the Assyrian language "the land of Gugu" is rendered as *Ma-gugu,* just as "the land of Zamua" is rendered as *Ma-zamua.** The Hebrew spelling of *Magugu* is "Magog," and thus, "Magog" simply means "the land of Gog."

When *Ezekiel 38:2* refers to Gog from the land of Magog, as the chief prince of Meshech and Tubal, scripture refers to a specific geographic area – Asia Minor; to a specific time period– when Magog (Lydia), Meshech, and Tubal were coexistent; and to a specific ruler – Gog (Gyges to the Greeks) – who led the defensive efforts of Magog, Meshech and Tubal against invading Gomer (the Cimmerians). Again, the archeological record shows that Magog, Meshech, Tubal, Togarmah and Gomer all ended long before population centers began to develop in Russia. Thus, archeologists know that **none** of the nations referred to in *Ezekiel 38-39* had anything to do with the origins of modern Russia.

---

*   In Akkadian the word *mat* means "land." From *mat-gugga* we get *ma guggu* which means "land of guggu."

Who will be the end times Magog and why?

The historical nations of Magog, Meshech, and Tubal over which Gog was a "prince" as referred to in *Ezekiel 38-39* served as "historical types" of the antichrist and the multi-national confederacy the antichrist is to rule over during the tribulation period. The antichrist's empire is to participate in the great end times "Battle of Armageddon." The Bible uses this historical Gog as a type or picture of the future antichrist in the same way that the Bible uses King David as a type or picture of Jesus' Second Coming as a king and a shepherd to rule over all of God's people (*Ezekiel 34:23* and *Ezekiel 37:22-24*).*

More is revealed by examining the Sumerian language, that was used by the Assyrians as a sacred language, much like Latin has been even until the present used by the Roman Catholic Church. In Sumerian the word *gug* (Gog) means "darkness." In Gog from the land of Magog, the chief prince of Meshech and Tubal from *Ezekiel 38:2*, refers to the chief prince of darkness from the land of darkness, as another type of reference to the antichrist and the lands of his kingdom.† "Gog and Magog" are also referred to in *Revelation*

---

*   *Ezekiel 38:17* gives added insight about Gog being a "type" of the antichrist when it asks, "Art thou he of whom I have spoken in old time by my servants the prophets of Israel . . .? And it shall come to pass at the same time when Gog shall come against the land of Israel, saith the Lord God, that my fury shall come up in my face." Space does not permit, but there are indeed passages in the Old Testament before the *Book of Ezekiel,* where the antichrist is spoken about. There are no passages where Russia or its originators are spoken about, per se.

†   In the chief prince Gog from Magog leading Meshech, Tubal, and Gomer during the end times, we have in ancient terms "the chief prince of darkness from the land of darkness leading men that "caused their terror

20:8, as opposing the saints of Jesus the Christ at the end of the Millennium. Here "Gog and Magog" (Gog and the land of Gog) in their most basic meaning, are taken to be types or references to any "antichrist and the land of this antichrist"; just as *I John 2:18, 22, 4:3,* and *II John 7* say "there are many antichrists."

The *Book of Daniel* and the *Book of Revelation (Daniel 2:32-45, 7:7-28, Revelation 13:1-18,* and *17:7-18)* symbolically speak of a succession of ancient and historical empires. The apostle John in *Revelation 17:10-11* wrote,

> And there are seven kings: five are fallen, and one is, and the other is not yet come; and when he cometh, he must continue a (relatively) short space. And the beast that was, and is not, even he is the eighth, and is of the seven, and goeth into perdition (destruction).
>
> *Revelation 17:10-11*

It is from this succession of empires that the ultimate empire of the antichrist arises. At the time John wrote the *Book of Revelation,* the reigning empire referred to as "one is," was the Roman Empire. Many years later, it would be replaced by the Holy Roman Empire. According to this Scripture, the end times empire of the antichrist will be a third and final expression of the Roman Empire (Roman Empire – Holy Roman Empire – and a modern terminal Roman Empire) growing out of the present day multi-national umbrella, the European Economic Community or EU. Although initially, the antichrist may only be a leader within the European Economic in the land of the living" (*Ezekiel 32:26*) and "creatures from hell" (Assyrian term for Gomer).

Community, eventually this economic/political conglomeration will become the antichrist's empire (the eighth empire of *Revelation 17:11*).*

Beyond Gog having led a multi-national force or empire, as the antichrist is to do, how does *Ezekiel 38-39's* use of the historical leader Gog of Magog (Lydia) as a type of the antichrist, relate to a revived Roman Empire? How does the ancient Magog or Lydia connect to a revived Roman Empire that is to invade Israel during the end times? Not many people know that the roots of Roman culture and legal authority historically come from Magog or Lydia.†

Rome's relationship to Magog

Recall that *Gog* is the Hebrew spelling of the name of the militant leader from western Asia Minor who was known as Gyges of Lydia to the Greeks. Migrants from Lydia called the Etruscans took control of the struggling local population of central Italy and put them on the fast track to becoming the powerful Roman Empire. The *Penguin Dictionary of Archeology* says "the influence

---

* The first empire in this succession of empires from *Revelation 17:10 – 11* is the Akkadian empire, the second is the Assyrian Empire, the third is the Neo-Babylonian Empire, the fourth is the Persian Empire, the fifth is the Greek Empire, the sixth is the ancient Roman Empire, the seventh is the Holy Roman Empire, and the eighth is the antichrist's Empire. The identity of the third through sixth empires is symbolically communicated in *Daniel 2:32-45* and *7:7-28*. The Holy Roman Empire is referred to symbolically in *Revelation 13:1-18* as a beast that recovered from a wound. The ancient Roman Empire is the *"one is"* from *Revelation 17:10*, since the *Book of Revelation* was written around 90 AD during the time of the Roman Empire and its system is still around!

† Roman legal authority called *imperium* is also known as Justinian Law. The word "justice" echoes Justinian Law.

of the Etruscans on Roman civilization was enormous." Grahame Clark in *World Prehistory* states "Roman civilization was deeply indebted to the Etruscans." [18]   In the 5th century BC the famed Greek historian Herodotus, who is called the Father of History, told how the Etruscans came to Italy as a result of a migration from Lydia. Herodotus in *The History (1.94)* wrote that the Lydians set sail across the Mediterranean and colonized Etruria (Tuscany) because of a severe eighteen year famine in Lydia.[19]   To relieve the strain upon their food supply, half of the Lydian residents left with the King's son, and called themselves the Etruscans after his son, Tyrrenius.

Lydian oriental customs were at the heart of the Etruscan culture of Italy. According to Plutarch, Livy and Vergil, when Romulus (twin of Remus) founded Rome in 753 BC, he sent for men of Etruria (Tuscany) who directed him in all sacred ceremonies, ordinances and religious rites according to the *Disciplinia Etrusca.*[20]   These same ceremonial rites were used when Rome began establishing its own colonies in the fourth century BC and later.[21]   By 600 BC the Etruscans took control of Rome and became the ruling aristocracy in a new society with an economy based on trade and commerce. The Etruscans were responsible for the urbanization of Rome, as well as it's army and its economic advancement. They made Rome the leading state in Latium. *(See Illustration E)*

Between 753 and 509 BC Rome was ruled by seven kings; the last three were Etruscan. Then in 509 BC the Romans revolted and overthrew the monarchy and established Rome as a Republic. The Etruscan king on the throne at this time was a tyrant called Tarquinius Superbus. During the Etruscan rule of Rome, the king served as commander of the army, judge and priest. The Etruscans built temples, introduced the worship of the mother goddess and even began the wearing of the purple embroidered toga. The eagle or falcon was a symbol of Lydia, which the Etruscans also used, and in time the eagle headed scepter became a symbol of Rome. The symbol of Rome's political power was the "fasces" (a tightly bound bundle of rods enclosing a double edged ax – see the 1916-1945 US Mercury dime reverse) also came from the Etruscans. Thus, both of the favored symbols of the Roman Empire, the fasces and the eagle were of Lydian origin. From fancy pottery to oriental style burial tombs, the culture of Lydia was lifted and replanted just to the north of the city of Rome. As Gog's descendents became the Kings of Lydia, their Etruscan cousins became the Kings of Rome (*Herodotus 6:22*), and they even used the word "*Etruria*" as a synonym for Italy.

Now we can see the genius in using Gog and Magog as types of the antichrist and the final expression of the Roman Empire. Gog from Magog represents the rightful legal heir to the throne of the Roman Empire, since it was an Etruscan who last sat on the throne of Rome, and the Etruscan right to rule traces back to Lydia, whose eponym during Biblical times was Magog which means "the land of Gog."

Recognition that Herodotus was correct about the Etruscans' origins and that the Etruscans were the true source of Roman culture and Roman legal authority *(imperium)* is something that was not confirmed until recently. On April 3, 2007, the *New York Times* ran a feature story entitled "DNA Boosts Herodotus' Account of Etruscans as Migrants to Italy." The *Times*' report gave new findings that support Herodotus' specific account of the Etruscan migration to Italy from Lydia. The Times reported three new and independent studies by geneticists from different universities in Italy and Spain that support Herodotus' account of the Etruscans migration to Italy from Lydia.[22] One study cited was based on bone samples taken from ancient Etruscan tombs in the Tuscan area of Italy. It showed that based on DNA evidence, Etruscans were genetically different from the general population of modern Italy. Another study published in the April 2007 issue of the "American Journal of Human Genetics," "supports a direct and rather recent genetic input from the Near East, a scenario in agreement with the Lydian origin of the Etruscans."[23] According to this study, "the roots of Etruscans were in the ancient Lydian region of Anatolia."[24] *(See Illustration E)*

A third study has since shown that DNA from four unusual ancient breeds of cattle currently found in Tuscany (Etruria) is different from that of seven other breeds of cattle also currently found in Italy. The DNA of the Tuscan cattle has been related to cattle typically found in Anatolia (western Turkey) and the Near East; while the DNA from the other breeds of cattle found in Italy has been related to herds found in northern Europe. When people migrate by sea to a new land, it is not unusual for them to take their herds of livestock with them. The Etruscans having taken their cattle with them is consistent with Herodotus' early account of them taking "everything useful that would go aboard ship" and setting sail from the ancient port of Smyrna in Western Turkey (ancient Lydia).[25]

The *New York Times* article quoted a technical report by Dr. Ajmone-Marsan and other geneticists showing that evidence from DNA indicates "both humans and cattle reached Etruria from the Eastern Mediterranean by sea."[26] Dr. Anthony Tuck, an archeologist at the University of Massachusetts Center for Etruscan Studies, admits that "three clear genetic threads linking a Tuscan population, human and bovine, to groups in the Near East is pretty compelling evidence."[27]

So what is the point to this exhaustive narrative about the Etruscans being from Lydia (whose eponym was Magog which means "the land of Gog"), and the significant part the Etruscans played in early political Roman history? Well, it is safe to say there have been scores of books written by Christian authors in the last fifty years that speak of the end times Magog representing the present day country of Russia. Ancient historical writings and now genetic DNA studies not only prove conclusively that Russians are not the descendants of Magog, but that immigrants from Lydia to Rome called Etruscans helped to found Rome. It is interesting to note that the Bible weighed in on what has been a very controversial subject for over 200 years, i.e. as to who Magog was. Also the Bible's descriptions about these people from western Asia Minor and their neighbors (*Ezekiel 27:13-14,* and *32:26)* has been upheld.

Where does Armageddon take place?

Now, analysis of the Scriptures has it made clear that the battle of *Ezekiel 38-39* and the battle of Armageddon are one and the same. We have looked at which end time country or political conglomerate Magog represents. The next question is where this battle may take place? *Ezekiel 38-39* repeatedly refers to a battle in the mountains, and Armageddon literally means a "mountain rendezvous of troops." Analysis of Scripture reveals a specific mountain location; it is clear that this great battle would be at Mount Hermon, **not** at the hill of Megiddo as popularly thought.

As noted earlier, in Hebrew names often give information about the place being named. In Hebrew *chermon* (word #2768 in

Strong's Concordance) means "sacred" or "sanctuary" or "asylum."[28] So as "Armageddon" (in the context of the armies of the world that are to gather in the mountains) should be translated as a "mountain rendezvous of troops" (#4023 in *Strong's Concordance*); "Mount Hermon" (in the context of the Jews fleeing these invading armies) should be translated as "sacred mountain" or "mountain asylum." This makes Mount Hermon a sacred mountain, a mountain of asylum or sanctuary. So in the Hebrew words *har maggedon* and *chermon* we have a "mountain rendezvous of troops" taking place at the "mountain asylum" to which Jews have fled. (The Jews taking refuge on rugged Mount Hermon, which is full of caves and hiding places [*Song of Solomon 4:8*] is not unlike the Afghanis who took refuge in the mountains after Russia invaded the country in 1979.)

Mount Hermon is a massive mountain on Israel's northern border with Lebanon. The highest of Mount Hermon's three distinctive peaks is called Mount Zion (Sion) (*Deuteronomy 4:48, 3:8, 9* and *Psalm 125:1* and *133:3*).* This is why scripture sometimes refers to this battle as taking place at Mount Zion or Zion (Sion).† For example, *Isaiah 34:1-10* tells how the "indignation of the Lord" will be upon all nations and their armies to utterly destroy them,

---

*   The peak Mount Zion on Mount Hermon should not be confused with the hill called Mount Zion in the Southwest part of Jerusalem; or "Zion" ("fortress") which is sometimes used as a reference to Jerusalem or as a reference to Israel. Mount Hermon and Mount Zion are over 2 miles high and snow usually covers them all year round, so dews are common. *Psalm 133:3* refers to the dew of Hermon and Zion. *Psalm 125:1* refers to the massiveness of Mount Zion and says that it "cannot be removed."

†   Some argue that Mount Zion (Sion) is just an alternate name for Mount Hermon. Zion (Sion) lends its name to a stream and a large fault on the southeast flank of Mount Hermon. Aryeh E. Shimron, "Early Cretaceous Tectonism on Mount Hermon," *Israel Geological Society, Annual Meeting Ramot, 1989*, p. 141.

and "the mountains shall be melted with their blood," on the "day of the Lord's vengeance" and "the controversy of *Zion*, and the streams ('torrents' -Interlinear) thereof shall be turned into pitch . . . ." "The streams (torrents) thereof" being turned into pitch tells us this is the Mount Zion of well watered Mount Hermon which has a number of streams on it, not the small hill in the southwest part of arid Jerusalem that is also called Mount Zion.* The streams thereof (including the N. Sion, Banias, N. Avar, and N. Guvta) being turned into *pitch* (as a reference to molten rock or lava flowing down the streambeds) speaks of a volcanic eruption to take place on the "day of the Lord's vengeance." [29] This volcanic eruption is taken up later in the chapter. *(See Illustration H)*

Under the Biblical principles of *Ecclesiastes 1:9 and 3:15* where there is "no new thing under the sun," and "that which is to be hath already been" there is Biblical precedent for the armies of the nations coming to fight against Israel and their defeat in the mountains of Israel, specifically at Hermon and Zion. On the other hand, there is no precedent for Israel having won a victory at Megiddo.† It was "under Hermon" that God came to the rescue when Joshua won a great victory against a multinational force (Amorites, Hittites, Perizzites, Jebusites, and Hivites) gathered together against Israel in the large valley south of Mount Hermon *(Joshua 11:1-6)*. *Joshua 11:3* and *7, 8* tell how the Lord "delivered" the armies gathered together into the hand of Israel who chased and smote them from the valley

---

*   *Revelation 14:1* tells how Jesus is to stand on Mount Zion (Sion) with 144,000 saints. This should be the Mount Zion on Hermon, not the hill in Jerusalem. *Joel 2:30-32* tells of cometary events then says that "in Mount Zion (Hermon) and in Jerusalem shall be deliverance."

†   In fact, a great loss was suffered at Megiddo. Josiah, one of Judah's exceptional kings, died in a battle between the Egyptians and the Assyrians at Megiddo (ca. 609 BC).

ARMAGEDDON, THE SEALS AND A ROCK 349

of Mizpah at the foot of Mount Hermon "unto great Zidon" (Zion), which as noted earlier is Mount Hermon's highest peak.

The Winepress running with blood

Additional information indicating that the great end times battle takes place at Mount Hermon and Zion comes from *Revelation 14:19-20*. *Revelation 14:19-20* refers to "the great winepress of the wrath of God," describing that on the day of judgment, the harvest in this "winepress" will be trodden, so that blood will come out of the winepress, even to the height of the horses' bridles, by the space of a thousand and six hundred furlongs. *Revelation 19:15* explains that Jesus on the day of his return, comes dressed for battle to "treadeth the winepress of the fierceness and wrath of Almighty God." Now, if the battle is to take place in the mountains *(Ezekiel 39:17-19)* on Mount Hermon, then the winepress should be nearby. The physical expression (representation) of this "winepress" must be the Hulah Basin that lies below Mount Hermon. *(Joel 3:11-14* tells of the press being full and multitudes being in the valley of decision on "the day of the Lord.") *(See Illustration H)*

Moving down from Mount Hermon, one encounters the upper valley of the Jordan River, and the Hulah Basin. The Hulah Basin is a 5 km x 15 km x 1 km depression that the Jordan River flows through.* Ancient lava flows once blocked the Jordan Valley in the Hulah area and dammed up the river within this depressed area creating a rough bowl shaped area. However, in time the Jordan River cut through the lava dam and continued its rapidly

---

*   This basin is also referred to as the Damascus Basin.

descending journey to the south. (In Hebrew the world *"jordan"* means "descends.") Since ancient times, a small lake called Hulah Lake ("the waters of Merom" *Joshua 11:5 and 7)* has been fed by the waters that drain into this basin, which is likened to a winepress in *Revelation 14*. From Hulah Lake, the Jordan River continues to flow south until it enters then flows into the Sea of Galilee. After it emerges from the Sea of Galilee it continues on until it ultimately reaches the Dead Sea.

In addition to the streams that flow down snow capped Mount Hermon there is a second way for blood spilled on Mount Hermon to get into the winepress. Because of its geology (limestone with hollowed pockets from water percolation, known as a **karst**), Mount Hermon has significant underground drainage, and the basin that lies at the foot of Mount Hermon would in part drain away the blood if vast quantities of it were spilled on the mountain. In fact, a portion of the rainfall and snow melt that occur on Mount Hermon seeps through the ground, emerging from a spring at the foot of Mount Hermon. The waters that flow out of this spring and the streams that run down Mount Hermon constitute the headwaters of the Jordan River. Jordan River water is the lifeblood of Israel, and it begins at Mount Hermon, passes through the basin or winepress, and heads south, eventually emptying into the Dead Sea, the lowest place on the surface of our planet.

When grapes are crushed in a winepress, the released liquid drains out of a pipe at the bottom of the winepress. The section of the Jordan River that emerges from the basin or winepress below Mount Hermon can be likened to this drainage pipe. So, from this winepress like basin, blood and water would then flow out into the Jordan River, one of the very few waterways in Israel whose waters can reach the height of a horse's bridle, as also called for in *Revelation 14:19-20.*

Revelation 14:19-20 says that these blood stained waters are to flow for a total distance of 1600 Hebrew furlongs or about 184 miles.* The distance from the spring at Mount Hermon and the beginning of the Jordan River down to the Dead Sea, and then continuing southeast *up* into the River Zered, with headwaters near Bozrah; is about 1600 Hebrew furlongs. While the River Zered flows from southeast to northwest, an earthquake or impact generated river wave could easily cause water to go up the Zered and flow the entire 1600 furlong length of this interconnected water system. Or, blood could enter the River Zered near the mountain area of Bozrah and flow northwest into the Dead Sea. So, one way or another, blood could run as high as the horses bridles for a total distance of 1600 furlongs. The **only** interconnected water system in Israel that could let blood flow for a distance of 1600 Hebrew furlongs is the Jordan River/Dead Sea/River Zered track. Along this 1600 furlong drainage track, we have the battle in the mountains at Hermon, the "winepress," that is, the bowl shaped Hulah Basin, south of the mountains, the blood running south and high as the horses bridles down the Jordan River into the Dead Sea; and the slaughter in the mountains near Bozrah discharging blood stained water running down the River Zered into the Dead Sea. These parallels are haunting. *(See Illustration H)*

Now, *Revelation 19:13* has Jesus appearing, "clothed with a vesture dipped in blood," and *Revelation 19:15* tells how he "treadeth the winepress." *Isaiah 34:6* says that "the Lord hath a sacrifice at Bozrah," *Isaiah 63:1-6* asks who is coming from Bozrah with

* A Hebrew "furlong" is approximately 606 feet and it is equivalent to a Roman "stadion." A Hebrew furlong is 1/8 of a Roman mile which is 4,854 feet. (606 x 1600 = 969,600 ÷ 5,280 = 183.7 miles)

garments like he has trodden in the winepress, and *Jeremiah 49:22* tells of the Lord flying like an eagle over Bozrah. *Revelation 14:1* tells of Jesus appearing on Mt. Zion with 144,000 faithful Jews *(Revelation 7:1-8)* who have taken refuge there.* (Recall, that Mount Hermon can mean "sacred mountain" or "mountain asylum.") The Mt. Zion of *Revelation 14:1* would be the highest peak of Mount Hermon *(Deuteronomy 4:48)*, not the small hill in Jerusalem, incapable of accommodating 144,000 people.

It seems that at the main battle site of Mt. Zion on Mount Hermon, Jesus trods the winepress, the bowl like area at the foot of the mountains, where the invading armies have assembled. From Mt. Zion at Mount Hermon, Jesus is to then head south to the mountainous area near Bozrah and effect another bloody slaughter. This is why *Isaiah 63:1-5* and *Jeremiah 49:22* tell of Jesus flying like an eagle coming from Bozrah with garments already stained with blood as if he had trodden the winepress. Presumably, he is seen flying like an eagle on his way to effect a third bloody slaughter. (*Isaiah 31:4-5* and *Jeremiah 48:40* may apply here). This time at a smaller winepress, called the "Valley of Jehosphaphat" (*Joel 3:11-15*), which seems to be in or around Jerusalem. The "multitudes in the valley of decision" referred to in *Joel 3:14* may be a reference to the much larger valley below Mount Hermon where the main battle takes place.† These

*   Isaiah 31:4-5 NIV says "so the Lord Almighty will come down to do battle on Mount Zion and on the heights." *Isaiah 31:4-5* in the *Septuagint* says, "so the Lord of hosts shall descend to fight upon the Mt. Zion (Mount Hermon), even upon her mountains."

†   The location of the Valley of Jehoshaphat is not certain. It has been equated with the Valley of Barachah, south of Bethlehem (*II Chronicles 20:26* and with several valleys in and around Jerusalem. Tradition associates it with the Valley of Hinnom in Jerusalem. Since *jehoshaphat* means "Yahweh judges," and a location called the "Valley of Decision," is also referred to, the use of the term *jehosphaphat* may simply be a refer-

sites and topography seem to fit the scene painted by Scripture.

Again, considering *Ecclesiastes 1:9* and *3:15*, where that which is to be has already been; it is interesting to note that Jehoshaphat, like Joshua, won a great victory against an invading multinational force (the children of Ammon, Moab and Mount Seir), when the Lord once again fought for Judah. So, in the case of Joshua's great victory at Hermon/Zion (*Joshua 11*) and Jehoshaphat's great victory (*II Chronicles 20*), we have historical precedent for the battle of Armageddon, where Yahweh will judge in a valley of decision by again fighting for Israel.

The Seven Seals Connection to The Seven Trumpets and Seven Vials

As we have seen, *Revelation* tells about catastrophic events to take place during the end times. In particular, the Seven Trumpets focus on how seven separate impact episodes will affect the Earth, while the Seven Vials focus on how these *same* seven impact episodes will affect humanity. In essence, *Revelation* gives two witnesses for each round or episode of cometary impact, just as the Bible teaches that two or more witnesses are necessary to establish a matter (*Deuteronomy 19:15,* and *Genesis 41:32*) beyond dispute.

Understanding the disasters of the Seven Trumpets and Vials of *Revelation* is essential for understanding the main thrust of the end times, even though *Revelation* calls for other catastrophic events to take place during the end times. Before the Trumpets and Vials, *Revelation* describes a series of frightening events called the Seven Seals.[30] There has been considerable confusion about

---

ence to the judgment God brings about in this valley.

the relationship between the Seven Seals, the Seven Trumpets and the Seven Vials. It is beyond the scope of this book to discuss the Seven Seals in detail, but it is important to at least understand the basic relationship between the Seven Seals, Seven Trumpets and the Seven Vials. This results in a more complete and clearer picture of the catastrophes the Bible prophesies will happen during the end times. In doing this we go from the relatively straight forward catastrophic physical events of the Seven Trumpets and Seven Vials to the somewhat more complex societal, physical, religious, political and economic events of the Seven Seals.

Just as understanding how the people of the ancient Near East and the authors of the Bible wrote about comet activity allows us to recognize the parallel relationships between the Seven Trumpets and Seven Vials; similar understanding also enables us to recognize the correct relationship of the Seals to the Trumpets and Vials. Based on knowledge of how the people of the ancient Near East wrote about cometary activity, we see that one of the Seven Seals clearly describes the Earth being bombarded by comets. Naturally one wonders how the bombardment event described in this particular Seal relates to the other impact events described in *Revelation*. For the most part, the Seven Seals (*Revelation 6:1-17* and *8:1-5*) describe terrible geopolitical, religious, and economic events and conditions. The first four Seals tell of four horsemen, the "so-called four horsemen of the apocalypse" riding forth and bringing false religion, war, famine, and a deadly plague, causing death to the Earth.* The Fifth Seal

---

*   The seemingly innocent *white horse* of the *First Seal* brings a *"false religion."* The blood colored *red horse* of the *Second Seal* uses the "false religion" of the First Seal to bring world-wide political changes that in turn brings *"war."* The ominous *black horse* of the *Third Seal* uses the devastation of the *"war"* of the *Second Seal* to bring food shortages and *"famine."* The *sickly pale* or *pale green horse* of the *Fourth Seal* uses the

concerns martyrs who are to be killed for their belief in "the word of God" and their refusal to accept the "false religion" of the tribulation period (*Revelation 6:9-11, 7:14, Matthew 24:21,* and *Jeremiah 30:7*).[*] The Sixth Seal then clearly describes an event where the Earth is bombarded by comets.

Now, the obvious question . . . is the cometary bombardment event described in the Sixth Seal in some way related to one of the seven separate cometary bombardment events described in the Seven Trumpets and Vials? The answer is yes! If the Seven Seals (*Revelation 6:1-17* and *8:1-5*) give an overview of the major events of the seven year tribulation period, there should be some overlap between one or more of the Seals and one or more of the Seven Trumpets and Vials; and indeed there is. The cometary activity of the Sixth Seal relates to the cometary activity indicated for the battle of Armageddon that occurs at the end of the Sixth Vial and the beginning of the cometary activity described in the Seventh Trumpet and Vial.[†]

malnutrition and famine" of the *Third Seal* and a *plague* (NIV, NAS) that causes widespread death; while the sword of religions and political persecution continues to bring death. This sequence brings to mind the geo-political changes, famine, and plague (1918) that followed World War I.

[*] In *Matthew 24:21,* Jesus calls this a time of "great tribulation" and says that there has never been, nor shall there ever again be a time such as this. In *Jeremiah 30:7* this same unique time is referred to as "the time of Jacob's trouble."

[†] The cometary activity at the **end** of the Sixth Vial occurs *after* the cometary impacts called for in the beginning of the Sixth Trumpet and Vial. Recall that the Sixth Trumpet tells of the loosing of four comets "which are bound for the great river Euphrates" (which kills the third part of mankind). Then the Sixth Vial compliments this by telling how "the great river Euphrates" is struck and "dried up that the way of the kings of the east might be prepared." Finally, the Sixth Vial tells how *all* the kings of the earth will be gathered together to "the battle of *that great day of God Almighty,*" "into a place called in the Hebrew tongue *Har-mageddon,*" which means a mountain gathering together of troops.

Earlier in this chapter it was shown that the battle of Armageddon and the battle of *Ezekiel 38-39* are the *same* final end times battle when God shows the fierceness of his wrath *(Revelation 6:17, 11:18, 16:19, 19:15, Ezekiel 38:19,* and *Isaiah 34:2 NAS)*. As explained earlier, this battle is to involve the kings of the Earth gathering together for a battle in the mountains and comets (bringing great hailstones, fire and brimstone) falling to the Earth, including a large impact that causes a great earthquake and raises dust into the atmosphere on "the great day of his (God's) wrath" *(Revelation 6:17)*. The Sixth Seal as well as the Seventh Trumpet and Seventh Vial all involve these same elements: the kings of the Earth gathering together for a battle in the mountains, comets falling to the Earth *(Revelation 16:17-21)*, and a large impact causing a great earthquake and raising dust into the atmosphere (sun blocked and moon turning red).

These events occur on a day the Sixth Seal calls the great day of God's wrath, the day of which the Seventh Trumpet *(Revelation 11:18)* says, "thy wrath is come," and the Seventh Vial *(Revelation 16:19)* says God will show "the fierceness of His wrath" (also see *Isaiah 34:1-4* and *2:10-22)*.[*] The Sixth Seal *(Revelation 6:13)* tells about comets coming in to bombard the Earth, "the stars of heaven fell unto the earth" (see also *Isaiah 34:4* and *Matthew 24:29* which also tell of the stars falling from heaven on this day) while *Luke 21:25* says the sea and waves will roar. The Sixth Seal also tells how these

---

[*] The Seventh Vial which tells of the "fierceness of (God's) wrath" *(Revelation 16:19)* seems to give details about the Battle of Armageddon, and like the Sixth Seal, the Seventh Vial talks about a comet impact that is so large that it causes the mountains and islands to move on the day that God *shows* "the fierceness of His wrath." Like the Seventh Trumpet and *Ezekiel 38-39*, the Seventh Vial also tells about the comets bringing "great hail."

impacts cause "a great earthquake," with dust from the impacts rising up into the atmosphere so that the "sun became black... and the moon became as blood" with the shock waves from the impacts causing the atmosphere to roll back ("the heaven [atmosphere] departed as a scroll when it is rolled together" – *Revelation 6:14; Isaiah 34:4* says 'the heavens shall be rolled together as a scroll'). Some of the later impacts are so large that "every mountain and island were moved. .." *(Revelation 6:12-17).**

> And I beheld when he had opened the sixth seal, and, lo, there was a great earthquake; and the sun became black as sackcloth of hair, and the moon became as blood; 13) And the **stars of heaven fell unto the earth,** even as a fig tree casteth her untimely figs, when she is shaken of a mighty wind. 14) And the heaven (atmosphere) departed as a scroll when it is rolled together; and every mountain and island were moved out of their places. 15) And the kings of the earth, and the great men, and the rich men, and the chief captains, and the mighty men, and every bondman, and every free man, hid themselves in the dens and in the rocks of the mountains; 16) And said to the mountains and rocks, Fall on us, and hide us from the face of him that sitteth on the throne, and from the wrath of the Lamb. 17) For the **great day of his wrath** is come, and who shall be able to stand.
>
> *Revelation 6:12-17*

---

*   Also see *Revelation 16:18-20.*

Since the Sixth Seal covers the same events called for at the end of the Sixth Vial and the beginning of the Seventh Vial, it is clear that the Seven Seals give a general overview of the major events of the seven year tribulation period, including the time period of the Seven Trumpets and Seven Vials. As already noted, the first four Seals tell of the first few years (3-1/2 years?) of the tribulation period begin with political and religious upheaval, the subsequent move to war, then worldwide famine, and finally worldwide plague and death. On the other hand, the Seven Trumpets and Vials concern tell of the last few years of the tribulation period (3-1/2 years?) which are to be characterized by seven rounds or episodes of comet impacts.

In addition to the Sixth Seal covering the same events described in part of the Sixth and Seventh Vials, the tragedy of massive deaths across the world related in the Fourth Seal *may partially* be caused by one or more of the impacts given by the earlier Trumpets and Vials. Note that the Fourth Seal tells about the fourth part of the Earth being subject to death. The First Seal spoke of the white horse bringing religious deception; the red horse of the Second Seal bringing war, the black horse of the Third Seal bringing famine, and the pale [or light green] horse of the Fourth Seal bringing plague (NIV, NAS) and death.

It is possible that the plague and death brought by the pale horse of the Fourth Seal is partly the result of comet impact. This plague or pestilence may be from the Fifth Trumpet and Vial's impact that opens up a deep pit or crater, where the dust raised by the impact contains "locusts," which are used to symbolize the pathogens that may be borne by a comet [a controversial idea]. The Fifth Trumpet tells how this bacteria torments men, and the Fifth Vial tells how

ARMAGEDDON, THE SEALS AND A ROCK

men "gnawed their tongues" because of the pains from their "sores."*
The Sixth Trumpet tells of comets **bound** to strike the "great river Euphrates," while the Sixth Vial tells how the water of the "great river Euphrates was dried up, that the kings of the east and their armies could pass and gather together with the other kings of the Earth in the mountains of Israel at a place called *Har-mageddon*.

By the Sixth Seal, it is easy to recognize that the "Sequence" of Seven Seals covers some of the same events given by the Sixth Vial's ending and the beginning of the Seventh Trumpet and Vial (see illustration below). More specifically, this is the period of the great end times battle called Armageddon, the same one as the battle described in *Ezekiel 38-39*.

First Seal   4th   5th   6th   7th

Seventh Seal (Silence in heaven - *Revelation 8:1* before impacts begin, also see *Matthew 24:36*)

First Trumpet   5th   6th   7th

Seventh Trumpet (impacts, earthquake, great hail, nations judged)

---

*   To be more definite about the match-up between the Fourth Seal and the Fifth Trumpet and Vial, more scriptural integration is necessary.

```
              First Vial   5th      6th     7th
              |_____|_____|_____|_____|
```

Seventh Vial (impacts, earthquake, great hail, nations judged)

```
Trumpets                    5      6      7
                            /      /      /
   SEALS  1     2    3    4    5    6    7
                            \      \      \
Vials                       5      6      7
```

ARMAGEDDON

Ezekiel 38-39 and Armageddon (at start of final impacts, earthquake, great hail, volcanic eruption, nations judged).

In other words, the last three sequences of prophesied events given in the Seven Seals, the Seven Trumpets and the Seven Vials all come to an end at *about the same* time during the last part of the tribulation period, and all involve the same basic events!

The Sixth Seal (*Revelation 6:12-17*) tells of comets that will bombard the Earth and cause a "great earthquake" with mountains moving. Dust will be rising into the atmosphere so that it blackens the sun and reddens the moon. This will occur while the "kings of the earth," gathered in the mountains of Israel, ask the mountains "to hide" them "for the great day of his (the Lamb's) *wrath* is come" (also see *Isaiah 2:19-22*). God's victory is clear when *Revelation 6:17* asks "who shall be able to stand?"

The Seventh Trumpet (*Revelation 11:15-19*) also speaks of comet bombardment and an impact generated earthquake, since it is associated with "great hail," or ice blocks that come from the ice of comets breaking up. Then the Seventh Trumpet tells of the **nations** "being angry." This equates to the Sixth Seal referring to the "kings of the earth" as a target. The Seventh Trumpet explains that the "nations were angry," because the Lord's "wrath is come. . . that they (the nations) should be judged." The nations of the Earth being angry and the Lord's wrath coming equates to the Sixth Seal telling how "the great day of His (the Lambs) wrath is come." As shown by the Sixth Seal asking "who shall be able to stand?" Victory is made clear, because the Seventh Trumpet (*Revelation 11:18)* refers to the Lord destroying "them which destroy the earth," and says, "The Kingdoms of this world are become the kingdoms of the Lord" *(Revelation 11:15).*

Like the Sixth Seal and the Seventh Trumpet, the end of the Sixth Vial and start of the Seventh Vial also tell of nations gathering and a comet bombardment producing a "great earthquake." This earthquake is to be the largest one that mankind has experienced, the largest since mankind has been on the Earth. This earthquake's magnitude is so great that it is to cause all the cities of the nations to fall, and its intensity moves islands and mountains. As we learned in Chapter 2, the planet Earth by itself cannot produce an earthquake of this magnitude. Only a large cosmic body such as a 100 mile diameter comet hitting the Earth could produce this kind of force. Just like the Seventh Trumpet, the Seventh Vial tells of "great hail" to come from the ice of comets breaking up. This great hail is to fall upon men and every icy stone is to weigh about 100 pounds. The Seventh Vial, like the Sixth Seal and the Seventh Trumpet, also tells of God's wrath and the "fierceness of his wrath" being poured out upon the empire of the antichrist, which traces its religious and political roots back to the Akkadian Empire of Sargon, which involved a marriage of religion and government ("church and state") in the land of Sumer and Akkad, that came to be called Babylonia. (This was the empire that built the tower of Babylon. What unified this empire was the worship of the "Queen of Heaven," which the Bible calls "Mystery Babylon.")

It is important to recognize that the Sixth Seal, the Seventh Trumpet, the Sixth and Seventh Vials and *Ezekiel 38-39* have a number of different prophesied elements in common (*II Timothy 3:16*). These elements include comet bombardment of the Earth, an impact generated "great earthquake," mountains being moved, and "great hail" from comets that fall. As well, the "kings of the earth" and the armies of the nations gather together for battle in the mountains of Israel, and God's wrath and judgment come upon the armies of the nations, resulting in God's victory on a set day.

As noted, the authors of most popular prophecy books have not recognized that the Seven Vials are actually second witnesses of the same seven impacts described in the Seven Trumpets. Instead, it is generally thought that there are 14 separate events, the Seven Trumpets being followed by the Seven Vials. Many of these same authors also believe that all of the Seven Seals precede the Seven Trumpets and the Seven Vials to give a total of 21 separate disastrous events. The main problem with this interpretation is that the Seven Seals tell of war, famine, and death, leading up to the Kings of the Earth hiding in the mountains as comets and cometary debris fall to the Earth and decimate the planet. Now, if the Seven Seals destroy most of the world's population and decimate the planet, who and what would be left for the Seven Trumpets to destroy? Who and what would be left for the Seven Vials to destroy? The inability to correctly identify the actual "cometary" cause of the Sixth Seal, and the comet impacts of (the events described in) the Seven Trumpets and the Seven Vials made it difficult for most prophecy book authors to ultimately recognize the obvious connections among the Seals, Trumpets, and Vials given by the author in these pages.

The Great and Terrible Day of the Lord

As we have shown the reader, the Sixth Seal, the Seventh Trumpet, the Sixth and Seventh Vials and *Ezekiel 38-39* have a number of different prophesied elements in common, culminating with the appearance of the Lord. One would expect numerous scriptural references to *such* a day of importance! Many have believed that the Sixth Seal, the Seventh Trumpet, the Sixth and Seventh Vial and *Ezekiel 38-39* speak of different dates. But, how many days can be called "**the day of the Lord**," and how many Second Comings actually occur? Can men shake at the Lord's *presence* when he reveals himself to the nations during the battle of *Ezekiel 38-39*, and then at another time hide from the *face* of the Lamb on the "Great Day of His Wrath" as called for in the Sixth Seal? What then of the Lord's wrath and judgment upon the nations as called for in the Seventh Trumpet and the fierce wrath upon the world called for in the Seventh Vial? How many times can God's wrath *utterly destroy* the armies of the nations? After the armies of the nations are destroyed, whence do nations get new armies so that God's wrath completely destroys them again, and again?

We have the "fire of (God's) wrath" upon the armies of the nations on "*the day* whereof I (God) have spoken." (*Ezekiel 38:19 and 39:8*), which seem to be the same day that the Sixth Seal calls the "great day of his wrath." (*Ezekiel 39:8* indicates that God has previously spoken of this day when it says "this is the day whereof I have spoken." There are a number of instances in the Bible where God has spoken of this day before the time of Ezekiel. For example, the prophet Isaiah [ca. 740-680 BC] came before the prophet Ezekiel [ca. 592-570 BC], and *Isaiah 2:11-22* and *Isaiah 13:3-13* speak of this day, the "day of the Lord" and his wrath where the Earth shakes terribly and moves out of her place.") [31]

On this day, *Ezekiel 38:20* exclaims "all men that are upon the face of the earth shall shake at my (God's) presence," and *Ezekiel 38:23* reveals saying that on this day God will "be known in the eyes of many nations!" The Sixth Seal tells how the "kings of the Earth" will ask the mountains to hide them from the "face" and the "wrath of the Lamb" (*Revelation 6:15-17*), just as *Isaiah 2:19-22* tells of men hiding in the rocks and caves when the Lord "ariseth to shake terribly the Earth."

Revelation 19:11-15 describes Jesus returning astride a white horse to make war, judge and smite the nations while clothed in a garment having been dipped in blood. *Revelation 19:19* tells of the beast who is the antichrist and his "armies gathered together to make war against him (Jesus) that sat on the horse." *Revelation 19:17-18* directs all fowl that fly to "come and gather yourselves together unto the *supper* of the great God. That ye may eat the flesh of kings . . . and the flesh of all men" after the slaughter that takes place on this day. This is consistent with *Matthew 24:28* where Jesus warns how the vultures will gather together to feed on the bodies of those who die in this battle. Jesus says, "Wherever the corpse is, there the vultures will gather" (*Matthew 24:28 NAS*).

The *"supper"* of men's flesh offered to the fowl that fly in the sky of *Revelation 19* seems to also be the *"sacrifice"* spoken of in *Ezekiel 39:4* and *17-19*. *Ezekiel 39:4* and *17-19 say:*

> "Thou shalt fall upon the **mountains** of Israel, thou and all thy bands, and the people that is with thee: I will give thee unto the *ravenous birds* of every sort . . . (17) . . . Speak unto every *feathered fowl,* and to every beast of the field . . . gather yourselves on every side to my **sacrifice** that I do **sacrifice** for you, even a great **sacrifice** upon the mountains of Israel, that ye may eat flesh and drink blood. (18) Ye shall eat the flesh of the mighty and drink the blood of the princes of the earth . . . (19) And ye shall eat fat till ye be full, and drink blood till ye be drunken, of my **sacrifice** which I have **sacrificed** for you.
>
> Ezekiel 39:4 and 17-19

Zephaniah 1:7-8 talks about this sacrifice, when the Lord is present on "the day of the Lord"* *Zephaniah 1:7-8* says, "Hold thy peace at the **presence** of the Lord God for **the day of the Lord i**s at hand: for the Lord hath prepared a **sacrifice**, he hath bid his guests. And it shall came to pass in the day of the Lord's **sacrifice**, that I will punish the princes and the King's children . . . ." It is interesting to note that

---

*   Recall that *Ezekiel 38-39* also tells of the Lord being present (*Ezekiel 38:20*) when this sacrifice takes place (*Ezekiel 39:4 and 17-19*).

Israel's unique location at the junction of three continents has made it a major crossroad for bird migration. Over 500 million "migratory birds, almost twenty-five percent of the world's migratory birds, fly over northern Israel on migrations traversing Europe, Western Asia and Africa. Large concentrations of birds of prey including vultures, buzzards, eagles, and hawks migrate over Israel twice per year. [32]

Revelation 14:1 describes Jesus standing on Mt. Zion (Sion), the highest peak of Mount Hermon. *Revelation 19:15* says that Jesus is the one who treads the "winepress," which we learned is located below the aforementioned Mt. Zion. This is the location of the battle of *Ezekiel 38-39* and the battle of Armageddon (which are one in the same), so that the blood comes out of the winepress as high as the horses bridles down the Jordan River and into the Dead Sea and then up the River Zered for a distance of 1600 furlongs as called for in *Revelation 14:14-20*. Earlier in this chapter it was explained that, there is another winepress and slaughter at Bozrah near the River Zered (*Isaiah 63:1-4* and *34:1-8).* *Isaiah 63:4* says that this slaughter takes place on "the day of vengeance," a day of great woe also referred to in *Isaiah 61:2*. *Isaiah 34:1-8* speaks of the "day of the Lord's vengeance" in the context of the same events found in the Sixth Seal, the Seventh Trumpet and the Sixth and Seventh Vials. *Isaiah 34:1-8* tells of the "host of heaven," that is, the comets of heaven falling, the indignation of the Lord being upon the armies of the nations, so that "he hath utterly destroyed them and delivered them to the slaughter," and including "a sacrifice in Bozrah," where "the mountains shall be melted with their blood." [*] *(See Illustration H)*

---

[*]   *Isaiah 2:10-22* tells of men "hiding in the mountains when the Lord will ariseth to shake terribly the Earth on **the day of the Lord.**" Using different descriptions for the comets, *Isaiah 13:1-13* tells of the Lord mustering the host (comets), His sanctified ones (those set apart – *Job 38:22-*

ARMAGEDDON, THE SEALS AND A ROCK

As shown a few pages before, *Revelation 19* essentially describes Jesus appearing on a white horse clothed in a bloody garment to make war and smite the kings of the Earth and the armies of the nations, which have gathered together. It foretells of Jesus coming to tread the "winepress" and providing a supper or sacrifice so that fowl may eat the flesh of Kings along with the flesh of all men on earth. The horrors of *Revelation 19* are clearly consistent with the events called for in the Sixth Seal, the Seventh Trumpet, the Sixth, Seventh Vials, and *Ezekiel 38-39*. The day of Jesus' Second Coming and his judging the nations is the day of His wrath, the day of vengeance, and the day of the Lord; these days are all one in the same.

In all of this, a great number of Scriptures fit together and work together, supporting each other, giving a more detailed picture of the events to take place at the end of the tribulation period. Contrary to what most end times prophecy books have said, the Scriptures do not support modern theories of nuclear events or nuclear warfare. However, as we have seen, there are scores of Scriptures describing comet caused disasters. Similarly, the inability to identify the cometary causes of the major Old Testament catastrophes (discussed in Chapter 4 and 5) has made it difficult for authors of prophecy books to recognize that God used comets as the weapons of His wrath in the past, and that He will use comets again to bring about the disastrous events prophesied in the book of *Revelation*.*

---

23), the weapons of His wrath, who came from a far place, the edge of heaven (the Oort Cloud), and the kingdoms of the nations gathered in the mountains as destruction from the Almighty on **the day of the Lord**, a day of fierce anger, with the sun being darkened from the dust raised by the impacts, with an impact so large that it shakes the Earth out of her place.

*   Recall, that in Greek the word *dis-aster* literally means "bad star," where the word "star" *(aster)* can refer to any luminous body in the sky

On the day of God's wrath, God is to take vengeance on the nations of the world that have become part of the antichrist's empire and have invaded Israel. On this day, God is to use comets, the sanctified weapons of His wrath to fight for and save Israel, just as comets served as His weapons of choice in the past. God is to destroy the forces of the end times empire of the antichrist with comets, just as he destroyed the Near East's first empire (the Akkadian Empire) with a comet (recall the Amarah Crater in Iraq). This is the day when the Lord says, "gather the nations that I may assemble the kingdoms to pour upon them mine indignation, even all my fierce anger: for all the earth shall be devoured with the fire of my jealousy" (*Zephaniah 3:8*).*

Of this day, in *Matthew 24:29-30*, Jesus says, "Immediately after the tribulation of those days shall the sun be darkened, and the moon shall not give her light, and the stars (comets) shall fall from heaven, and the powers (objects) of the heavens shall be shaken: (30) And then shall appear the **sign of the Son of Man** in heaven; and then shall all the tribes of the earth mourn, and they shall see the Son of Man coming in the clouds of heaven with power and great glory." In Cecil B. DeMille's movie *The Ten Commandments,* the scene that portrays the parting of the Red Sea is startling, but it pales in comparison to what is foretold in *Matthew 24:29-30*, where light is to suddenly shine out of darkness (*II Corinthians 3:6*).

---

including comets.

*   It is interesting to note that *Joel 3:11-15* first tells the heathen to assemble themselves and then asks the Lord to "cause thy **mighty ones** to come down." The "mighty ones" referred to in *Joel 3:11* are comets; just as *Isaiah 13:3* also refers to the Lord's comets as "mighty ones."

ARMAGEDDON, THE SEALS AND A ROCK

The Stone Cut Without Hands

Now, in addition to the use of comets, at the battle of Armageddon (*Revelation 16:14-16* and *Zephaniah 3:8*) there is even more Scriptural information describing the great end times battle of Armageddon that takes place at Mount Hermon in northern Israel. This additional information comes from the *Book of Daniel*. *Daniel 2:34-35* and *45* tell about a "stone . . . cut out of the mountain without hands that smote the kingdoms of the world." These kingdoms relate to the armies of the nations of the world being destroyed at the battle of Armageddon on the day of the Lord. But how can a stone be cut out of a mountain without hands, and how could the stone then destroy armies?

To geologists a stone cut out of a mountain with no hands that can destroy armies describes a ***nuée ardente*** which involves the side of a volcanic mountain exploding outward (horizontal eruption) instead of the top of a volcanic mountain exploding upward (vertical eruption). The French term *"nuée ardente"* refers to the "glowing cloud" (*nuée* – cloud, *ardente* - glowing) of material that hugs and rolls down the mountain slope after a lateral volcanic eruption. This cloud of fiery material contains gases, ash, lava, and rock fragments of all sizes known as pyroclastics, some the size of a house, that are shot out by the explosive eruption. Not only is the rocky material plugging an old volcanic caldera or crater blown out or "cut out," but a large part of the side of the mountain is also "cut out" and incorporated into the glowing cloud of material that begins avalanching down the mountain slope. This glowing cloud can reach temperatures up to 2000° F and move at speeds that are over 300 miles per hour. This fast moving cloud would destroy everything in its path, even melting and twisting metal and glass. This is what happened at Mount Pelee on the Caribbean Island of Martinique on May 8, 1902.

On this day of woe, a *nuée ardente* rolled down the mountain, crossed the Blanche River Valley, and spread out laterally before it hit the seaside city of St. Pierre. In just a few fiery minutes the superheated volcanic debris left the entire city a smoldering (ruin), killing all but two of its 30,000 inhabitants. Those who escaped the blast and incineration by incandescent ash, died from intense heat, while still others died miserably of asphyxiation by inhaling extremely hot volcanic dust and sulfuric gases.

The May 18, 1980 eruption of Mount St. Helen's in Washington also involved an explosive *nuée ardente*. After a series of warnings, the north face of the mountain was suddenly shattered by a horizontal blast that released a high velocity cloud of superheated debris. The energy released by this lateral "blast" was equivalent to 500 Hiroshima size atomic bombs or ten million tons of TNT.[33] The debris rolled down Mount St. Helens in less than a minute and then spread out in a fan shaped (140 degree) pattern. Four foot in diameter trees up to fifteen miles away were knocked down. The deadly explosive cloud turned 232 square miles of once beautiful timberland into a dull gray ash covered lunar like wasteland virtually devoid of life.[34]

When the air finally cleared, the entire north face and top of the mountain was blown away. It is fair to say that the blown away north face represents "a stone cut out of the mountain without hands." Where the north face once stood an amphitheater like crater with 2,000 foot high walls was created. This striking amphitheater looks down upon the widespread devastation below. The crater measures 1.2 miles wide and 2.4 miles long. And the 10,500 foot high mountain was instantly made 1,313 feet lower.[35] The cloud of superheated ash and debris reached a temperature of 500° F and traveled at 230 miles per hour. Its powerful ash laden blast killed people as far away as sixteen miles from the mountain, and filled valleys with pumice and ash as far as 17 miles away. The resulting avalanche of pyroclastic, mud and debris swept away everything in its path from bulldozers to bridges as it moved down the mountain and spread out in a fan shaped pattern. [36]

Is there any physical evidence for Mount Hermon being capable of a volcanic eruption producing a *nuée ardente*? Mount Hermon is basically an uplifted fault block of limestone where there have been several lava flows evidenced by a number of volcanic dykes. The remnants of several basalt cones on the flanks of Mount Hermon have also been found. Basalt cones are evidence of past eruptive craters. In particular, half way up the mountain on its southeastern flank there is a large depression, a "collapsed caldera" that is approximately 2.5 miles long by 1.9 miles wide, equivalent to Mount St. Helen's crater in Washington. [37] This ancient caldera is called the Newe Ativ Caldera, and it contains a volcanic crater called the Guvta Crater.* Israel's Geological Survey study and mapping of

---

* A "caldera" is produced by the collapse or rapid subsidence of a magma chamber's when a great volume of magma is removed by an eruption. A caldera collapse often contains the volcanic cone or crater that erupted. The collapsed volcanic caldera on Mount Hermon contains a complex of magnetic and volcanic materials including a small basalt cone that pierced through the Jurassic limestone at Newe Ativ; and a volcanic crater at the core of the complex called the *"Guvta Eruptive Breccia Crater."* Evidence for the *Guvta Crater* comes from a magnetic rock unit called the Guvta Eruptive Breccia, which resulted from an explosive volcanic eruption – Aryeh E. Shimron, "Early Cretaceous Tectonism on Mount Hermon," *Israel Geological Society, Annual Meeting, Ramot, 1989,* p. 141. Also as above "Early Cretaceous Magmatism," 1989, pp. 1-5 and 10. On page 10 the author asks, "Is the Guvta Breccia the throat of an Early Cretaceous volcano?"

this area refers to it as the "Newe Ativ Volcanic Complex," and the "Newe Ativ Magmatic (Caldera) Complex." [38] *(See Illustration H)*

A few miles up the mountain slope from the large Newe Ativ Caldera with its small cone and eruptive crater, there is a large ancient basaltic cone that cuts through the limestone formation at a place called Majdal Shams. This large prominent cone seems related to the small cone and eruptive crater that the Survey's geologist Aryeh Shimron found in the large Newe Ativ Caldera, and to another nearby but smaller cone at Nahal Nimrod. [39] There is also a small volcanic crater at Mas'ade, less than one kilometer to the southeast of the Newe Ativ Caldera.* These cones and craters, together with others in the area, indicate a large, central magma chamber feeding a number of cones and craters. These collectively characterize a classic circumstance of lateral eruptions releasing incandescent clouds, where the biggest eruptions occur not at the main cones and craters, but at one of the lower cones or craters, a parasitic lateral vent. This geomorphology duplicates what happened at Mount St. Helens in Washington in 1980 and at Pelee on Martinique in 1902. [40]

There are also a number of faults on Mount Hermon that produced earthquakes, some of which likely accompanied past volcanic eruptions. The eruptions of Mount Hermon in the past were responsible for a portion of the lava streams and ejected pyroclastic rocks that are found on the east side of the Jordan River. There are also a number of extinct volcanic craters in the Hauran, the plain

---

* Personal communication from Aryeh E. Shimron, based on his "Preliminary Geological Map of the SE Hermon Range" in *New Geological Data and K-Ar Geochronology*, as above, Fig. 1-2, p. 7. In a letter to the author herein dated November 29, 1989, Shimron wrote, "some Pleistocene hunters got caught having a barbecue near the crater (at Mas'ade) and got well cooked during the eruption. What was left of their bones was found some years ago in some ashes.".

ARMAGEDDON, THE SEALS AND A ROCK

to the south of Mount Hermon. This is not surprising since Mount Hermon sits at the northern end of the 4,000 mile long Great Rift that begins in Africa. This geologically active, relatively young rift (twenty five million years old) still produces new volcanic craters that spew tremendous quantities of ash and lava. Thus, Mount Hermon is capable of a volcanic eruption, an eruption that produces a lateral blow out called a *nuée ardente*.

The fact that Mount Hermon has experienced this type of volcanic eruption in the past fulfills *Ecclesiastes 1:9* and *3:15* which say, "that which is to be hath already been." Interestingly, since Mount Hermon ceased volcanic activity long before literate man occupied the area, and since these prophecies were written long before scientists had the knowledge to interpret the evidence from past volcanic events, it is doubtful that *Ezekiel* or *Daniel* had any idea that snow capped Mount Hermon had a volcanic history and equally doubtful that they knew what a *nuée ardente* is.

The Newe Ativ Caldera on the southeastern flank of Mount Hermon lies near the head of a stream called the Banias, located about four and one half miles to the east of Caesarea Phillippi along the head waters of the Jordan.[*] If this caldera or the large Magdal Shams basalt cone above this caldera were to be the location of a lateral volcanic eruption on the day of the Lord, then any *nuée ardente,* or burning cloud released would race down the valley and slam into Caesarea Phillippi (the Banias Valley already contains basalt related to the Magdal Shams basalt cone). Caesarea Phillippi is where the antichrist and the kings of the Earth and their vast armies are most likely to be camped as they prepare to lay siege to Mount Hermon to

---

[*] The *el Liddani* meets the *Nahr Banias* above Caesarea Phillippi. Caesarea Phillippi lies about three miles to the east of Dan.

exterminate the Jews who have taken refuge on the mountain. (In Hebrew the word *hermon* means "sanctuary.") [41]

As discussed earlier, *Isaiah 34:1-10* tells how the "indignation of the Lord" will be upon all nations and their armies to utterly destroy them during "the controversy of Zion." It then speaks of a volcanic eruption when it states that "the streams thereof shall be turned into pitch (***lava***), and the dust thereof into brimstone, and the land thereof shall become burning pitch (***lava***)." The word twice translated as "pitch" in this verse, the Hebrew word *zeh-feth* (word #2203 in *Strong's Concordance*) is better translated as "lava," since it comes from a root meaning to "liquefy" as in molten rock. The cloud released down the Banias Valley would travel from east to west, and like the ash clouds released by both Pelee and Mount St. Helens, there would definitely be lightning associated with pizoelectric discharge by the cloud. *Matthew 24:27-28* says, "For as the lightning cometh out of the east, and shineth even unto the west; so shall also the coming of the Son of Man be. For wheresoever the carcass is there will the eagles be gathered together."*

When *Ezekiel 39:6* says God "will send fire on Magog," in addition to the comets that are to fall, this could also designate the burning cloud or fireball released by a *nuée ardente*. *Zechariah 14:12* seems to describe the effect of a burning cloud up to 2,000 degrees Fahrenheit, moving at up to 300 to 400 miles per hour, would have on people. It says, "And this shall be the plague wherewith the Lord will smite all the people that have fought against Jerusalem. Their flesh shall consume away while they stand upon their feet, and their eyes shall consume away in their holes, and their tongue shall consume

---

*   The word "carcass" indicates the carnage that will result from this event.

away in their mouth." Many prophecy authors have believed that this Scripture is describing the effects of a nuclear bomb being exploded at Armageddon, but the intense burning wind from a *nuée ardente* would easily cause the very same effects. When God tells Gog, a type of the antichrist in *Ezekiel 39:3* that, "I will smite thy bow out of thy left hand and will cause thine arrows to fall out of thy right hand," it seems God is telling the antichrist that his armies would not even be able to fire a shot or detonate a bomb when they come against Him.

Interpreting *Zechariah 14:12* as a description of a volcanic disaster rather than a nuclear blast makes this verse compatible with a number of other Scriptures that call for and/or describe a volcanic eruption and an earthquake that will occur on the day of the Lord. In *This Shaking Earth* (1978) physicist John Gribbin, in discussing the explosive eruption of Mount Pelée in 1902 *nuée ardente*, says,

> The phenomenon which produces this kind of hot avalanche (*nuée ardente*) is familiar today in another even more grim context, as the **'base surge'** which rolls away from the site of a nuclear explosion at ground level. The destruction caused by a *nuée ardente* which hits an inhabited region can only be compared with the destruction caused by a nuclear explosion . . . Even in bare outline the story of the destruction of St. Pierre reads more like the horrors of war than any natural disaster. [42]

Since Scripture reveals that comets are to come in and impact the Earth on the day of the Lord, it is no surprise that Scripture also indicates that a volcanic eruption will also occur on the day of the Lord. Just as large cosmic impacts can trigger mega earthquakes, large cosmic impacts can punch into or through the Earth's crust, triggering massive volcanic eruptions on one or both sides of Earth. A cosmic impact can trigger a volcanic eruption by raising the pressure and temperature in the mantle underlying Earth's crust. In particular, this could initiate and/or strengthen plumes of hot magma that rise up within the mantle and circulate toward the lower crust in concentrated streamers or plumes. A mantle plume is not unlike a rising glob of material in a 1960's style decorating lava lamp. Dr. Dallas Abbot of Columbia University's Lamont-Doherty Earth Observation says, "The general idea is that plumes are strengthened by impacts."[43] At a Geological Society of American meeting "she showed a correlation between the timing of purported superplumes and large impact events . . ."[44] Dr. Abbott and Ann Isley of State University New York at Oswego found a correlation between meteorite and comet impacts and an increase in volcanic activity. Ten major episodes of impacts were found to correlate with nine major episodes of volcanism.[45] They also found that "two prominent lulls in impact activity matched up with periods of decreased volcanism."[46] Dr. Abbott said, "Large impacts generate large earthquakes . . . These earthquakes can trigger volcanic eruptions. If the earthquake is large enough to do more, it could make the eruption more intense by allowing more magma to escape."[47]

There is growing support in the scientific community for the idea that volcanic eruptions can be set off by enormous cosmic impacts. For example, Adrian Jones and David Price from University College London say Abbott's work backs up their recent computer simulations. These models suggest that meteorites larger than ten kilometers across occasionally punch right through the Earth's crust causing huge volcanic eruptions on both sides of the Earth, simultaneously. "A large impact has the ability to cause instant melting where it hits, creating its own impact plume in the mantle and resulting in a massive surge of molten ejecta or lava spilling out."[48] Some scientists believe that the Hawaiian Islands were born of a cosmic impact that created a plume or hot spot in the mantle below this island chain (see Chapter 7). *(See Illustration I)*

In Chapter 7 we learned that continental areas of extensive ancient volcanism such as India's Deccan Plains with its step like layers of basalt and the Siberian Traps (Trappen for steps) may be the result of large cosmic impacts. Recall the very large impact that struck the moon and produced the extensive lava flows referred to as the "Man in the Moon." Now some scientists believe that the dinosaurs did not all die from a cosmic impact alone calling for a combination of factors to explain the destruction. In particular, Jones says the dinosaurs probably all died from a "double whammy," consisting of a cosmic impact followed by devastating volcanic eruptions. [49]

The Timing of the Events

Revelation 16:14-16 tells of the armies of the world being gathered together "to the battle of that great day of God almighty . . . into a place called in the Hebrew **Armageddon**."* In *Zephaniah 3:8* the Lord adds to the picture and says, "gather the nations that I may assemble the kingdoms to pour upon them my indignation,

---

*   This great day of battle is also referred to in *Revelation 11:18* which talks about the Seventh Trumpet and tells how the nations were angry that God's wrath is come that they should be judged.

even my fierce anger: for all the Earth shall be devoured with the fire of my jealousy." Whereas *Revelation 16:14-16* tells how this battle is to occur at a place called in the Hebrew **Armageddon**, remember that in Hebrew **Armageddon** literally means "mountain rendezvous of troops." The troops or armies of the nations that are to gather together for a battle at Mount Hermon (the "mountain of asylum") on this great day is consistent with *Ezekiel 38-39*, which describes a battle in the mountains where the "Lord God" is present, fights for Israel against the invading multinational force, and "saves" Israel. This is also consistent with *Revelation 14:1-5* which tells of Jesus being seen on Mount Zion ("mountain fortress"), the highest peak of Mount Hermon with 144,000 redeemed Jews, who no doubt fled to the mountain to hide from the invading armies.

While the "Lord God" comes to fight for Israel and "save" Israel at the Battle Armageddon, the author does not want to give the impression that only Jews will be "saved" during the tribulation period. Both before and during the Battle of Armageddon, a great number of Gentile believers from all nations, cultures and tongues are to be "saved." Among the believers who are "saved," some will be martyred (*Revelation 6:9-11* and *7:9-14*), some will escape all the things that come to pass (*Luke 21:36*), and some will be kept from the hour of trial (*Revelation 3:10 NIV*). In *Matthew 24:22* (*Mark 13:20*) Jesus says that except the days of the "great tribulation" were shortened "there should no flesh be saved: (but for the elects' sake those days shall be shortened.") *I Thessalonians 4:15-17* talks about believers being "caught up" or "raptured" up into "the clouds to meet the Lord in the air" at Christ's coming.* The "rapture" is a

---

* The word translated "caught" is from the Greek word *harpazo* (word #726 in *Strong's Concordance*). The word translated "up" in *I Thessalonians 4:17* is from the Greek word *anō* (word #507 in *Strong's Con-*

ARMAGEDDON, THE SEALS AND A ROCK

controversial doctrine among Christians. There is much debate on whether believers will be "raptured" at the beginning, middle or end of the tribulation period. Somehow, believers will either be removed or protected during the tribulation period. Believers being protected during the catastrophic tribulation period is not unlike Moses and the Israelites being protected during the ten plagues of the Exodus.

Scripture says the battle of Armageddon will take place in northern Israel at Mount Hermon. It is in the expansive open area of the Hulah Basin below Mount Hermon where the armies of the antichrist will camp, and where "the great *winepress* of the wrath of God" is located (*Revelation 14:19-20)*. This is the "winepress" spoken of when scripture says Jesus will "treadeth the *winepress* of the fierceness and wrath of Almighty God" (*Revelation 19:15*). This is the "press" that *Joel 3:13-14* says will be full with "multitudes in the valley of decision," the valley below Mount Hermon on "the day of the Lord." The "day of the Lord," is the day *Isaiah 13:2-13* says that the Lord will destroy the kingdoms of the nations gathered together, and the day *Zephaniah 1:7-8* says that the Lord will "punish the princes and the kings' children" and prepare a "sacrifice."* In *Matthew 24:28 cordance)*. In Latin Bibles such as the *Vulgate*, the Greek word *harpazo* ("caught up") is translated as *raptus*, which in English is rendered as the word "rapture."

*   The phrase the "day of the Lord is used twenty-one times in the Old Testament (*Isaiah 2:12; 13:6, 9; Jeremiah 46:10; Ezekiel 13:5, 30:3; Joel 1:15, 2:1, 11, 31; Amos 5:18, 20; Obadiah 15, Zephaniah 1:7, 14; Zechariah 14:1;* and *Malachi 4:5)*. The phrase "the day of the Lord is used four times in the New Testament (*Acts 2:20; II Thessalonians 2:2; and II Peter 3:10* and *12)* and referred to in comparable terms in *Revelation 6:17* and *16:14*. On the "day of the Lord" *Isaiah 13:2-13* says God will punish the world" with "wrath and fierce anger" and bring "destruction " on the "kingdoms of nations gathered together." On the "day of the Lord" *Zephaniah 1:7-8* says the Lord God will be present and "will punish the armies and the King's children" and prepare a "sacrifice" and "bid his guests" in the "day of the

Jesus makes reference to this sacrifice.

The Bible tells us the Jews of Israel will take refuge at several places in the wilderness to escape annihilation by the antichrist. Yet it is on Mt. Zion, the highest peak of Mount Hermon (the "sacred mountain" or "mountain asylum"), where Jesus is to appear astride a white horse coming to deliver or save 144,000 "redeemed" Jews who have taken refuge on the Mount (*Revelation 14:1-5, 12:13-17,* and *Isaiah 16:1-5*).* The 144,000 believing Jews will be surrounded by the armies of the antichrist. *(See Illustration H)*

Based on Scripture, the antichrist's plan is to annihilate all the Jews and eliminate the need and basis for Jesus returning to the Earth *(Revelation 12:13-17)*. This desire or intention to eliminate Jews is not new to either the Scriptures nor the historical record. As early back as the Exodus story, Pharaoh sought to enslave, overwork and kill all male infants. Even after releasing the Hebrew nation, Pharaoh and his army pursued them to destroy them (*Exodus 15:9*). In the story of Esther, "Haman sought to destroy all the Jews" of the Persian Empire "in one day" (*Esther 3:6, 10* and *13*). The Holy Roman Empire in their medieval crusades to recover the Holy Land

Lord's sacrifice." *Ezekiel 39:4* and *17-20* also make reference to the sacrifice of flesh and blood to be prepared by the Lord, which will include the blood of princes. On the "day of the Lord," *Joel 2:1-11* says to sound the alarm in (His) holy mountain (Mount Hermon) and to blow the trumpet in Zion for the "day of the Lord is great and very terrible and who can abide it." *Joel 3:13-14* refers to the day of the Lord and to the winepress in the valley below Mount Hermon where the armies of the antichrist are to camp and says, "Put ye in the sickle, for the harvest is ripe: come, get you down: for the press is full, the vats overflow; for their wickedness is great. (14) Multitudes, multitudes in the valley of decision: for **the day of the Lord** is near in the valley of decision."

* The Jews taking refuge on rugged Mount Hermon which is full of caves and hiding places is not unlike the Afghanis who took refuge in the mountains after the Russians invaded the country in 1979.

ARMAGEDDON, THE SEALS AND A ROCK

from both the Turks and the Jews for the Roman Catholic Church sought to kill many Jews, and Adolph Hitler of the Germany's Third Reich (empire) sought to annihilate all the Jews of Europe and then the world.

The promise of Messiah's coming as a "conquering" Messiah to save the Jews was made to the Jews of the Old Testament. Jesus is to fight against the multinational armies of the antichrist, just as "the Lord fought for Israel" against Pharaoh (*Exodus 14:14*). "The Lord cast down great stones from heaven" and helped Joshua defeat the armies of five kings gathered against Joshua at Gibeon as recorded in *Joshua 10:1-14* and *42*. In *Isaiah 35:4* the prophet tells Israel to "Be strong and fear not: behold, your God will come with a vengeance . . . he will come and save you." Just as *Jeremiah 46:10* calls "the day of the Lord God of hosts, a day of vengeance," *Isaiah 34:8* refers to "the day of the Lord's vengeance," and *Isaiah 61:1-2* refers to "the day of vengeance of our God."* If there are no faithful Jews for Jesus to save, then there is no express need for Jesus to return. The name "Jesus" is the Greek rendering of His name in Hebrew. In Hebrew Jesus' name is *Yeshuah,* and in Hebrew *yeshuah* (#3442 or #3444 in *Strong's Concordance*) means "he will save."† If Jesus does not return to save the Jews and others who accept Him as Savior, and defeat the armies of the antichrist, then the antichrist could complete his plan of empire and rule the world from David's temple in Jerusalem. *Isaiah 14:12-14* and *Ezekiel 28:2, 14-19* tell how satan wanted to be like God

---

* In *Luke 4:18-19* Jesus quotes from *Isaiah 61:1-2*, but leaves out the words regarding "the day of vengeance of our God." His appearance at the battle of Armageddon constitutes "the day of vengeance."

† The Hebrew word *yehoshua* (#3091 in *Strong's Concordance*) which means "the Lord is salvation " also comes into play. In the Greek *yehoshua* is rendered as *"Joshua."*

and sit on God's throne in heaven and rule, which led to satan being thrown out of heaven. Through the antichrist, satan seeks to sit on God's throne on Earth and rule Earth. *II Thessalonians 2:4* tells how the antichrist will exalt himself above the God of the Bible and seek to be worshipped as God while sitting on God's throne in the rebuilt temple of God in Jerusalem.* *Daniel 8:25* tells how the antichrist will "stand up against the Prince of Princes (Jesus), but he shall be broken without hand ('without human agency' – NAS)."

In *Deuteronomy 4:29-31* the God of the Bible promises to save the children of Israel during the "tribulation" of the "latter days" ('the time of Jacob's trouble' – *Jeremiah 30:7*), and He says He will not let them be destroyed. In *Deuteronomy 7:19 (NAS)*, God promises to do the same type of terrifying and destructive "signs and wonders," as during the Exodus, upon "all the peoples of whom you (Israel) art afraid." *Micah 7:15* says, "According to the days of thy coming out of the land of Egypt will I shew unto him marvelous things ('wonders' in the NIV)." From these verses we can see that during the "tribulation" of the "latter days," when Israel is "afraid" of the antichrist and his vast armies, the God of the Bible will again use comets to show terrifying "signs and wonders" that bring death and destruction upon the empire of the antichrist, just as He also used comets to destroy the kings and their armies in Joshua's battle (*Joshua 10*) and destroy the empire that was building the temple tower at Babylon (see previous chapter).† *Zechariah 12:8-9* says, "In

---

*   *Isaiah 14:13-14* tells how satan wanted to be like God and exalt his throne above God's throne in heaven. After being thrown out of heaven, satan then planned to rule from God's throne on Earth in Jerusalem. In other words, satan planned to rule on Earth and let God rule in heaven. By indwelling the antichrist, satan could come to rule the Earth.

†   As *Deuteronomy 7:19 (NAS)* tells how the God of the Bible used "signs and wonders" from cometary phenomena to save Israel during the

that day shall the Lord defend the inhabitants of Jerusalem . . . and it shall come to pass in *that day* that I will seek to destroy all the nations that come against Jerusalem." So, at the last minute, when all seems lost, the Messiah (**Maschiach ben David**) will appear astride a white horse to deliver and save the children of Israel as promised. *Romans 11:25-27 NIV* says,

> Israel has experienced a hardening in part until the full number of the Gentiles has come in. And so **all Israel shall be saved**, as it is written (quoting *Isaiah 59:20*): 'The *deliverer* (redeemer) will come (to) Zion, he will turn godlessness away from Jacob: And this is my covenant with them when I take away their sins.
>
> *Romans 11:25-27 NIV*

Exodus and would again do so to all the peoples coming to destroy Israel; it is interesting to note that the following verse, *Deuteronomy 7:20* seems to refer to cometary debri as "hornets." *Deuteronomy 7:20 (NAS)* says "Moreover, the Lord your God will send the *hornet (hornets* in *Septuagint)* against them, until those who are left and hide themselves from you perish." The Hebrew word *tsir-aw* (#6880 in *Strong's Concordance*) which is translated as "hornet" in *Deuteronomy 7:20 (KJ, NAS, NIV,* and *Septuagint)* comes from the Hebrew word *tsah-rah* (#6879 in *Strong's Concordance*) which means "to scourge or strike." As "hornets" constitute a swarm of flying objects that scourge or strike, *maybe* the word "hornets" is a figurative term for a swarm of cometary debris that scourges or strikes. This Hebrew word *tsir-aw* (#6880) has been translated as "hornets" and seems to be used in a similar way in *Exodus 23:28* and *Joshua 24:12,* and *8.* As cometary phenomena seem to be figuratively related to "hornets" in *Deuteronomy 7:20,* cometary phenomena is figuratively related to "locusts" in *Revelation 9:3-12.* The second part of *Deuteronomy 7:20* speaks of men "hiding" themselves; and in this regard relates to *Isaiah 2:10, 19,* and *21 and Revelation 6:15-17,* which speaks of men hiding themselves in the context of cometary phenomena.

The "great day of God Almighty" of *Revelation 16:14* is also the "great day of His (Jesus) wrath" of *Revelation 6:13*. On the "great day of His wrath" *Revelation 6:13:17* says that the stars of heaven, that is, comets will fall unto the earth and the kings of the Earth will seek to hide "from the face of him that sitteth on the throne and from the wrath of the lamb," just as *Matthew 24:29-30* says that the stars of heaven will fall on the day of Jesus' return. *II Peter 3:10-13* says that on the "day of the Lord" the entire Earth and the heavens (atmosphere) will be "on fire," which speaks of a mega impact that engulfs or baptizes the entire planet in flames.

Based on the catastrophic events of the Sixth Seal, the Seventh Trumpet, and the Seventh Vial, this great day of battle starts with stars falling from heaven, that is, comets ("hairy stars") and/or comet fragments coming in to impact the Earth (*Matthew 24:29, Revelation 6:13* and *Isaiah 34:4*). *Isaiah 13:3-9* and *13* say:

> 3) I have commanded my sanctified ones (comets), I have also called my mighty ones (comets) for mine anger, even them that rejoice in my highness. 4) The noise of a multitude in the mountains, like as of a great people, a tumultuous noise of the **kingdoms of nations gathered together**: the Lord of hosts mustereth the *host* (comets) of the battle. 5) They come from a far (place), from the end of heaven (the Oort Cloud), even the Lord, and the weapons of his indignation (comets), to destroy the whole land. 6) Howl ye; for **the day of the Lord** is at hand; it shall come as a destruction from the Almighty. 7) Therefore shall all hands be faint, and every man's heart shall melt:

8) And they shall be afraid: pangs and sorrows shall take hold of them; they shall be in pain as a woman that travaileth; they shall be amazed one at another; their faces shall be as flames. 9) Behold, **the day of the Lord** cometh, cruel both with wrath and fierce anger, to lay the land desolate: and he shall destroy the sinners thereof out of it . . . 13) Therefore I will **shake the heavens, and the earth** shall remove out of her place, in the wrath of the Lord of hosts, and in the **day of his fierce anger.**

Isaiah 13:3-9 and 13

Initially, these impacts will cause a "great earthquake" (*Revelation 16:18 and 11:19*), so intense that every mountain and island will be moved out of their place (*Revelation 6:14*) with "the sea and waves roaring," according to *Luke 21:25*. There will also be volcanic eruptions. A volcanic eruption, a ***nuée ardente*** or lateral eruption will occur on Mount Hermon, where the armies of the antichrist are camped in pursuit of the Jews who have taken refuge there. The lateral blow out of Mount Hermon would represent a rock or "stone . . . cut out of the mountain without hands," that smote the kingdoms of the world (*Daniel 2:45*).*

Picture the multinational forces of the antichrist moving up the Banias Valley on the southeast flank of Mount Hermon. The Newe Ativ Volcanic Complex sits at the top of this valley. An eruption in the Newe Ativ Caldera would send a very hot and glowing cloud of

---

*   *Zechariah 9:15* says, "The Lord of hosts shall defend them; and they shall devour and subdue with ***sling stones*** . . . ."

ash down the valley, and according to *Zechariah 14:12* the flesh of the troops moving up the valley would "consume away while they stand upon their feet and their eyes (would) consume away in their holes, and their tongues (would) consume away in their mouth." We know that these verses in *Zechariah* apply to a volcanic eruption because of the type of consumption described.

The fiery blast from a volcanic eruption (or pyroclastic flow) can reach temperatures upwards of 2000° F, and is capable of causing the specific effects on men described in *Zechariah 14:12*. In contrast, the fiery blast from a cometary impact could reach temperatures of many thousands of degrees, and people caught down range from an impact would be instantly and completely vaporized. Further, *Isaiah 34:8-9* seemingly makes reference to this lateral volcanic eruption on the "day of the Lord." First, *Isaiah 34:1-3* says,

> Come near, ye nations to hear; and hearken, ye people: let the earth hear, and all that is therein; the world and all things that come forth of it. For the indignation of the Lord is upon all nations, and his fury upon all their armies; he hath utterly destroyed them, he hath delivered them to the slaughter. Their slain also shall be cast out, and their stink shall come up out of their carcasses, and the ***mountains*** shall be melted with their blood.
>
> *Isaiah 34:1-3*

Then *Isaiah 34:8-9* says,

> For it is the **day of the Lord's vengeance** and the year of recompenses for the controversy of Zion. (9) And the streams thereof (of Zion) shall be turned into pitch (**lava** - *zeh-feth*, word #2203 in *Strong's Concordance*, a root meaning to *"liquefy"* as in molten rock or **lava**), and the dust thereof into brimstone (volcanic eruptions release brimstone or burning sulfur), and the land thereof shall become burning pitch (**lava**, word #2203 in *Strong's Concordance*).
>
> *Isaiah 34:8-9*

The troops that are not killed by the burning cloud will then be struck down by a "plague" of "great hail" that involves an "exceeding great" fall of one hundred pound stones of cometary ice (*Revelation 16:21* and *11:19*).* This is not unlike (*Ecclesiastes 1:9* and

---

*   *Zechariah 14:13* tells how in this "great tumult" the invading troops will turn on each other, as "they shall lay hold every one on the hand of his neighbor, and his hand shall rise up against the hand of his neighbor ('they will attack each other' – NIV). *Ezekiel 38:21* which tells about this same battle tells how "every man's sword shall be against his brother." Ironically, as the word **Har-mageddon** as used in *Revelation 16:16* should be translated as "mountain rendezvous of troops;" it could also be translated as a "mountain gathering of troops to cut one selves." This is because the Hebrew word *megiddo* (#4023 in *Strong's Concordance)* that is translated as "rendezvous" comes from the Hebrew word *gaw-dad* (#1413 in *Strong's Concordance*), which means "to crowd" or "gash" as if by pressing into, or "gather selves together in troops" or "cut selves." In this the word **Har-mageddon** more completely describes the battle of *Ezekiel 38-39*. Also there is precedent for when *Ezekiel 38:21 and Zechariah 14:13* say that the invading troops will turn on each other. *Judges 7:22* speaks of Gideon's great victory

3:15) the time when "great stones from heaven" helped Joshua win a victory over the five kings of the Amorites (the king of Jerusalem, Hebron, Jarmuth, Lachish and Eglon) that were gathered against him (*Joshua 10:3-14*). *Joshua 10:11* says "the Lord cast down great stones from heaven . . . (and) they were more which died with (cometary) **hailstones** than they who the children of Israel slew with the sword." *Joshua 10:42* explains that "all these kings and their land did Joshua take at one time, because the Lord God of Israel fought for Israel."

Ezekiel 38:19-23, talks about the great end times battle in Israel, and reflects the sequence of events that has been suggested here. This "Sequence" includes a cosmic impact that causes an earthquake with great shaking, a volcanic eruption that blows out the side of a mountain and sends a fiery cloud of rock, ash, and burning sulfur on the antichrist's troops, and then one hundred pound hailstones and burning sulfur from cometary ice raining down on the troops. *Ezekiel 38:19-23* says:

> Surely in that day there shall be ***great shaking*** in the land of Israel . . . (20) all the men that are upon the face of the earth (land) shall shake at my ***presence,*** and the ***mountains shall be thrown*** down, and the steep places shall fall . . . (22) I will rain upon him and upon his bands ('troops' – NAS, NIV), and upon the many people ('nations – NIV) that are with him . . .

in battle when "the Lord set every man's sword against his fellow." *I Samuel 14:20* tells how Israel's King Saul won a battle against the Philistines and "every man's sword was against his fellow and there was a very great discomfiture." *II Chronicles 20:17-23* tells how King Jehoshaphat of Judah won a great battle when the Lord sent confusion against the invading forces of Ammon, Moab, and Mount Seir and they all helped to destroy one another.

great hailstones, fire, and brimstone (burning sulfur). (23) Thus, I will magnify myself and sanctify myself; and I will be known in the eyes of many nations and they shall know that I am the Lord.

Ezekiel 38:19-23

Ezekiel 39:4-5 adds:

Thou shalt fall upon the mountains of Israel, thou and all thy bands, and the people that is with thee: I will give thee unto the ravenous birds of every sort, and to the beasts of the field to be devoured . . . (6) And I will send a fire on **Magog** (the antichrist and the armies of the lands or nations that are part of his empire).

Ezekiel 39:4-5

After the initial impact or impacts, the burning volcanic cloud, and the cometary hailstones, it seems that at least one much bigger and more penetrating impact is to lead to the Earth being rocked in space (*Isaiah 13:13* and *Isaiah 24:18-23*). This bigger impact is to also lead to the cities of the nations falling due to the earthquake of unprecedented size (bigger than any since men have been on the Earth - *Revelation 16:18-19*) this impact causes. *Jeremiah 30:23-24* and *Jeremiah 25:31-32* (also *Jeremiah 23:19-20*) may relate to the Earth being rocked in space, and the great shock wave and winds that would occur when they say:

Behold, the *whirlwind* of the Lord goeth forth with fury, a continuing *whirlwind*: it shall fall with pain upon the head of the wicked. The fierce anger of the Lord shall not return, until he have done it, and until he have performed the intents of his heart: in the *latter days* ye shall consider (understand) it.

Jeremiah 30:23-24

A noise shall come even to the ends of then earth; for the Lord hath a *controversy with the nations*, he will plead with all flesh; he will give them that are wicked to the sword, saith the Lord. Thus saith the Lord of hosts, Behold, evil shall go forth from nation to nation, and a great *whirlwind* shall be raised up from the coasts of the earth.

Jeremiah 25:31-32

II Peter 3:10-13 tells how such a mega impact on "the day of the Lord" would temporarily engulf the entire planet in flames so that everything is burned up, including the atmosphere and the oceans; where the "elements shall melt with fervent heat" (also see *Isaiah 66:15-16*). As discussed in Chapter Seven regarding the Seventh Trumpet and Vial, a crust penetrating mega impact could raise the temperature of the Earth's interior and also cause vast amounts of molten rock to rise up from the Earth's mantle and flood out onto the Earth's surface. From all of this activity, the Earth's relatively thin crust could come to melt so that "every island fled away and the mountains were not found" (*Revelation 16:19*). While all the mountains and islands would have been shaken and moved at the start of the day, now with the much bigger impact the mountains and islands, and all of the Earth's crust would literally melt away leaving the Earth with a molten surface, a sea of molten rock, not unlike the Moon's Mares. As science has come to learn of continental drift and then of plate tectonics; the reality of catastrophic melting of the Earth's crust represents the geological **"END GAME,"** a game we have seen played out on planet Venus. No one and nothing would survive as the Earth undergoes a "Baptism of Fire" (*Matthew 3:11-12,* also see *Matthew 13:40-42,* and *49-50*). In the Old Testament fire was used for cleansing and such a Baptism of Fire would cleanse both heaven (the atmosphere) and the Earth. [50] After this cleansing, when all cools down again, there would be a new crust, a new Earth, and a new heaven (atmosphere) *(II Peter 3:10-13).*

Before this very big impact that comes to engulf the Earth in a sort of **"Baptism of Fire,"** there is a "most spectacular" event that takes place in the heavens for all mankind to see. This event will be taken up in the next chapter.

*   *   *   *

We have looked at the battle of Armageddon, the description of the winepress running with blood, how the Seven Seals relate to the Seven Trumpets and Vials, God's response to the multinational attack of Israel, and what happens at the return of Jesus. We have put forth explanations for the events that the Bible warns of and have seen how they are scientifically possible. We have also seen how the Bible provides important scientific information that modern researchers in geology, archeology and astronomy have only recently been able to recognize. A better understanding of the disastrous events that Scripture says will befall the nations and their armies on the day of Jesus' return could help us integrate the timing and relationship of these events to each other.[*]

---

[*] It must be remembered that the "**Sequence**" of events given here is not absolute and only a possibility. The thrust of this book is on the interpretation of the physical events prophesized by Scripture. Theological arguments are not in the domain of this book.

TEN

THE SIGN OF THE SON OF MAN IN HEAVEN

I shall see him, but now: I shall behold him, but not nigh: there shall come a Star out of Jacob and a Scepter shall rise out of Israel, and shall smite . . . and destroy all . . . .

Numbers 24:17

Immediately after the tribulation of those days shall the sun be darkened, and the moon shall not give her light, and the stars shall fall from heaven, and the powers (objects) of the heavens shall be shaken: (30) And then shall appear the **sign of the Son of Man in heaven**; and then shall all the tribes of the earth mourn, and they shall see the Son of Man coming in the clouds of heaven with power and great glory.

Jesus in *Matthew 24:29-30*

As we have seen in the preceding chapters, the Bible contains a great deal of scientifically sound information about comets ranging from their origin, composition, behavior, and what happens during and after various types of impacts. They have even been called the weapons of God's wrath. The Bible says they have an appointed time, destination and purpose. Could the above passage from the *Book of Matthew* be about a specific comet? What is this "sign of the Son of Man in heaven" that is to appear on the day of Jesus' second coming? Based on several Scriptures, the **"sign of the Son of Man in heaven"** seems to be a star, not a comet!

A "star" heralded Jesus' first appearance at the time of his birth. Remember that the Star of Bethlehem was a **"*sign*"** Scripture says appeared before Jesus' first coming (*Matthew 2:2-10*). Could the appearance of a star, be the sign in heaven that is to occur before the Son of Man's, (Jesus') second coming? In this context, the word "star" is not being used to designate a comet, a planet or a planetary conjunction, but rather a self luminous celestial body like our sun. This is because the testimony of the wise men who saw the Star of Bethlehem over a period of weeks or more spoke of an object whose position against the celestial background was relatively fixed. A comet or planet would be seen moving against this celestial background.[*] The Old Testament also seems to reflect this theme of a star of Israel – a future King to come when *Numbers 24:17* speaks of a "Star out of Jacob." The phrase a "Star out of Jacob," also means a "Star out of Israel," since *Genesis 32:28* tells how "Jacob" was given a new name

---

[*] To the ancient Greeks the word "star" (*aster*, word #792 in *Strong's Concordance*) was a non-specific term that was used to designate any of the luminous bodies seen in the heavens: comets, meteors, planets and the sun-like bodies we now call stars.

and would thereafter be called "Israel."* In an ancient Near Eastern manner of speaking, the "Star out of Jacob," and thereby, the "Star out of Israel," implies that a future King of Israel will come and be represented by a star in the heavens just as the Star of Bethlehem was perceived by the wise men to announce the arrival of the King of the Jews (*Matthew 2:2*).†

Amidst all that the Bible has to say about comets and comet bombardment, how does a "star" fit into the whole scheme of things? Scientists now know that comets coming from space to bombard the Earth originate from the huge cloud of comets at the outer reaches of the solar system (*Job 38:22-23*). The scientists know that these comets are displaced from this cloud and are driven to the Earth by a slow moving dwarf star passing through this cloud. NASA's Donald Yeoman's in *Comets: A Chronological History* writes, "Showers of comets might be expected to strike the Earth as a result of a star passing near the solar system's inner comet cloud...."[1] Studies have shown that the movements of a very small type of star called a dwarf star passing through the comet cloud is the ultimate cause of comets being unleashed to travel in and bombard the inner planets of our solar system including the impacts on Earth. These studies showed clustering of the points of origin for many of the comets in their samples. This clustering indicated that groups of comets were driven into the inner solar system a few million years ago by **the passage of**

---

*   In *Jeremiah 30:7* the "great tribulation" period of *Matthew 24:21, 29* (also see *Deuteronomy 4:30*) is referred to as the "time of Jacob's trouble."

†   This is not unlike the Sumerians and Akkadians referring to Inanna or Ishtar, the so-called "Queen of Heaven" (*Jeremiah 44:17-19* and *25*) as "the Radiant Star" or "the Lone Star." *Inanna – Queen of Heaven and Earth*, Diane Wolkstein & Samuel Noah Kramer, Harper and Row, 1983, pp. 93, 101 and 105.

a single slow moving dwarf star through the reservoir of comets that exists at the edge of the solar system. [2]

One particular study conducted by German astronomer Ludwig Biermann and his colleagues, identified an unknown dwarf star that dislodged 17 comets. [3] Investigators identified "one conspicuous star track" marked by 17 comet origin points falling along a straight line covering about ninety degrees of the sky. [4] In this distribution we can see the ancient footprints in space of an *unknown* dwarf star that once moved along the edge of the solar system, at a distance of only 0.63 light years (40,000 Au) away from Earth, which is about 1,000 times farther away than our most distant "planet," Pluto. To put this in perspective, the closest *known* star to the Earth, Proxima Centauri is 4.22 light years away, seven times farther than the area through which this unknown star passed.[*] At the very least, these studies provide direct evidence that a single dwarf star has in fact driven a number of comets into the inner solar system in the past, and could do so again or may indeed *have already* done so.

Not only have astronomers learned that the movements of a dwarf star through the comet cloud at the end of heaven can drive comets in to bombard the Earth, but some astronomers suspect that one star in particular has periodically done this. They believe that the star is an as yet undiscovered dark or extremely faint companion star to our Sun. In 1984, award winning astrophysicist, Richard Muller from the University of California, Berkeley, led one of two separate

---

[*] This distance to Proxima Centauri is equivalent to 7,000 times the diameter of the solar system. "Evidently stars are very sparsely distributed in space: the average distance between stars is about five light years." *The Cambridge Encyclopaedia of Astronomy*, Simon Mitton, editor, Crown Publishers, NY 1978, p. 301.

scientific teams. Working independently of each other, these two teams simultaneously proposed that our local star, the Sun, like *most other* stars in the universe has a companion! These are called "Binary star" systems.

This proposed companion is a *dark* dwarf star that is popularly referred to as "Nemesis." [5] They say that Nemesis has not yet been seen and discovered because it is "dark," meaning that it gives off very little light. Some scientists say that it is not simply faint but actually "sub-luminous" meaning this type of star is relatively cool and does not give off enough radiant energy to be clearly seen via electromagnetic spectral emission in the visible light band. These scientists suspect that about every 28 million years, as Nemesis moves along its orbit, it passes through the comet reservoir at the edge of the solar system. Recall that this reservoir is the distant holding pen or lockup for *several trillion* comets called the Oort Cloud. The Oort Cloud at the edge of the solar system is spherical in shape, and encloses the *entire* solar system like a giant womb. These scientists believe that, when Nemesis passes through this comet reservoir, it sends a storm of comets hurtling away from and some toward the inner solar system, a few of which hit the Earth, and cause periodic sudden mass extinction on the Earth. They also suspect that when some of the comets pass through the asteroid belt that lies between Jupiter and Mars, they may in turn cause some of these asteroids to also fly into the Earth.

The worldwide media loved the Nemesis hypothesis when it was made public in 1984. Feature articles were written in a great number of popular magazines. For example, the late but noted science writer Dr. Isaac Asimov, former professor of biochemistry at Boston University School of Medicine, was quick to hail Nemesis as the possible culprit behind the extinction of the dinosaurs and other episodes of extinction. He wrote about these "great dyings" and said, "it begins to appear that they are not unpredictable accidents at all."[6]

While this hypothesis with Nemesis passing through the Oort Cloud every 28 million years is far from being established, the important thing is that a number of astronomers suspect that there is an undiscovered dark second star in our solar system. In his book *Nemesis: The Death Star - The Story of a Scientific Revolution*, Dr. Muller tells how he came to "reconcile" established facts and believe that there had to be a *second star in our solar system* causing the Earth to be periodically bombarded by comets. Muller wrote, "If there was a star orbiting the Sun, we would have to change our entire theory on the origin of the solar system. The star would have been a decisive factor in the evolution of life, due to its periodic 'shotgunning' of asteroids (and comets) at the Earth. The asteroid (or comet) that killed the dinosaurs had been the bullet, but the solar companion was the murderer."[7]

Dr. Muller called this sought after second star in our solar system a "murderer" in the sense of it being the instigator of the cosmic bombardment that killed the dinosaurs. This raises an interesting Scriptural question. *John 5:22* says, "For the Father judgeth no man, but hath committed all *judgment* unto the Son." If the sign of the Son of Man in heaven indeed involves a star, could this same star be the trigger for *Revelation's* Seven Trumpets and Seven Vial bombardments and judgments of the Earth at the end times? Note that *Numbers 24:17* says, "I shall see him, but now: I shall behold him, but not nigh: there shall come a Star out of Jacob and a Scepter shall rise out of Israel, and shall smite . . . and destroy all . . . ."

Beyond the Nemesis theory, some astronomers are now searching the sky for small sources of light and infrared radiation, which might indicate the location of a secretive companion star to our Sun.* The likely companion star to our sun may be an infrared emitting brown dwarf or a red dwarf, types of stars that are **much** smaller, cooler, and fainter than our sun, but have sufficient mass to affect comet movement. Astronomers Clark Chapman and David Morrison in *Cosmic Catastrophes* wrote, "Our hypothetical Death Star could be either a faint **red dwarf** or a still fainter **infrared dwarf**."[8] Red dwarfs have masses that range from forty percent to just eight percent of the mass of the Sun, and diameters as small as one-seventh the diameter of our Sun. They emit little light, sometimes less than 1/10,000 that of the Sun. [9] Red dwarfs are the most common star type in the Galaxy, and twenty-one of the thirty nearest stars to the Sun are red dwarfs.

Interestingly, the Bible appears to provide specific information about this mystery star. There are verses that indicate this star may not be readily seen until a specific time, that it has a cloud that conceals it, and it produces heat. In the Bible's story of the Star of Bethlehem there are indications that it is a red dwarf and even more specifically, the story indicates it is a special type of red dwarf residing in a hydrogen cloud. In terms of discovering the location of this companion star, it is important to know red dwarf stars residing in clouds can suddenly erupt or and brighten the cloud that enshrouds

---

* Several new astronomical surveys (Pon-STARS, LSST) are planned for 2008 to 2010. In particular the *"Wide Field Infrared Survey Explorer"* or NASA funded satellite mission for 2010 will carry on infrared sensitive telescope. Note that instead of giving off radiation in the visible light spectrum, an infrared dwarf would only give off radiation of invisible infrared wavelengths.

and conceals it. But whether a red dwarf, eruptive red dwarf or brown dwarf, this so-called "lost star" likely exists.

Astronomers call a cloud enshrouded eruptive red dwarf a "flare star."* More simply, a flare star is a red dwarf that may undergo very sudden short term changes in brightness. Most dim red dwarfs are also flare stars. Since flare stars can suddenly brighten as much as six magnitudes and go from invisible to visible and then back to invisible in minutes, a "wise man" who knew both where and when to look could see an otherwise dark star when it flares unveiling itself.† For example, on April 25, 2008 the tiny "red dwarf star known as EV Lacertae, unleashed a mega flare packing the power

---

* On December 7, 1947 the American astronomer K. G. Carpenter *accidentally* photographed UV Ceti while it was flaring. UV Ceti increased in brightness by a factor of 12 and returned to normal within a three minute time span. The word didn't get out to the astronomical community until 1948, and the next flare star discovered, AD Leo, was in 1949. G. A. Gurzadyan, *Flare Stars*, Pergamon Press, New York, 1980, p. 1. A different account of the discovery of flare stars is given in *Burnham's Celestial Handbook* by Robert Burnham, Dover Publication, New York, 1966, pp. 641 and 642, and in *Variable Stars* by Michel Petit, John Wiley & Sons, New York, 1982, p. 156. In these two books Willem J. Luyten of the University of Minnesota is credited with first discovering the flaring of the UV Ceti system (L726-8) in 1948, and this discovery being announced by Harvard Observatory in April 1949. The fainter member of this pair of stars called the UV Ceti system has been called "Luyten's Flare Star" and it is the classic example of the type.

† The flares of a flare star (stellar flares) are unlike the more familiar flares of our Sun (solar flares). The flares of a flare star are of a different nature and infinitely greater magnitude than the flare that our sun undergoes. "A stellar flare is a grandiose event, truly stellar in its scale, an event that, each time it appears, encompasses huge volumes of space around the star and develops with fabulous large speed"— G. A. Gurzadyan, *Flare Stars*, Pergamon Press, New York, 1980, p. 334. In an article called "The Atmosphere of M Dwarfs" which appears in *The M-Type Stars* – NASA (J. Johnson and F. Querci editors) 1986.

of thousands of solar flares.[10] The flare was seen by Russian and American satellites. To those who could see the star in the night sky, the flare would have been visible to the naked eye. Rachel Osten of NASA's Goddard Space Center said, "Here's a small cool star that shot off a monster flare."[11] It is interesting to note that Proxima Centauri, the closest known star to the Earth (4.22 light years away), and many other known stars closest to the Earth are red dwarf flare stars. The possibility of being able to find an unknown flare star in the Earth's stellar neighborhood is demonstrated by the discovery in 2003 of a faint red dwarf, a probable flare star, that at first appeared to represent the third closest star system to our own. "Our new stellar neighbor is a pleasant surprise, since we weren't looking for it," wrote Bonnard Teegarden, an astrophysicist with NASA's Goddard Flight Center.[12] Teegarden and his colleagues detected the unknown star while searching a database of sky survey observations. "It was while going through the database that researchers discovered the dim red dwarf, which shines about 300,000 times fainter than the Sun. It's faintness has veiled it from astronomers until now. . . ."[13]

The new star called "Teegarden's Star" (SO 025300.5 + 16528) is in the constellation Aries and it is too faint (17.47 magnitude) to be seen without a telescope. Calculations show that the new star is somewhat farther away than initially thought, but astronomers currently only know of six other stars with proper motions greater than the new star.* "Proper motion" reflects distance from the Earth. Astronomers think that many other dim red dwarfs that exist within

---

* "Teegarden's Star," Sol Station, *www.solstation.com/stars/SO 02530.htm*. The star closest to the Earth, the flare star Proxima Centauri is 4.22 light years from the Earth; and the second nearest star to the Earth, the probable flare star called Barnard's Star is about 6.0 light years from the Earth; and this new star, a dim red dwarf is about 12.4 light years from the Earth.

twenty light years of the Earth have also been overlooked, since their count on stellar population surveys is much lower than would be expected. [14] Bonnard Teegarden, the discoverer of the new star said, "since the survey only covered a band of the sky (about twenty-five degrees in declination), it is entirely possible that other faint objects remain to be discovered." [15]

The Bible appears to give information about this special type of flaring red dwarf star in a number of different places. *Presumptions* not speculations, must be made that some of these Scriptures talk about this star, but these presumptions are not unreasonable, considering that the Bible contains an impressive body of scientifically sound information about comets. Since the Bible gives information about comet origins, composition, behavior, and what happens during and after the various types of impacts, it would not be surprising that the Bible also gives information about the type of star that drives comets to the Earth.

Psalm 19:4 says, "he hath set a tabernacle for the sun," and since a "tabernacle" is a "dwelling place" (Interlinear) this **suggests** that this dark dwarf, or lost star is part of the solar system. *Psalm 19:6* says, "His going forth is from the end of the heaven ('heavens' – Interlinear, NAS, NIV) and his circuit (*'orbit'* – Interlinear Bible) unto the ends of it: and there is nothing hid from the **heat** thereof ('from His heat' – Interlinear)." This suggests that the star travels at the end of the solar system (heaven).* Thus, it would pass through the inner part of the Oort Cloud which begins at the end of heaven, that

---

*   Recall that in the Bible, when used in regard to astronomy the term "heaven" pertains to the "solar system," and the term "heaven of heavens" pertains to the "greater solar system" which includes both the solar system and the Oort Cloud. So in regard to astronomy, the term "heavens" indicates both "heaven" and the "heaven of heavens."

is, at the end of the solar system, and it could drive comets from the Oort Cloud to the Earth.* When *Psalm 19:6* also says that "nothing is hid from the heat thereof or from His heat," it **suggests** that this far travelling star is an infrared emitter, as infrared radiation is the heat given off by a body. Infrared radiation is a function of temperature, and this is how night vision work by making heat waves visible.

A flare star, that is, an eruptive red dwarf star enshrouded in a hydrogen cloud, is an infrared emitter, because the dim light it gives off is mostly blocked and converted to additional heat by the thick and opaque hydrogen cloud that conceals it. *II Peter 1:19* may apply here since it talks about "a light that shineth in a dark place" until the day comes when the "day star" arises. "A light that shineth in a dark place" is an accurate description of a flare star, a star that shines inside the cloud that covers and conceals it. Finally, it must be noted that most flare stars are dim eruptive red dwarfs and that all flare stars are infrared emitters. Interestingly, astronomers believe that the companion star to the Sun is a red dwarf or an infrared (brown) dwarf.[16]

Job 22:14 seems to refer to the thick cloud that conceals this star as it moves along its orbit in the heavens, that is, in "heaven" and the "heaven of heavens," as references to the "solar system" and the "greater solar system." *Job 22:14* says, "Thick clouds are (or a cloud is) a **covering** to him, that he seeth (is seen) not; and he walketh in the circuit (orbit) of heaven ('the heavens' – Interlinear)" (also see *Job 26:8-9*). *Job 9:11* also seems to tell how this star hidden by a cloud goes by unseen. *Job 9:11* says, "Lo, he goeth by me and I see

---

*   *Isaiah 13:4-5* tells how "the Lord of hosts mustereth the host (comets) of the battle. They came from a far (place), from the **end of heaven** even the Lord, and the weapons (**comets**) of his indignation to destroy the whole land."

him not: he passeth on also but I perceive him not ('but I do not see Him' – Interlinear)." *Job 37:21* says, "And now men see not the bright light which is in the clouds (cloud)," **may** also pertain. *Job 37:15* asks, "Dost thou know when God . . . caused the light of his cloud shine?" *could* be speaking of this dwarf star erupting or flaring and causing the cloud that conceals it to shine and be seen.

As every translation is an interpretation, the better one understands what a Scripture is describing, the better the translation. Understanding how the ancient Sumerians, Akkadians and Babylonian spoke about astronomical phenomena provides a context in which we can determine that the *Book of Job's* chapters 37 and 38 give a great deal of information about comets and dwarf stars. For example, in the literature of the ancient Near East, and in the Bible the word "cloud" can designate an "atmospheric cloud," a "cometary cloud" or a "stellar cloud."* Context is the key in deciding what type of "cloud" is being referenced.

These two chapters from the *Book of Job* contain multiple references and details about comets and stars. So what does *Job 37:6* mean when it talks about the "small rain" and the "great rain?" What does *Job 37:12* mean when it says, "And it is turned round about by his counsels: that they may do whatsoever he commandeth them upon the face of the world in the earth?" What does *Job 37:13* and *15* mean when they say, "He causeth it to come, whether for correction, or for his land, or for mercy"; "when God . . . caused the light of his cloud to shine?" What does *Job 37:16* mean when it asks, "Dost thou know the balancings of the clouds?" What is *Job 37:21* communicating when it says, " "And men now see not the bright

---

*   An outgassing comet is enclosed in a cloud, which astronomers call a "coma." A star can be enclosed in a hydrogen cloud, which astronomers also call a "nebula."

light which is in the clouds (cloud)?"

From *Job 38* we have, "When sang together the stars of the morning" (*Job 38:7 Interlinear*), and "when (God) made the cloud the garment thereof and thick darkness a swaddling band for it" (*Job 38:9*). "Have you entered the storehouses" of the snow, or have you seen the storehouses of the hail (*Job 38:22*), or **"Who has put wisdom in the dark cloud? Or who has given understanding to the meteor?"** (*Job 38:36 Amplified based on Revised Version margin*). Three simple sentences from *Job 38* provide amazing information about the vast number of comets, their composition of water and how they are formed: "Who can number the clouds in wisdom? Or who can stay the bottles ('water jars' – NAS, NIV – that is, comets) of heaven" and "When the dust groweth into hardness and the clods cleave fast together? [how comets, asteroids and planets are made] (*Job 38:38*). Unfortunately, a critical study of *Job 37* and *38* is beyond the scope of this book. Suffice it to say for now, the Bible appears to give information about an eruptive red dwarf star concealed by a thick cloud (a flare star) that drives comets from the comet reservoir at the end of heaven (the storehouses of snow and hail we call the Oort Cloud) into the inner solar system to bombard the Earth.

Only in the last fifty years has scientific knowledge of comets caught up to the specific information contained in the Bible about comets. If we propose to do an analysis of what the Bible has to say about dwarf stars, the Bible should not be rejected as an accurate source of information because of a long history of inaccurate translations and misinterpretations. Both critics and advocates of the Bible should be careful about prejudging the scientific information contained in it. While the Bible is not a science book, the scientific knowledge that it mentions is accurate. The truth of the matter is that

Biblical passages with scientific implications have not been studied to uncover their full meaning. Can anyone familiar with scientists' efforts to find a second star in our solar system, a companion star to our sun, not wonder if the Bible also contains information about this star?

A growing number of astronomers and astrophysicists now believe that a dwarf star of some type passing through the comet reservoir at the end of our solar system can trigger a storm of comets to hurtle toward our inner solar system. They speculate that some of these dislodged comets have struck the Earth, causing periodic sudden mass extinction. A number of astronomers also suspect that there is an as yet undiscovered second star in our solar system, a dark dwarf star or eruptive red dwarf (a flare stare) and that this may be the dwarf star responsible for some of Earth's catastrophic cosmic history.

There are Scriptures that now clearly describe cosmic objects enshrouded by clouds, objects that shine in dark places, and objects that unseen by we Earthlings. Could the Star out of Jacob, a Scepter out of Israel, the star seen only by the wise men at Jesus' birth, be the sign in heaven that appears with Jesus' return? Could this star be connected to the objectives the returning Messiah is to fulfill?

ELEVEN

What the Bible Knew First

But thou, O Daniel shut up the words, and *seal* the book, even to the time of the end: many shall run to and fro and **knowledge shall be increased** .... And he said, Go thy way Daniel: for the words are closed up and *sealed till the time of the end.*

Daniel 12:4 and 9

Behold the storm (***cometary***) of the Lord has gone forth in wrath, Even a whirling tempest (***cometary***); It will swirl down on the head of the wicked. The anger of the Lord will not turn back, until He has performed and carried out the purposes of His heart. In the ***last days*** you will clearly understand it.

Jeremiah 23:19-20 NAS

> Ask now about the former days, long before your time, from the day God created man on the earth; ask from one end of the heavens to the other. Has anything so great as this ever happened, or has anything like it ever been heard of?
>
> *Deuteronomy 4:32 NIV*

Millions of people have wondered if the Bible is the Word of God and if what it says within its pages is true? Can we discover scientific facts in the Bible that may help us in determining the truth? What if the themes of "catastrophe" and "prophecy" are pieces of a mosaic, found throughout the Bible's 66 books (written over 1,500 years by a couple dozen different people) that reveal a specific message? Do these themes contain scientific facts that may help us understand what the Bible is saying? How are we to respond to the reality that the Bible contains complex scientific descriptions of events that our brightest scientists have only begun to recognize in the last half century? This important information recorded throughout the pages of the Bible has often been simply overlooked or blatantly misinterpreted. *Daniel 12:9* says that certain prophecies were "closed up and sealed till the time of the end." Significant discoveries and advances in the fields of archeology, geology and astronomy over the last sixty years will not allow us to dismiss *Daniel 12:4*, which speaks of "increased knowledge" in the latter days. A growing number of scientists have recognized and acknowledged that certain Biblical catastrophes appear to be descriptions of cometary events. By virtue of the "increased knowledge" of today (*Daniel 12:4*), the mystery of Old Testament catastrophes and Old and New Testament prophecies for the future can now finally be correctly interpreted, and "clearly understood" (*Jeremiah 23:19-20 NAS*). We are now able to relate the connections between catastrophe and prophecies for the future to the full body of Scripture. When the Bible's Scriptures about the heavens are properly understood, instead of being considered a hodge-podge of fanciful and poetic verses, we find quite a surprise, a surprise that ultimately speaks to the scientific credibility of the Bible. [1]

The surprise is that when the correct definition of certain key ancient words are used in conjunction with recent astronomical knowledge, a large body of internally consistent and detailed *scientific information* about comets emerges. This Biblical information includes descriptions and scientific information about comets – their origin, composition, behavior, and impact effects, and it includes a history of cometary impacts on Earth over the last 7,000 years. We see that the Bible contains a comprehensive body of accurate scientific data on these subjects. This amazing truth should make us ask, "How did the ancient work we now call the Bible come to contain scientific observations that mankind would not know until the last sixty years?"

In our 21st century some of this scientific data about comets is consistent with verifiable evidence about actual historical events, and some is contained within prophecies about events that are to occur in the future. These prophecies provide new opportunities for scientific verification of the astronomical information given in the Bible. In a sense, a "fulfilled prophecy" is a powerful form of evidence for establishing scientific knowledge. Most importantly, the Bible contains a comprehensive textbook full of detailed scientific information about comets that was recorded at least 1,500 years before the scientific community learned of this information.

This gives the Bible priority on a great number of scientific points of information about comets. These points, or what we will call here, **Bible Firsts**, will be reviewed in this chapter.

A number of prophecies written in the books of the Old Testament and the New Testament could not have been correctly understood and interpreted before just five to forty years ago, since the specific astronomical and geological knowledge needed was previously unavailable. For example, before this new wealth of information about cometary activity was discovered, it was nearly impossible to see how the "plagues" of the **Book of Exodus**, and the "end times" catastrophes of the Seven Trumpets and the Seven Vials in the **Book of Revelation** could be related to each other. Since comets actually streak across the sky to strike the atmosphere and the Earth, they logically become the "signs and wonders in the heavens and in the Earth" referred to in **Daniel 6:27** and prophesied in other parts of the Bible (*Joel 2:30, Micah 7:15 NIV,* and *Luke 21:11*). Their impacts would inevitably bring the "blood (death) and fire and pillars of smoke" (*Joel 2:30-31*) that have been prophesied to occur during the end times.

There are a number of scientific points about comets and stars that the Bible first reveals. These instances of scientific "priority" recorded in the Bible include:

- the different types of phenomena that can take place when a comet or asteroid impacts the Earth;
- how a comet can explode upon impacting the Earth's atmosphere;
- how a comet can cause a tsunami driven flood after hitting the ocean;
- how cometary bombardment of the Earth can involve multiple

impacts during a relatively short period of time;
- a reservoir of comets exists at the edge of our solar system, which astronomers call the Oort Cloud;
- comets mainly consist of frozen water (ice) covered by a rocky crust;
- comets were the main source of the water in the Earth's oceans;
- how comets are a source of water for the Earth;
- comets have "mouths" or vents in their crust through which they outgas;
- how cosmic impacts can introduce very harmful chemicals to the Earth;
- how comets can carry bacteria capable of causing disease;
- how the scientific response to comets or asteroids found on a collision course with Earth encompasses plans for shooting down these "messengers of God."

Based on these astronomical points and facts found first in the Bible, the Bible establishes a certain authority about comets and comet bombardment. *These facts demand serious review of the Bible's prophecies about future comet bombardments, and provides us with scientific reasons to believe that the Bible's prophecies are valid predictions about Earth's future.* If the trend of the last forty years continues, new scientific discoveries and "increased knowledge" in the years to come will make scientific information and prophecies from the Bible even more compelling and convincing for those who do believe, as well as for those who in the past have not believed in the God of the Bible.

1) *The different types of phenomena that can that take place when a comet or asteroid impacts the Earth.*

Several books in both the Old and New Testament provide specific details about what happens when a comet strikes Earth (for example, *Isaiah 13:5-13; 24:18-20, 34:4; Ezekiel 38:19-22; Joel 2:30-31, Revelation 6:12-14; 8:6-12; 9:1-19;* and *16:2-21*). Scientists have recognized the different effects of cosmic impacts by studying the effects of massive nuclear detonations. Only after entering the Nuclear Age did scientists recognize that a majority of craters on the Earth and Moon were caused by massive explosive impacts of comets or asteroids.* Only after scientists witnessed the specific events of nuclear explosions did they learn of the thermal pulse, fierce winds, base surge, towering mushroom cloud, and dust that fills Earth's atmosphere and blackens the sky after massive explosions. Scientists now know that comet impacts produce these same types of phenomena. The comparison between nuclear explosions and cosmic explosions allowed scientists to recognize that an explosion resulting from the impact of a six to ten mile wide comet or asteroid could release energy in excess of the simultaneous detonation of *five billion* atom bombs.

Only cosmic impacts could produce catastrophes of the magnitude and extent called for in the Bible. It is now clear that when the Bible describes these catastrophes, it refers to the different types of phenomena caused by cosmic impact. For example, when the Bible speaks of such events as stars (comets) from heaven falling to Earth; and an earthquake of unprecedented magnitude and size,

---
* For example, many of the craters on the moon that were once classified as volcanic came to be reclassified as impact craters almost over night.

where the sea and waves roar, every mountain and island is shaken, the cities of the nations fall, the Bible characterizes a level of shaking that can **only** be produced by a cosmic impact. Scientists have recently learned that an enormous earthquake can actually displace the Earth's axis and rate of rotation, and that a massive cosmic impact would rock the entire planet in space. Nevertheless, the *Book of Isaiah* (*Isaiah 24:18-20* and *13:13*), written around 680 BC, tells how a massive comet impact can cause the foundations of the Earth to shake, the Earth to reel to and fro, and be moved in space like a cottage. When the Bible tells of a great pit being opened (impact crater), with smoke and dust arising out of this pit (impact ejecta) like that of a great furnace (debris laden mushroom cloud), and the air departing as when a scroll is rolled together, with scorching winds (base surge after impact), it is now clear that the Bible is giving accurate observations about what happens when a truly massive impact occurs. For in the "fire, smoke and brimstone" and "one hundred pound hailstones" the Bible refers to, science describes observations about comets that are outgassing and breaking up. When the Bible talks about the islands and mountains suddenly disappearing we even have information about mega cosmic impacts, "planetary scale impacts" that can actually blast away a large part of the Earth's crust and/or release enough heat inside of the Earth to cause the crust to melt. Now that knowledge has increased and we know that crustal melting, crustal removal and planetary disruption are a function of impact size, we need only look at Mars to see that a large part of a planet's crust can be blasted away. And one need only look at Venus to see that a planet's crust can be melted and all of the surface features lost before a new crust is formed. *(See Illust. I)*

Remember that astronomer John Lewis in his book *Rain of*

Iron and Ice refers to *Revelation 9:1-2* and asks:

> Why should a star, falling to Earth, open a great pit and fill the air with enough smoke to darken the Sun? The central theme is clear and unambiguous: the events described in *Revelation* are of astronomical origin, and describe real physical events, not mere portents or symbols. Did John (the author of the *Book of Revelation*) somehow know more about impact phenomena than any scientist before the present decade? [2]

Lewis' recognition of the Bible's scientific accuracy is consistent with that of other astronomers, such as Clube and Napier, who have studied comet and asteroid impacts. Clube and Napier recognize passages in *Revelation* and the Old Testament that clearly and accurately describe impact phenomena exactly like the phenomena scientists have discovered in the last forty years. [3] As we can see now, *long before astronomers learned of the various types of catastrophic effects that take place when a comet or asteroid impacts the Earth, the Bible has undeniable descriptions of impact phenomena.* (For more information, see Chapter 2.)

2) How a comet can explode upon hitting the Earth's atmosphere.

In addition to giving information about the catastrophic effects that take place when a comet strikes Earth's surface, the Bible makes it clear that atmospheric impacts can occur, and gives information about the catastrophic effects unique to atmospheric cometary impacts or airbursts. Only in the last twenty years or so have scientists learned that some comets can be loosely consolidated and that these comets or their fragments can explode and disintegrate when they leave the vacuum of space and impact the (relatively dense) gaseous envelope that covers the Earth (the atmosphere).

Airburst explosions were not recognized until after the 1908 Tunguska Event in Siberia had been intensely studied by scientists. The Tunguska event is now believed to be the result of an airburst explosion high in the atmosphere of a small asteroid or cometary fragment approximately 100 meters in diameter. Unlike cometary impacts on the surface of the Earth that leave craters, atmospheric impact explosions do not produce craters, but otherwise level structures, trees and the like below at ground zero! And these insights mark the centennial or 100th anniversary of Tunguska.

Astonishingly enough, data indicates that cosmic airburst explosions are not all that rare. Recently, military satellites have recorded a number of airburst explosions high in the Earth's atmosphere. Astronomers suspect as many as eight major explosions occur per year with the combined energy that is released equivalent to that of one Hiroshima atomic bomb. [4] Fortunately, it is believed that up until now, these explosions have all occurred above unpopulated areas. This is possible because oceans cover seventy percent of Earth's surface area, and 46% of the remaining land mass is still wilderness.

Paul Chodas, an astronomer with the Jet Propulsion Laboratory in California studies military satellite records. He says, "They have seen a number of explosions which can be traced to large meteors or small asteroids entering the Earth's atmosphere and completely exploding at an altitude of perhaps thirty or forty kilometers." [5] Chodas notes that in the fall of 1990 a satellite detected a one kiloton blast high above the Pacific Ocean. As noted in Chapter 6, this explosion was at first thought to be caused by a land based missile, since the Gulf War was about to begin. A few years later, Jay Tate, an astronomer with the United Kingdom's Spaceguard project, reported that a British satellite passing over southeast Greenland on December 9, 1997 picked up an airburst explosion over the ice fields below that produced a flash bright enough to light the night sky. [6] Tate also states that a village 35 miles from the explosion felt the blast in the form of hurricane like winds. [7]

In addition to scientists' recent recognition of comet airburst explosions in Earth's atmosphere, they found these explosions are often characterized by certain atmospheric aftereffects, such as great winds, a fireball, and great heat. Airbursts produce intense heat and thermal radiation from the fireball capable of scorching people and animals at great distances. As noted in Chapter 4, the destruction of Sodom and Gomorrah, where fire and brimstone (burning sulfur) rained from heaven (*Genesis 19:24* and *Luke 17:29*), is an accurate description of airburst explosions. Chapter 6 deals with the Fourth Trumpet and Fourth Vial found in *Revelation*, and specifically tells of an airburst explosion of a comet, that describe the same types of airburst explosive effects now recognized by scientists. *As we can see now, long before astronomers learned that comets can explode when impacting the Earth's atmosphere and understood what the effects such explosive impacts can produce, the Bible had already accurately described these phenomena.*

3) How a comet can cause a tsunami driven flood after hitting the ocean.

Chapter 4 recalls the Bible's account of Noah's Flood as given in *Genesis 7* telling how a comet impact caused a tsunami and how this tsunami produced the Great Flood. A more accurate translation for *Genesis 7:11* reads:

> In the six hundredth year of Noah's life, in the second month, the seventeenth day of the month, the same day were all the **sources** of the great primeval ocean (**comets**) broken up, and those ***that lie in wait in heaven*** (**comets**) were **loosed**.
>
> *Genesis 7:11 – retranslation*

While the Sumerians, Akkadians, and Babylonians also give accounts of the Flood, only the Biblical account correctly refers to comets as the source of the water in the Earth's oceans, while indicating that the break up of incoming comets caused the Flood. Only in the last sixty years have scientists learned that comets are the main source of Earth's ocean water, and that ocean impacts by comets can cause mega tsunamis with mountain high waves capable of suddenly flooding the land.* In addition to saying that comets breaking up caused the Flood, the Bible reiterates that comets were the cause of the destruction when *Genesis 7:11* says comets which "lie in wait in heaven" were "loosed." Comets that lie in wait in heaven being loosed, as in the clearer translation presented here says, indicates

---

* It must be emphasized that while comets are the main source of the Earth's water, it is not the water that they bring that caused the Flood, but the waves created when they hit the ocean that caused the Flood. Calling comets "the sources of the deep" in *Genesis 7:11* is just a clever way of referring to comets. Also see item #7 here – "How comets were the main source of the water in the Earth's oceans."

some of the comets stored in the heavens in the Oort Cloud moved out from this storehouse toward the inner solar system to impact the Earth and bring about the Flood. Although astronomers did not discover the Oort Cloud of comets until 1950, *Job 38:22-23 (NIV)* tells of the snow and hail, that is, the frozen waters we call comets are stored and reserved "for times of trouble, for days of battle and war." Noah's Flood was certainly a "time of trouble." *(See Illust. F)*

Loosed comets being broken up suggests their disintegration after entering the Earth's atmosphere and impacting the ocean.* Further, Chapter 4 details physical events of the Great Flood given in the Bible that are consistent with an impact driven tsunami flood. For example, only the towering waves of a tsunami could quickly cover the mountains, and yet, after pulling back, only leave the land flooded to a depth of 22-1/2 feet as stated in *Genesis 7:20*. Obviously in a tsunami, the tsunami waves reach heights flood water could never reach. During the deadly 2004 Indonesian tsunami, the destructive waves' run up was higher than thirty feet in some places, while average floodwaters only rose several feet.

In 450 BC "the Greek historian Thucydides wrote in *History of the Peloponnesian Wars* his speculation about the causes of tsunamis. He argued that it could only be explained as a consequences of ocean earthquakes and could see no other possible causes for the phenomena." [8] Geologists now know that most tsunamis occur by sudden motion of the ocean floor, and that a tsunami can be caused

---

* Note that a more accurate translation for *Job 38:31* asks, "Cans't thou bind the chains (NAS, *Tanakh* – "cords") of the ***stored aways*** (not Pleiades, #3598 from the same as #3558 - 'to store away', referring to the comets of *Job 38:22-23*), or loose the bands of the ***brutish*** (not Orion, #3685 as if a ***burly one***, same as #3684 from #3688 'to be foolish' as synonymous with 'brutish' - see *Brown-Driver-Briggs,* p. 492)." Also see the discussion of Noah's flood in Chapter 4.

by earthquake, explosive volcanic eruption, landslide, and even cosmic impact.* While over 160 impact craters have been found on the land, only recently were impact craters found under the ocean. No one had spent much time looking for them down below. Up until recently, scientists did not even know how to look for craters in the deep ocean. Chapter 4 explains the new techniques based on satellite technology that allow geoscientists to locate craters left by ocean impacts, and associated deposits that have been carried onto the land by the mega tsunamis generated by these ocean impacts, so-called tsunamites.

Most notably, these geoscientists just recently discovered an eighteen mile wide underwater impact crater that is called Burckle Crater. Burckle Crater lies in the middle of the Indian Ocean under 12,500 feet of water. Tsunami deposits or tsunamites from this impact have recently been found in Africa, India and Australia. Given that the land of Noah, and the Sumerians as well as the Akkadians is located down range from Burckle Crater, it is easy to see how a towering tsunami wave moved up the Tigris-Euphrates Valley like a huge rising hooded cobra until it stopped at the mountains of Ararat in Turkey (*Genesis 8:4*). It has been estimated that Burckle Crater is today 4,500 to 5,000 years old, which is around the time the Bible gives for Noah's Flood.

The recently discovered Burckle Crater on the Indian Ocean floor is the proverbial "smoking gun" for cosmic impact causing the <u>Flood and it is</u> proof positive that a comet impact can cause tsunamis

---

* Note that like *Genesis 7* and the Flood, the Second Trumpet and Vial of the *Book of Revelation* (*Revelation 8:8-9* and *16:3*) also call for a cometary ocean impact. *Genesis 8:8* says "a great mountain burning with fire was cast into the sea . . . ." While powerful and deadly seismic sea waves are indicated, no landfall is called for. Regarding the Second Trumpet and Vial see, Chapter 6.

with consequent floods. The Burckle Crater and related tsunami deposits also provide evidence that the better translation of *Genesis 7:11* presented here is correct! In this translation we have comets being "loosed" and then "breaking up" to ultimately impact the ocean and cause a tsunami and the Flood. **As we can see now, long before astronomers and geoscientists learned how comets can cause a tsunami and a flood, the Bible describes this type of phenomena.**

4) How cometary bombardment of the Earth can involve multiple impacts during a relatively short period of time.

The Bible's *Book of Revelation* speaks of multiple cometary bombardments that impact Earth during the relatively short period of 42 months. Long before scientists discovered the reality of multiple cometary bombardments, the Bible described them. The Bible makes the point for the possibility of rapidly successive multiple cosmic impacts most clearly when it delineates a series of prophecies in the Seven Trumpets and the Seven Vials. In Chapters 6 and 7, the Seven Trumpets and Seven Vials refer to seven episodes of cosmic bombardment that occur during a three and a half year period of time. In addition, each of the seven episodes of bombardment spoken of in the *Book of Revelation* is also spoken about in some small way by one or more books of the Old Testament. The Bible makes it quite clear that bombardment of the Earth can involve multiple impacts during a relatively short period of time.

On the other hand, up until only twenty years ago, scientists believed that bombardments of the Earth only occurred singly, and that these bombardments were separated by very long periods of time. However, this is not the position of astronomers today. Astronomers

Clube and Napier, quoted in Chapter 2, wrote that bombardments of the Earth by comets and asteroids might be "concentrated into brief catastrophic periods in which multiple impacts occur."[9] Multiple impacts can occur as a result of several different comets coming in during the same time period, or as the result of one larger comet breaking up into several smaller bodies as it approaches the Sun and passes nearby the Earth.

Scientists have looked for a single impact crater that caused the extinction of the dinosaurs 65 million years ago. They actually found not one but a number of craters in different parts of the world. Analyses have shown that these craters occurred about the same time, indicating multiple impacts. In *Fire on Earth* authors John and Mary Gribbin ask,

> Were the dinosaurs actually wiped out by a ***triple*** whammy? It looks very likely. Indeed this may be a conservative assessment, since there are several other features of about the right age (but not yet dated accurately) in a line linking these sites. [10]

From this, it appears possible that more than three impacts were involved in the dinosaurs' demise. Scientists now believe that the 112 mile wide crater off of Mexico's Yucatan Peninsula, a 22 mile wide crater near Mansion, Iowa, a 180 mile wide crater in the Caribbean off Columbia's coast, and a sixty-eight mile wide crater at Popigai in Siberia were all formed at this time.[11] It is a wonder that not all forms of life were destroyed at that time. In December of 1990, University of Arizona scientist Bill Boynton, one of those doing research on the crater in the Caribbean off Columbia's coast said, "We think we have found the smoking gun, but there may be other guns and maybe a firing squad." [12] In July of 1991 an article entitled, "At Last, the Smoking Gun" appeared in *Time* magazine confirming the connection of **both** the crater off of Mexico's Yucatan Peninsula (the Chicxulub Crater) and the crater near Manson, Iowa to the extinction of the dinosaurs 65 million years ago.[*]

In addition to all of the "smoking gun" craters associated with the dinosaurs' demise, proof positive for multiple impacts during a

---

[*] Leon Jaroff, "At Last, the Smoking Gun," *Time,* July 1, 1991, p. 61. (Further proof was found when ocean core samples taken in February of 1997 contained "proof positive of the impact" off of the Yucatan Peninsula in the Gulf of Mexico 65 million years ago. [Paul Recer, "Dinosaur Death Reports Claim Asteroid did it." The Associated Press, *Tucson Citizen*, February 17, 1997, p. 6A.] These samples were taken from 370 feet beneath the seabed in the Atlantic Ocean 300 miles off the coast of Northern Florida in about 8,500 feet of water. The samples had the "unmistakable signature of an asteroid impact," including brownish clay called the "fireball layer" because it contains the "vaporized remains of the asteroid itself." [Recer, "Dinosaur Death Reports" 1997, p. 6A] Just below the fireball layer is a layer with small green pebbles believed to be ocean floor material that was instantly melted by the enormous impact. Just above the fireball layer is a layer of gray clay, whose "almost total absence of fossils speaks of a nearly dead world." This layer of clay is called the "dead zone." Dr. Robert W. Corell, assistant director for Geosciences of the National Science Foundation "called the core samples the strongest evidence yet that an asteroid impact caused the dinosaurs' extinction." [Recer, "Dinosaur Death Reports" 1997, p. 6A] Dr. Corell said, "In my view this really nails it down." [William J. Broad, "New Proof of Asteroids Devastation," *New York Times*, Science Section, February 18, 1997] Although the main asteroid impact probably occurred off Mexico's Yucatan Peninsula in the Gulf of Mexico, Richard D. Morris, leader of the successful international ocean drilling expedition, reasoned that the Gulf of Mexico would be too disturbed by the impact to contain clear layers of debris reflecting the event. Norris chose to drill 1,000 miles to the east, off the east coast of Florida, reasoning that waves from the impact would have washed completely across Florida and left clear and comprehensive layers of debris reflecting the event there. This evidence was exactly what he found.)

short time span comes from the 21 separate large comet fragments from the breakup of Comet Shoemaker-Levy. This comet struck Jupiter during July of 1994, as the world watched on TV and through small telescopes. The multiple impacts of Comet Shoemaker-Levy left a chain of craters on Jupiter. Direct observation of these multiple impacts on Jupiter (1994) gave astronomers insight for assessing the images of Ganymede, the largest of Jupiter's moons, that were sent back by NASA's Spacecraft *Galileo* a few years later. These images, taken in 1996-1997, show a chain of 13 craters on Ganymede as evidence of a comet breaking up and producing multiple impacts over a relatively short period of time. [13] *As we can see now, long before astronomers learned that cometary or asteroid bombardment of the Earth can involve multiple impacts during a relatively short period of time, the Bible made repeated references to a period of multiple impacts that is to occur during the end times.*

5) *A reservoir of comets exists at the edge of our solar system, which astronomers call the Oort Cloud.*

> *Nehemiah 9:6* says,
> Thou, even thou, art Lord alone; thou has made **heaven** (the **solar system**), the **heaven of heavens** (the **Oort Cloud**), with all their **host** (**comets**), the Earth, and all things that are therein ...

> *Job 38:22-23 NIV* says,
> Have you entered into the **storehouses of the snow?** Or seen the **storehouses of the hail** (the **Oort Cloud**), which I reserve for times of trouble, for the days of

war and battle?

Isaiah 13:3-5 says,
 I have commanded my **sanctified ones** (***comets***), I have also called my **mighty ones** (***comets***) for mine anger . . . the **Lord of hosts** mustereth the **host of the battle**. They came from a far (place), from the **end of heaven** (the **Oort Cloud**), even the Lord and the **weapons of his indignation** (**'wrath'** in NIV) to destroy the whole land.

The Bible repeatedly refers to a reservoir or cloud of icy comets at the outer edge of the solar system. The above verses are just a few that do so. However, few people have understood the meaning of these verses. *Job 38:22-23* (*NIV*) calls the cometary reservoir at the end of heaven "the storehouse of the snow" or the "storehouse of the hail" (astronomers sometimes refer to comets as "dirty snowballs"), where "snow" and "hail" refer to the comets' cradle that holds them in "reserve for times of trouble; for days of war and battle." This "storehouse" or reservoir is an important part of the greater solar system that astronomers did not discover until 1948-50.* This reservoir or cloud of comets begins at the edge of the solar system, and extending out about one-third the distance to the nearest star. Astronomers call this reservoir or "storehouse" the Oort Cloud. Comets in the Oort Cloud complex represent material

---

* The Hebrew word *owtsar* (#214 in *Strong's Concordance*) which is translated as "treasures" in the KJV and "storehouses" in the NAS and NIV means "depository armory, storehouse or treasure house" from the root *atsar* (#686 in *Strong's Concordance*) which means "to store or lay up." For example, *owstar* is translated as "storehouse" in *Malachi 3:10* and "storehouses" in *I Chronicles 27:25 and Psalm 33:7.*

from a primordial disk of material around the young Sun believed to be left over or "reserved" from when the solar systems' planets came into being. *(See Illustration A)*

The Oort Cloud is a huge spherical shell of comets that surrounds or encases our solar system. It is a ***thick*** shell of comets with an inner and outer boundary containing as many as several trillion comets. The Oort Cloud was first proposed in 1950 by Dutch astronomer Jan Oort. Within this Oort Cloud is the Kuiper Belt, first proposed in 1951 by Canadian astronomer Gerard Kuiper. The Kuiper Belt is a flattened disk of comets that contains anywhere from 100 million to a few billion comets. The Kuiper Belt begins between Neptune and Pluto, and extends and expands outward to the inner part of the Oort Cloud. In 1981 astronomer Jack G. Hills proposed the theory of an "inner Oort Cloud" which represents the densest part of the Oort Cloud and gives it a dense inner core. To get an idea of its relative size, picture the Sun and the planets of the solar system out to the planet Pluto as occupying the 3/8" hole of an old fashioned 33-1/3 LP record. The comets of the Kuiper Belt would occupy the flat disk's full width of the vinyl LP record. Then imagine this LP record being placed in the center of a domed professional football stadium. The trillions of comets of the outer Oort Cloud would <u>extend out from</u> beyond the LP record to fill the stadium!*

---

* Some astronomers believe that the International Astronomy Satellite or IRAS launched in 1983 picked up a background flux of far-infrared radiation coming from the comets of the inner core of the Oort Cloud - Bailey, Clube and Napier, *The Origin of Comets*, Pergamon Press, New York, 1990, pp. 296 and 303. Since 1992, the Hubble Space Telescope has taken direct photographs of over 40 different Kuiper Belt comets in orbit around the Sun. These are objects that up until now were too small and faint to be directly photographed - Sam Flamsteed, "Where Comets Come From," *Discover Magazine*, November 1995, pp. 83-90. And "Icy mini-planet found in outskirts of solar system," Associated Press, *Tucson Citizen*,

Long before scientific theories or photographs from space telescopes, the Bible referred to the Oort Cloud of comets in *four* different places. Bible verse calls the Oort Cloud "the storehouse of ice and snow" (*Job 38:22-23*), "the waters that are above heaven" (*Psalm 148:3-5*), "the heaven of heavens with all their host (comets)" (*Nehemiah 9:6*) and "the waters which were above the firmament or expanse" (*Genesis 1:6-8*). Not only does the Bible correctly speak of the existence of the Oort Cloud, it also correctly locates the Oort Cloud above heaven or above the solar system proper, beginning at the edge of the solar system.

The Bible states that this storehouse or cloud of comets contains a countless number of comets (*Jeremiah 33:22* says, "the host of heaven cannot be numbered"), and modern astronomers agree. They believe the Oort Cloud contains **trillions** of comets and that this cloud of comets was in place before the solar system and planets were formed.

Psalm 148:3-5 sings out, "Praise ye him, Sun and Moon: praise him, all ye stars of light. Praise him, ye ***heavens of heavens, and ye waters that be above the heavens.*** Let them praise the name of the Lord: for he commanded, and they were created." These verses thus enumerate the objects of our greater solar system. "Heaven" contains the objects of the solar system: the Sun, the Moon, and the "stars of light" (including all the luminous bodies of the heavens, such as planets, comets, and asteroids). The phrase "ye waters that be above the heavens," should read "***ye waters that be above heaven***" and refers to comets, which in fact represent "frozen waters." The "frozen waters" or "waters that be above heaven" are located (stored) in the "heavens of heavens," which is also referred to in the *Psalm* June 4, 1997, p. 4A. And Anne Denogean, "Solar system gets bigger, with UA seniors' help," *Tucson Citizen*, June 5, 1997, pp. 1A and 6A.

above.

So, in *Psalm 148:4* "the waters that are above the heavens" (NAS) arguably refers to comets in the Oort Cloud. *Nehemiah 9:16* supports the distinction made in this passage from *Psalms 148* between "heaven" and the "heaven of heavens." Referring directly to the Oort Cloud comets, *Nehemiah 9:16* says, "Thou, even thou, art Lord alone; thou has made **heaven**, the *heaven of heavens*, with all their **host** (**comets**), the Earth, and all things that are therein . . . ." The phrase "heaven of heavens" is an idiom or term for the "highest heaven" (NAS, NIV, *Tanakh*). While our solar system is "heaven," having comets in it, the host of stored comets numbering in the trillions is located in the outer "heaven of heavens." The "heaven of heavens" or "highest heaven" is the same as the Oort Cloud comet reservoir, because *Psalm 148:4* associates the "heaven of heavens" with comets when it says "ye *heaven of heavens* and ye *waters that be above the heavens*."* "Heaven" and the "heaven of heavens" taken together constitute the "greater solar system." In *II Corinthians 12:2* Paul refers "to the third heaven," which may be the place where the Bible says God prepared His Throne (*Psalm 103:19, Psalm 11:4*).

"The waters which are above the firmament or expanse" is another-reference to the Oort Cloud of comets found in *Genesis 1:6-10*.

> And God said, Let there be a firmament (expanse - NAS, NIV, *Tanakh*) in the midst of the waters, and let it divide the waters from the waters. 7) And God

---

*   It is interesting to note that the Hebrew word *shamayim* which means "lofty" and is translated as "heaven" and "heavens" (word #8064 in *Strong's Concordance*) contains the Hebrew word *mayim* which means "water" (word #4325 in *Strong's Concordance*). Thus, the "heavens" can be conceived of as the place of "lofty waters."

made the firmament, (expanse - NAS, NIV, *Tanakh*) and divided the waters which were under the firmament (expanse - NAS, NIV, *Tanakh*) from the **waters which were above the firmament** (expanse - NAS, NIV, *Tanakh*): and it was so. 8) And **God called the firmament** (expanse - NAS, NIV, *Tanakh*) **Heaven**. And the evening and the morning were the second day. 9) [The subject now moves from the solar system to planet earth] And God said, Let the waters under (***within***) the heaven be gathered together unto one place (into its places - *Septuagint*), and let the dry land (ground - NIV) appear: and it was so. 10) And God called the dry land (ground - NIV) earth (land - NIV); and the gathering(s) together of the waters called he seas: and God saw that it was good.

Genesis 1:6-10

In *Genesis 1:6-8* God created the "firmament" or "expanse" (NAS, NIV, and *Tanakh*) which he called "heaven" (the solar system). This "heaven" (the solar system) was in the ***midst*** of the waters that would be divided, where the waters that were ***under*** heaven were divided from the waters that were ***above*** heaven. The Oort Cloud comets are both above and below and therefore enclose the solar system so that the solar system or "heaven" was indeed "in the ***midst*** of the waters." (Note that in "the waters that are above the heavens" from *Psalm 148:4* we have a reiteration of "the waters, which were above the firmament," where the "firmament" is another word for "heaven" from *Genesis 1:6-10.*)*

As *Genesis 1:6-8* tells of the formation of the solar system "in the midst of the waters" of the already existing Oort Cloud of frozen waters, we are also given the **correct** order of formation and relationship between the Oort Cloud and the planets of the solar system. *II Peter 3:5* refers to this same division of the frozen waters of the Oort Cloud, and then the gathering of the water on the Earth, when it says "... by the word of God the heavens were of old, and the Earth standing *out of the water* (that is *dry land*) and *in the water* (in the solar system, that is *within the frozen waters of the spherical Oort Cloud*)."

The Bible even asks: who created this cloud of comets? *Job 38:29* seems to refer to the origin of comets when in a "*doublet*," a Hebrew grammatical form, God asks Job, "Out of whose womb came the ice (of comets) and the hoary frost (water bowls or water basins - **comets**) of heaven (the solar system), who hath gendered it?"† *Job 37:10 (NIV)* seems to answer this question, "The breath of God produces ice, and the broad (expanse of - NAS) waters become frozen." Notice that the Bible does not just mention the Oort Cloud

---

* Some of the early Hebrews erroneously regarded the "firmament" or "expanse" to be a solid vault of heaven which supported "waters" above it. From *The New Brown-Driver-Briggs-Gesenius Hebrew-English Lexicon*, Hendrickson Publisher, Peabody, Massachusetts, 1979, p. 956.

† As explained in **NOTE FOUR**, the Hebrew word *kefore* (#3713 in *Strong's Concordance*) that is translated as "hoary frost" in *Job 38:29* can just as well be translated as "bowl" or "basin," as it is in *I Chronicles 28:17, Ezra 1:10* and *8:27*. Translating the word *kefore* as "bowl" or "basin," to give "bowls of heaven" or "basins of heaven" in *Job 38:29* would indicate a comet, just as eight verses later in *Job 38:37* comets are referred to as the "the bottles of heaven" or "the water jars of the heavens" (NAS). *Psalm 147:16-17*, like *Job 38:29*, also associates the "hoar frost" or "bowls" of heaven with the ice of comets, when it says in a *doublet* "... he scattereth the **basins** (#3713) like ashes. He casteth forth his *ice* (*comets*) like morsels."

in passing, but gives comprehensive and accurate information about the Oort Cloud.*

In *Genesis Chapter 1* the Bible says the Oort Cloud was created first, before the solar system was created. Scientists indeed believe that the Oort Cloud was already in place when the solar system and the planets came into being. John and Mary Gribbin describe the Oort Cloud saying, "Most astronomers agree that it was 'in the first place' - at the time the Solar System was born and the planets formed."[14] Astronomers Clark Chapman and David Morrison write, "... the nucleus of a new comet seems to be the most pristine remnant of the original matter from which the solar system was made...."[15] *As we can see now, long before astronomers proposed a reservoir of comets involving matter left over from the formation of the solar system that existed at the edge of the solar system and enclosed the solar system, the Bible made accurate, detailed and repeated reference to just such a reservoir of comets!*

6) Comets mainly consist of frozen water (ice) covered by a rocky crust.

---

* Another possible reference to the Oort Cloud is in *Job 38:31* where a corrected translation asks, "Canst thou bind the chains of the *stored aways* (the *comets* in the Oort Cloud), or loose the bands of the *brutish* (the *comets* to come)." Here *loosing* the bands of the stored away comets relates to the retranslation of *Genesis 7:11* presented in Chapter 4, which tells of a time when the comets "that lie in wait in heaven were *loosed,* the sources of the great deep were broken up" - *Job 38:31* better translation: "Cans't thou bind the chains (NAS, *Tanakh* - cords) of the *stored aways* (not Pleiades, #3598 from the same as #3558 - 'to store away', the comets of *Job 38:22-23*), or loose the bands of the *brutish* (not Orion, #3685 as if a *burly one,* same as #3684 from #3688 'to be foolish' as synonymous with 'brutish' - see *Brown-Driver-Briggs,* p. 492)." Also see the discussion of Noah's flood in Chapter 4.

As seen in the last section about the Oort Cloud, there are several Bible verses describing things in the heavens composed of frozen water that we now recognize as comets (*Job 38:22-23*, *Nehemiah 9:6*, *Psalm 148:3-5*, and *Genesis 1:6-8*). Scientists didn't know that all comets are mainly composed of water until 1950. It was then that Harvard astronomer Fred Whipple proposed what became known as the "icy conglomerate model" (an unconsolidated mixture of ice and snow and boulders – compare to *Job 38:22* and the "treasures of ice and snow") for an active comet's nucleus. [16] Today we know that comets are composed of a consolidated mixture of dust particles and ices, and that these ices consist of various frozen gases with water (H_2O) ice predominating.

Bible verses also state that cometary ice is covered by a ***dusty or rocky crust.*** This is another specific detail about comets astronomers did not discover until 1985-1986. Before this, most astronomers believed that an active comet was a loose or *unconsolidated* mixture of ***bright*** and highly reflective ice and boulders instead of what we now know as the facts. Accurate comet models finally have a ***consolidated*** body of dust laden, frozen water covered by a ***dark rocky crust.*** This crust insulates the ice of comets from heating by the Sun. Referring to the rock covered ice of comets, *Job 38:29-30* asks, "Out of whose womb came the ice and the ***water bowls of heaven*** (from word #3713 in *Strong's Concordance*, the Hebrew word *kefore* meaning 'water bowls' not 'hoary frost)? [Note how 'ice' and 'water bowls of heaven' constitute a Hebrew '***doublet***' about comets] . . . The **waters** are hid as with a ***stone***, and the face (presence) of the deep (mass of water) is frozen."* Here *Job 38:30* says that the frozen waters or ice of comets

---

*   In *Job 38:30* the Hebrew word translated as "face" (#6440 *pawneh*) is better translated as "presence"; and the Hebrew word translated as "deep" (#8415 *teh-home*) is better translated as "mass of water" or "water-

is *covered* by rock or stone, that is, a rocky crust. After *Job 38:30* makes reference to this rocky crust, *Job 38:37* asks "who can stay (*cast down*) the bottles of heaven."* Here, comets, which we now know are mostly composed of water, were accurately described as the "bottles of heaven" (KJV) or "water jars of the heavens" (NAS, NIV) which is strikingly similar to the reference for comets as the "water bowls of heaven" in *Job 38:29*. *(See Illustration A)*

Notice the consistency of subject in *Job 38:29-30* and *37* where the "ice of heaven," the "frozen waters . . . hid as with a stone," the "water bowls of heaven," and the "bottles or water jars of heaven" are *all* used as descriptive references to comets. Today, astronomers also characterize comets using comparable terms such as "chunks of ice," "chunks of water-snow," "iceballs," "frozen snowballs," "dirty snowballs," "icy snowballs," and "fire and ice."

A similar description of comets, *Psalm 147:16-17* says, "he scattereth the hoarfrost (water bowls) like ashes. He (the Lord) casteth forth his *ice* like *morsels*: who can stand before his cold?" These verses clearly refer to the water ice of comets. *Psalm 147:17 (NIV)* translates the word "ice" as "hail" and reads, "He hurls down his hail like pebbles. Who can withstand his *icy blast*?" Here comets that can be more than ten miles across are said to be but "pebbles to the Lord," to the Lord of Hosts, the Lord of Comets.

The melting of an icy comet occurs when it enters into the

---

supply." So, a better translated *Job 38:30* would read "The waters are *hid* or *covered* (#2244 *khaw-baw* to hide) by a stone, and the *presence* (#6440) of the *mass of water* (#8415) is frozen. *Job 37:10 NAS* also tells of cometary waters being frozen; "From the breadth of God, ice (of comets) is made, and the expanse of the (cometary) waters is frozen."

* The Hebrew word *shawkab* (#7901 in *Strong's Concordance*) translated as "stay" in this verse is a prime root meaning "to lie down," and it is more often translated to mean lay, lie down, or *cast down*.

inner solar system, where the warm Sun produces 1) the comet's outgassing jets, with gas and dust passing through holes in its rocky crust, 2) its surrounding cloud, and 3) its spread of tail debris, all of which are indicated in *Psalm 147:18*. This verse says,

> He sendeth out his word, and melteth them (the comets - the 'ice like morsels' from the preceding verse): he causeth his wind to blow (the solar wind or the comet's gas jet) and the waters flow (the waters in gaseous form that flow to make up the comet's cloud and tail or the waters that flow after impact).
>
> *Psalm 147:18*

Isaiah 34:4 also tells of comets melting when it says, "And all the host of heaven (comets) shall be dissolved . . ." (#4743 *maw-kak*, to melt, translated 'melt' in the *Septuagint*). **As we can see now, long before astronomers discovered that comets are mainly composed of frozen water covered by a rocky crust, the Bible clearly referred to this fact.**

7) Comets were the main source of the water in the Earth's oceans.

Our discussion of Noah's Flood in Chapter 4, shows that *Genesis 7:11* (when retranslated) referred to comets as the "sources of the great deep," a fact scientists didn't agree on until 1986 when a satellite greeted the return of Comet Halley. From that encounter, Comet Halley's chemical fingerprint was proven to closely match that of the Earth's oceans. *Genesis 7:11* (when retranslated) states, "in the six hundredth year of Noah's life, in the second month, the seventeenth day of the month, the same day were all the **sources** (*comets*) of the great deep (the oceans) broken up, and **those that lie in wait in heaven** (*comets*) were loosed." When *Genesis 7:11* (retranslated) then says "that those that lie in wait in heaven (comets) were loosed," it relates to *Job 38:22-23 (NIV)* which tells of the snow and hail, the frozen waters in comets that are stored and reserved for times of trouble, for days of battle and war. These are the comets of the Oort Cloud. In fact, scientists now believe that the comets of the Oort Cloud were the source of almost *all* the waters in the inner solar system. The recent 2006 Deep Impact space probe found the first direct evidence that comets contain enough pure ice to "have delivered life giving water to a primeval Earth." [17]

The Bible also indirectly indicates that comets were the sources of the water in the Earth's oceans when it says that comets are the sources of the waters in heaven, that is, the sources of the waters in the solar system. Recall, that *Job 38:29* refers to comets as "ice" and "hoary frost," where "hoary frost" is more properly translated as the "bowls of heaven," that is, "the bowls of the solar system" (see discussion in Chapter 6 regarding "bowls"). *Job 38:37* refers to comets as the "bottles of heaven" or the "water jars of the heavens" (NAS, NIV), meaning the "water jars of the solar system." *As we can see now, long before astronomers discovered that comets were the source of the water in the Earth's oceans, the Bible provides us with this fact.*

8) *Comets are a source of water for the Earth*.

We have seen that the Bible refers to comets as the "bottles of heaven" or as the "water jars of the heavens" in *Job 38:37 KJV, NAS* and *NIV*, and as the "bowls or basins of heaven" in *Job 38:29-30* and *Psalm 147:16-17*.* These translations are confirmed when we see that the Hebrew word *kefore* (#3713 in *Strong's Concordance*) translated as "hoarfrost" (KJV) is more accurately translated as "bowls" or "basins" based on context. The Bible indicates that not only are comets the source of water in Earth's oceans but they continue to supply the Earth with water. This is exactly what physicist Dr. Louis Frank's research indicates. Dr. Frank of Iowa University was a principal investigator for fourteen of NASA's space missions. *Time* magazine's report about Dr. Frank's presentation of his discovery at the American Geophysical Union's annual meeting in May of 1997, referred to this extraterrestrial water as "the gentle cosmic rain."[18] Dr. Frank says that pictures taken from orbiting satellites show a rain of "mini comets," "house sized chunks of water-snow that are vaporized above the Earth's atmosphere" every minute of the day to produce *"cometary water clouds."* [19]

Initially, when Dr. Frank's book *The Big Splash* came out, his data was ridiculed. However, years later with time lapse photos from NASA's Polar satellite, his discovery has been confirmed. Steve Maran of NASA told a CNN reporter, "It is obvious to us that there are dark spots in our satellite pictures. And these are *"incoming water bearing objects."* [20] A report in the August 1997 issue of *Sky and Telescope* called "Cosmic Rain," long time critic Thomas A. Donahue of Michigan University admitted that "The Polar satellite data (photographs) definitely demonstrates there are objects entering

---

*   Remember, comets are also referred to as "ice" in *Job 38:29* and *Psalm 147:17*.

the Earth's upper atmosphere that contain a lot of water."* Updates and scientific papers on Dr. Frank's work are posted on his website at the University of Iowa (http://smallcomets.physics.uiowa.edu/).

Dr. Frank's discovery of "cometary water clouds" is amazing! And it is even more incredible but true that the Bible spoke of it first! In the NASA scientist's statement about *"incoming water bearing objects"* we have the counterpart of *Job 38:37* asking "who can stay (#7901 *cast down*) the bottles (water jars – NAS, NIV) of heaven?" *As we can see now, long before astronomers discovered evidence that comets are still a source of water for the Earth, the Bible repeatedly referred to comets as just such a source of water.*

9) Comets have "mouths" or vents in their crust through which they outgas.

The Sixth Trumpet (*Revelation 9:17-19*) refers to the features of an active comet that is to enter our inner solar system and begin to outgas. Included in this passage is a statement that active comets have mouths or vents through which they outgas, a quintessential fact about active comets astronomers didn't know until 1986. *Revelation 9:17-19* says:

> And thus I saw the *horses* in the vision, and them that sat on them, having *breastplates of fire, and of jacinth, and brimstone*: and the *heads of the horses*

---

* Julie Wakefield, "Cosmic Rain," *Sky and Telescope*, August 1997, p. 29. Another former critic Robert R. Meir of the Naval Research Laboratory says, "the observation of both oxygen and hydroxyl (OH) emissions (in the trails in the ultraviolet photos taken over consecutive frames) constitutes the *smoking gun* for the water interpretation." Robert Naege, "Cosmic Rain of Mini-Comets," *Astronomy*, September 1997, pp. 24 and 26.

were as the **heads of lions**; and out of their **mouths issued fire and smoke and brimstone**. By these three was the third part of men killed, by the fire, and by the smoke, and by the brimstone, which issued out of their **mouths**. For their power is in their **mouths,** and in their **tails**: for their **tails** were like unto serpents, and had **heads**, and with them they do hurt.

Revelation 9:17-19

In this passage active comets are likened to an army of fast moving horsemen, just as they are in *Joel 2:3-4*. (*Joel 2:3* describes the "fire" and "flame" before and behind comets entering the atmosphere, then *Joel 2:4* says they have the appearance of horses and horsemen.) In the description of horses' heads like "the heads of lions," we have a description of the fuzzy lion mane like appearance of a comet's head after it begins to give off gas. Remember that the Greek term for comets, *kometes astares*, means "hairy stars." "Tails like serpents" that "had heads" is a good composite of an active comet's head and gaseous dusty tail. In the "mouths" issuing fire, smoke, and brimstone (burning sulfur), we have the quintessential illustration of active comets, in which they outgas through "mouths" or holes in their crust. This refers to the rocket like jets of "fire and smoke and brimstone" that come from the comets' mouths when they are warmed by the Sun. The water ice in a comet's nucleus begins to "melt," or rather "sublimate," and transform directly from a solid to a gas that in turn comes under the influence of the "solar wind" (*Psalm 147:17-18*). The comets' tails are extensions of materials jetted out of active comets mouths and can act somewhat like the rudder of a boat affecting the speed of a comet's travel through space (*Psalm 147:17-18*). The "gas jet" that comes out of a comet's mouth works much like a rocket thruster, and the comet's long tail works as an extension of this gas jet. We can see that *Revelation 9:17-19* is scientifically correct when, referring to comets, it says, "their power is in their mouths and in their tails." *(See Illustration A)*

Prior to 1986, astronomers thought that the comet's entire surface gave off gas and dust rather than just the 1% or 2% of a comet's surface. However, that changed when a spacecraft flew close to Comet Halley and sent back pictures of gas and dust jetting out of small "mouths" or vents on Comet Halley's surface. Writing about the rendezvous of Soviet, Japanese and European spacecraft with Halley's comet in 1986, astronomers M. E. Bailey, S. V. M. Clube and W. M. Napier in *The Origin of Comets* said, "From these we learned, for example, that the activity on the surface of a cometary nucleus is confined to a number of small areas (mouths) where gas and dust emerges in narrow jets before spreading into the coma and tail." [21] The late Harvard University astronomer Fred Whipple, who for fifty years led in the study of comets, was quoted in *Time* magazine as having said, "When I first realized about the jet action of comets, Boy! That was a thrill." [22]

Astronomers had previously suspected that an active comet had a solid nucleus but they believed that this nucleus consisted of **bright** ice. No one thought that this solid nucleus was **dark** or that the darkness came from the dark rocky crust enclosing a comet's ice. It was believed that asteroids as a group were darkish, rocky things, and comets as a group were bright, icy things. Dr. William Hartman, a senior scientist at the Planetary Science Institute in Tucson wrote, "A few comet researchers in the 1960's even claimed to have measurements of high comet reflectivity or albedo, said to be twenty to sixty percent or more. Comets were often visualized as being as bright as snow...."[23] In other words, even as late as a few months before the Vega probes and the Giotto probe encountered Comet Halley, the nineteenth century view was still in vogue. However, this impression turned out to be completely *wrong*. Vega and Giotto measured the reflection of Halley's nucleus at just four percent which is a very ***deep black***."[24] The reflectivity of a number of other cometary nuclei have also been measured at only a few percent confirming that comets and asteroids represent the darkest objects in the solar system. Once again, we see the Bible is correct about the physical characteristics of comets, since the Bible sometimes refers to both comets and asteroids as the "deep and hidden things ... (that) lie in darkness," the "deep (secret) things of darkness," and as "the stones of darkness" (*Daniel 2:22 NIV, Job 12:22 NIV,* and *28:3*).*

In *Comets* astronomer Donald Yeomans writes about "emission from discrete areas on the nucleus surface" and about "icy vents...issuing jets of gas and dust." These are other ways to describe what the Bible referred to first, and called "mouths" (*Revelation 9:17-19*).[25] It was only after Comet Machholz developed a "mouth" or vent in its crust exposing its ices and permitting outgassing that it went from being dark and inactive to being active and visible. Comet expert Zdenek Sekanina of the Jet Propulsion Laboratory in California "believes that until recently, Comet Machholz was a minute, dormant asteroid like object with a thick, dust laden crust much too faint (dark) to be discovered accidentally. Near the time of its 1986 perihelion (closest approach to the Sun) a ***vent (mouth)*** or

---

*   *Nahum 1:6* also seems to refer to comets and asteroids when it says "his fury is poured out like ***fire***, and the ***rocks*** are thrown down by Him."

active region opened in its crust, spewing out the gas and dust that created the cometary appearance first seen by Machholz in 1986."[26] As a result of the discovery that comets have vents or "mouths," Zdenek Sekanina also believes that Comet Encke's atypical tail, which points toward the Sun rather than away from it, can be explained. He said that this phenomenon that has long puzzled astronomers "can (now) be explained by gas and dust spewing forth from a single *vent* located at a high latitude on the comet's nucleus." [27]

Looking back to the Bible's referring to the "power" in a comet's tail, astronomers didn't learn that a comet's tail worked like the rudder of a boat, and affected the speed and direction of a comet's travel through space until the mid 1800's. It was not until these astronomers figured out Comet Encke's tail kept throwing off their carefully calculated orbits for this comet. In fact, it wasn't until 1577 and the studies of Tycho-Brahe and the calculations of Sir Isaac Newton, that European astronomers realized comets weren't just phenomena occurring in our own atmosphere but were denizens of outer space. One popular theory before the sixteenth century considered comets to be swamp gas that rose up into the atmosphere.

The detailed depiction of active comets in *Revelation 9:17* refers to the riders having "breastplates of fire and of jacinth (deep blue) and brimstone." This is a perfect description of an active comet's coma, which typically develops *after* it enters the inner solar system from deep space. This "breastplate" or "coma" represents an enormous round to elliptical shaped *cloud* of gas and dust which forms around the comet's body. Spectral analysis of this "breastplate" or "coma" or "cloud" shows that it indeed contains "brimstone" or sulfur gas, just as written of in *Revelation 9:17*.* As well, the color of a comet's cloud is indeed jacinth or deep blue. This deep blue color comes from the carbon monoxide molecules present which fluoresce. (Burning sulfur in oxygen can also color a comet's coma blue.) This cloud can have a small inner feature astronomers call a "false nucleus" or a "central condensation," which indeed has a "fiery" white star like appearance when set against the blue background of the cloud, just as it says in *Revelation 9:17*. The Bible has also given a correct description of the whirlwind like appearance an active comet can have when seen close up that is created by the gas spiraling out from the vents on its rotating nucleus. *As we can see now, long before astronomers learned about the vents or "mouths" of active comets, or that active comets outgas through these vents or "mouths," the Bible accurately gave precise information about comets' mouths along with additional accurate scientific information about active comets.*

10) How cosmic impacts can introduce very harmful chemicals to the Earth.

On September 15, 2007 a meteor was photographed as it streaked across the mid day sky over Peru. It made a fiery crash to Earth in southern Peru near Lake Titicaca and the Bolivian border. The meteorite, estimated to have been about ten feet in diameter before breaking up, left an impact crater that was 41 feet in diameter

---

* Actually, even when fireballs (brighter than -4 magnitude, which is the same as the planet Venus) produced by just 10-pound meteorites have come in to strike the earth, nearby "witnesses often describe the odor as sulfurous." Christopher Spratt, "It Came from Outer Space," *Astronomy*, February 1991, p. 68.

and 16 feet in depth. The impact caused a seismic shock equivalent to a 1.5 magnitude earthquake. A smoke column that lasted a few minutes rose from the crater and boiling water was seen in the crater.

There were also reports of a foul smelling toxic cloud that hugged the ground emerging from the crater. Regional authorities initially reported that people in 200 villages fell ill from the "mysterious gases" that came out of the small crater. "Scores of residents of the farming villages of Carancas began vomiting and complaining of headaches and dizziness after the space object struck the area."[28] Renan Ramirez of the Peruvian Nuclear Energy Institute said there was no radiation. Global news network Agence France-Presse reported that "Ramirez said the illnesses may have been triggered by sulfur, arsenic or other toxins that may have melted in the extreme heat produced by the meteorite strike."[29] Seven police officers guarding the crater were temporarily sickened. Preliminary lab tests were negative for bacteria and viruses. The director of the regional health ministry, Jorge Lopez said that despite wearing a mask while he approached the cordoned off crater, "the fumes irritated his nose and throat."[30]

The event drew worldwide interest and a number of astronomers immediately flew down to study the crater. Lab analysis of the meteorite and the acrid gas that emerged is still in progress. Initial analysis of the mineral composition of the meteorite by the Geological Institute of Peru (INGEMMET) found that it contained 15% kamacite, which *only* occurs naturally in meteorites, and 5% trolite, another mineral associated with meteorites. It seems that the temporary sickness that many people experienced was caused by vaporization of kamacites and trolites along with arsenic that naturally occurs in soils in this area. These processes also yield hydrogen sulfide gas.[31] The complex gas production was very irritating to the respiratory tract and the mucous membranes. While non-lethal at this level, this gas nevertheless caused panic and hysteria amongst the local people.

The Bible makes a connection between cosmic impacts and poisonous substances that can cause illness and death. In Chapter 6, the Third Trumpet (*Revelation 8:10-11*) tells of a comet hitting lakes and rivers, making the waters so poisonous that people who drank these waters die. In this last century, scientists have learned through spectral analysis that cometary material can contain poisonous chemical radicals such as cyanide and arsenic. Scientists have also learned that comet or asteroid impacts can produce great quantities of sulfuric acid, nitrous acid, and nitric acid. Cometary experts Clube and Napier cite several scientific studies on the subject, noting that it is now widely recognized that poisons introduced or produced by comet and asteroid impacts had important roles in the extinction of some animals on Earth.[32] Cometary expert John S. Lewis explains that beyond being toxic, the material introduced or produced by comet impact can be corrosive and can cause mutations, cancer and lung damage.[33] *As we can see now, long before astronomers learned that comets can introduce very harmful chemical substances to the Earth, substances that can even cause death, the Bible made this connection.*

11) How comets can carry bacteria capable of causing disease.

One significant difference between the Bible and science on

the subject of comets is whether comets can carry bacteria which could cause disease in men in the form of the infectious boils, painful sores and related skin diseases. These illnesses are all prophesied for the end times as a result of comet bombardment. Scientific thought has gone back and forth on the issue of comets and bacteria. It was once widely believed that the comets of 1664 and 1665 caused the London plague of 1665. Then scientists came to believe that pestilence brought by comets and/or their tails was nothing but an "old wives tale" that probably came from the Bible in the first place. Now, however, there is actually some scientific evidence for the connection between comets and bacteria based on discoveries made in the 1970's. In the 1970's astrophysicist Sir Fred Hoyle and several other world renown scientists, including Francis Crick (Nobel Prize for DNA), Leslie E. Orgel (Salk Institute for Biological Studies), and astronomer Chandra Wickramasinghe of Cardiff University in Wales began to think that some comets might at least carry bacteria like spores. This change in thinking occurred because it was discovered that comets, asteroids, and meteors can contain organic compounds, including tar like petroleum hydrocarbons, amino acids, and other materials considered to be essential for creating life on Earth.

Some scientists believe a comet can carry bacteria because the amino acids naturally found in comets might evolve into single celled organisms. It is also possible that a comet coming close to the Earth on its orbit could pick up microbes when it encountered bacteria or virus laden dust that was raised from the Earth into outer space by a cosmic impact. Amino acids are the building blocks of RNA and DNA, the proteins found in all living things on Earth. Based on the discoveries of the 1970's, Sir Fred Hoyle in a book entitled *Diseases From Space* speculated that microbe laden comet dust caused the widespread influenza epidemics that struck Europe during the Middle Ages and might even have caused more recent outbreaks of disease.* One study supported the concept of germs from space by noting the similarity between the spectra of a warm dust cloud in space with those of certain bacteria and with Hoyle's earlier comparative measurements. 34

Another study done by J. Mazio Greenburg, an astrophysicist at the University of Leiden in the Netherlands, concluded "a naked cell could survive in space for as many as ten million years if it is protected from radiation by a thin shell of ice." 35 On November 24, 2000 NASA told a *CNN* reporter that, "A recent discovery indicates that microbes can remain dormant for millions of years – enough time to travel from planet to planet (aboard comets and meteors)."36 An article in *Scientific American* even said, "Theories giving impacts a role in genesis are very trendy right now." It added that "Hoyle and a former student, the Sri Lankan astronomer, N. Chandra Wickramasinghe continue to promote this notion, even arguing that extraterrestrial microbes are the cause of influenza, AIDs, and other diseases." 37

"In 2000 an international team of scientists led by Javant Narlikar of India recovered micro organisms in the upper reaches of the atmosphere that could have originated from outer space."38 The team used a scientific balloon to collect this living bacteria from an altitude above ten miles. Commenting on the teams' work Wickramasinghe asked, "If we find microbes at great heights that are not contaminated from the ground, we have to wonder where they come from. One hundred tons of comet and meteor organic debris is

---

* A comet impact is not necessary. The Earth may just pass through a comet's tail to get a dose of bacteria-laden comet dust.

deposited in the atmosphere every day."[39] In 2000 Wickramasinghe and Sir Fred Hoyle published a paper on the WEB that reported that after looking at spectral data from the 1999 Leonid meteor shower they detected a "bacterial fingerprint" while the space rocks were travelling about fifty-one miles above the Earth.[40] It would seem that a height of 51 miles would rule out contamination from the ground.

A group of studies has shown some types of bacteria can survive in Antarctica, and thus could survive the cold of outer space. In Antarctica some microbes survive the extremely cold winters by actively freezing and going into a suspended state called "cryobiosis."[41] This relates to a new form of living bacteria. Richard Hoover, an astrobiologist at NASA's Marshall Space Flight Center in Alabama discovered these bacteria when he delved into Alaska's permafrost (a mixture of ice, soil and rock) near the town of Fox. Hoover was surprised when he looked under a microscope at a sample of these ice age micro bacteria and saw that they started swimming as soon as they thawed. They were apparently alive 30,000 years before that. Hoover said, "These bacteria that had just been thawed out of the ice . . . were swimming around. The instant the ice melted, they started swimming. They were alive, but they had been frozen for over 30,000 years."[42] Hoover also said that the bacteria (*carnobacterium pleistocenium*) were "instantly ready to eat and multiply."[43] NASA says that they believe "such creatures would be able to survive in a suspended state for millions of years." [44] NASA credits Hoover's find as "the first fully described validated species ever found alive in ancient ice.[45] NASA wonders if such microorganisms could have once flourished on Mars and survived in Mar's polar caps or in what appears to be a great frozen sea near the Martian equator.

On August 27, 2007 an international team lead by Eske Willerslev of the University of Copenhagen reported that they tested microbes from very deep in the permafrost of Canada, Siberia and Antarctica. They found active cells repairing DNA that survived by eating nutrients like nitrogen and phosphate in the permafroast. They concluded "ancient bacteria are able to survive nearly half a million years in harsh frozen conditions." [46] Comets, giant dirty ice balls traveling through the cold of space of course entail harsh frozen conditions with intense radiation and yet they provide nitrogen, phosphate, sulfur, and other nutrients.

This interest in bacteria surviving space travel was of course spurred on by NASA in August of 1996 when it was announced that they thought bacteria had been found in a meteorite from Mars. New studies of this meteorite have been conducted by Oregon State University professors. Their studies "revealed a series of microscopic tunnels that are similar in size, shape, and distribution to tracks left on Earth rocks by feeding bacteria." [47] Analysis done by the Oregon professors is not conclusive since they were not able to recover DNA, but nevertheless their findings are intriguing.

More than 2,000 years before scientists began to suspect a connection between comet activity and bacteria caused pestilence, the Bible connected pestilence with comet activity in a number of different places. In the discussion of the Fifth Trumpet (*Revelation 9:1-12*) and Fifth Vial (*Revelation 16:10-11*) of Chapter 7, a number of scriptures were given where the Bible associates pestilence with cometary activity (*Ezekiel 38:22, Habakkuk 3:5, I Chronicles 21:12-17, Isaiah 3:17* and *24, Revelation 16:2, Revelation 9:1-12,* and *Revelation 16:10-11*). In particular the Fifth Trumpet (*Revelation 9:1-12*) even seems to give a biologically sound physical description of a type of bacteria that it says will cause pestilence. This description was written over 1,600 years before the invention of the microscope and the subsequent discovery of bacteria, and before Louis Pasteur and others discovered that many diseases are caused by bacteria and other microorganisms.

While the connection between comets and bacteria has not yet been conclusively proven, there are prominent scientists seeking to prove the possibility. Some scientists are currently searching for evidence of bacteria in space, and many more are concerned about the possibility that one of own space probes might encounter some bacteriological life form and accidentally bring it back to infect planet Earth. [48] Recall Apollo 11's quarantine of astronauts visiting the Moon. *As we can see now, long before the invention of the microscope or before scientists began to theorize of a connection between comets and bacteria, the Bible made several references to such a connection and to the connection between bacteria and disease.* Obviously, credit for describing microorganisms before the invention of the microscope, constitutes another Bible first.

12) How the scientific response to comets or asteroids found on a collision course with Earth encompasses plans for shooting down these comets or asteroids.

It appears that the Bible anticipated mankind would one day attempt to shoot down comets on a collision course with the Earth. The general public was introduced to the idea of shooting down threatening cosmic bodies on June 18, 1991 when the *New York Times* ran a double featured section entitled "Asteroids, A Menace to Early Life, Could Still Destroy Earth." One of the two features were entitled "There's a 'Doomsday Rock,' But When Will it Strike?" and discussed the **inevitability** of an asteroid or comet one day bringing great devastation to the Earth. The article cautions us to take some sort of action before it is too late! A conclusion the *Times*' reporter William Broad said The American Institute of Aeronautics and Astronautics strongly agrees with. In a 1990 position paper about the threat of asteroids, this society of professional engineers said, "We would be derelict if we did nothing." [49] Mr. Broad quoted Dr. Edward Teller, the principal developer of the hydrogen bomb, when he gave a talk in Washington in 1989 where he suggested that if a large threatening extraterrestrial object could be observed while it was as much as a year away, this would allow enough time for us to send out a rocket "to meet it and give it a little sideways shove." [50]

The reporter also noted that Dr. Roderick Hyde from the Lawrence Livermore National Laboratory in California recommended the same type of action in a 1984 paper. In this paper Hyde advocates **nuclear** tipped rockets for either diverting or destroying an object discovered to be on a collision course with Earth. He wrote, "The human race should take out a cosmic collision insurance policy." [51] Buying such an insurance policy against a devastating impact is just what a committee of twenty-three leading U.S., Soviet, French, Australian and Indian scientists endorsed when they met in San Juan Capistrano, California at the First International Conference on Near Earth Asteroids three weeks after the *New York Times* article. [52]

Hollywood was quick to pick up on the issue and in 1998 released the films "*Deep Impact*" and "*Armageddon*." In "*Armageddon*" actor Bruce Willis plays an oil driller who leads a team of astronauts that land on an Earth bound asteroid in order to drill into it, bury and then detonate a nuclear bomb for destroying the asteroid. In

"*Deep Impact*" NASA scientists also try to destroy an Earth bound comet by drilling it and detonating a nuclear bomb on it.

After the release of these movies, cable TV's *Discovery Channel* and *History Channel* began running documentaries on comets and asteroids and their threat to the Earth. All of these documentaries cover strategies from the world's leading experts on how to shoot down or divert incoming cosmic bodies.* In 1998, Congress mandated NASA to find and track Near Earth Objects (NEOs) like asteroids. [53] Today NASA's Spaceguard Survey keeps constant watch on the night sky for asteroids as small as 140 meters in diameter; objects which could cause significant damage. NASA estimates that approximately 10,000 such asteroids, comets and meteors exist and so far they have found more than 4,000 of them. [54] In 2006, NASA called for ideas on diverting NEOs. These ideas were to be presented and discussed at the annual meeting of the American Association for the Advancement of Science. In particular, NASA was concerned with the newly discovered twenty-five million ton, 820 foot wide asteroid called *Apophis*, which is on a course likely passing dangerously close (within 22,000 miles) to the Earth on April 13, 2029 and April 13, 2036. [55] *Apophis* was the ancient Egyptian god of destruction.

The collision avoidance strategies discussed by NASA and other space agencies, to date involve for the most part sending out nuclear tipped rockets to either divert or destroy an object detected on a collision course with Earth. While this could be quite dangerous, scientists note that until the space age, society has had no defense at

---

* Some of these documentaries are: "*Comets: Prophets of Doom,*" "*Mega-Disasters: Comets,*" "*Mega Disasters: Asteroids,*" "*Cosmic Catastrophes,*" "*The Great Siberian Explosion,*" "*Fireballs from Space,*" "*Doomsday Asteroids,*" "*Menaces from the Sky,*" and "*Meteors: Fire in the Sky.*"

all against the threat of cosmic impactors. Incredibly, long before the nuclear age and the space age, it appears the Bible may have actually anticipated this strategy of mankind to counter the threat of cosmic impacts. *Daniel 8:9* refers to the antichrist as a "little horn" who grows in greatness, and *Daniel 8:10* says, "And it (the little horn -- the antichrist) waxed great (**reached up**), even to the **host of heaven**; and it **cast down** some of the **host** ('**starry host**' in NIV) and of the **stars** to the ground and stamped upon them." The NAS translation of *Daniel 8:10* says the antichrist "**caused** some of the **host** and some of the **stars** to fall to the earth . . . ." Note that the words "**host of heaven**," "**host**" and "**stars**" used in this verse are all three references to **comets**.*

As discussed in earlier chapters, during Biblical times the "host of heaven" were sky gods led by the "Queen of Heaven" and were the focus of idolatrous worship by most ancient Near Eastern

---

*   The term "**host of heaven**" is used 17 times in the Old Testament and twice in the New Testament. (For example, see *Deuteronomy 4:19, 17:3, II Kings 17:16, 23:5, Jeremiah 8:2, 19:13, Zephaniah 1:5* and *Acts 7:42*.) In all but three cases, which refer to the angels of heaven (*I King 22:19* [*II Chronicles 18:18*], *Nehemiah 9:6* and *Luke 2:13* - "heavenly host") the objects of the solar system, especially **comets**, are being referred to. Unfortunately, to date, translators have ignored the Bible's clear precedents for all of the key words involved in *Daniel 8:10*, and have instead taken the "**host of heaven**" to figuratively be "Israel or God's true people," and the "**stars**" to be "priests." It is interesting to note that these translators have provided no witnesses from Scripture to support these interpretations. Interestingly, *Revelation 12:3-4* tells how the tail of a great red dragon representing satan "drew the third part of the stars (comets) of heaven, and did cast them to the earth." Heaven or the solar system presently contains about 150 short period comets, and it is not hard to imagine how about 50 comets came to strike the earth during Sumerian and Babylonian times, and thereby influenced civilization and religion and brought about the worship of a pantheon of cometary gods.

heathen nations.* To the Sumerians and early Babylonians, the various gods of heaven, the "host of heaven," were mainly represented in heaven by very active comets that produced "signs and wonders in heaven" and whose attributes were later given over to the planets when these particular comets became inactive.†

So, in the "little horn," the antichrist reaching up to heaven and casting down some of the "host of heaven," some of the "stars," we have a picture of the antichrist sending a rocket up to heaven and shooting down some of the comets that are coming in to bombard the Earth. *Daniel 8:10* in effect says that the antichrist will implement NASA's strategy and send a rocket up to shoot down some of the first comets that are to come in during the Tribulation period. The Bible repeatedly tells of the comets that will come during the tribulation period, and in *Daniel 8:10*, the Bible says some of the comets will be shot down by the antichrist. **As we can see now, long before scientists decided to try and actually shoot down comets on a collision course with Earth, the Bible made references to mankind shooting down comets.**

* * * *

There are passages in the books of *Daniel*, *Ezekiel* and *Revelation* that anticipate the future affairs of mankind. According to *Daniel 2:40-45*, *7:7* and *23* and a number of other scriptures, during the end times, the antichrist will emerge as the leader of an end times

---

* The mother goddess Inanna, who became Ishtar, who became Ashtaroth, for example, see *I Samuel 7:3, Jeremiah 7:18, 44:17-19* and *25*.

† Regarding "signs and wonders" see *Daniel 6:27, Exodus 7:3, Deuteronomy 4:34, 6:22, 7:19, 26:8, 34:11, Nehemiah 9:10, Jeremiah 10:2, 32:20-21,* and *Acts 7:36*.

empire.*

Based on information we glean from Scripture, it is plausible that the antichrist might one day be a leader of the European Community. This individual may very well be in place specifically at a time in which an incoming comet would need to be shot down. It would not be surprising for the European Nuclear Commission and the European Space Agency to participate, since they may soon have the requisite experience. It was the European Space Agency, a consortium of 11 countries that sent up *Giotto*, the spacecraft that made the closest rendezvous with Comet Halley in 1985-1986. The ability to rendezvous with and then shoot down or divert a comet is not a foregone conclusion due to the erratic non-gravitational movements a comet makes as a result of the erratic eruptions of its gas jets. Among the fleet of spacecraft sent up to rendezvous with Comet Halley, the Russians sent up *Vega 1* and *Vega 2*, and the Japanese sent up *Suisei* and *Sakigake*, but it was the European Space Agency's *Giotto* that came the closest. "European scientists were able to reduce *Giotto's* targeting error to within approximately 100 kilometers." [56] The United States did not send up a spacecraft.

---

* The Bible indicates this empire will be a descendant of the ancient Roman Empire. Both the books of *Daniel* and *Revelation* speak of a future line of empires. The last empire is symbolized in the Bible as having three heads or three different reigns over the same or expanded geographical area. Historically, the first head of this third empire was known as the Roman Empire (27 BC – 476 AD), which was run by a dictator. The second head would arise in the same geographical area seeking to renew the extended power of the first Roman Empire. This second empire – the Holy Roman Empire begun by Charlamagne in 800 AD would last about 1,000 years ending in 1806. With the united economical power of the European Community (EC) begun in 1992 covering an even larger geographical area, we are left to wonder: could the European Community in fact prove to be the third and final expression of Roman Empire?

Shooting down a few incoming comets that were threatening to destroy the world would no doubt contribute to the growing renown that *Daniel 8:10-11* says the antichrist will come to enjoy.

> And it (the little horn - the antichrist) waxed great (reached up), even to the host of heaven (comets); and it cast down some of the host (comets) and of the stars (comets) to the ground, and stamped upon them. Yea, he magnified himself even to the Prince ('Commander' in NAS) of the host (***The Lord of the host, The Lord of the Comets***) and by him (the little horn - the antichrist) the daily sacrifice was taken away, and the place of His (God's) sanctuary (in the rebuilt Temple in Jerusalem) was cast down.
>
> *Daniel 8:10-11*

When the antichrist shoots down some comets targeting the Earth, possibly where other world leaders have failed, some may see the antichrist as ***"saving the world."*** The idea of someone "***saving the world***" by shooting down an incoming comet has now taken on theological significance.

A feature story in *Time* magazine called "Save the Earth" reflected on scientists calling for a system of telescopes and missiles to shoot down incoming comets. The *Times* author writes,

> Life on Earth may no longer have to wait helplessly for the next catastrophe…scientists at Lawrence Livermore National Laboratory declared that 'terrestrial life now

has a representative (mankind) capable of activity defending it from the bombardment after four eons of simply enduring it. [57]

In a book review entitled "Death from Above?" comet and asteroid astronomer Lucy McFadden writes, "With enough warning, we would then use the most effective technology of the time to divert an incoming projectile and **save our civilization** . . . Experts from all sides have made a collaborative, multinational effort to **save the world** from a chunk of rock or a ball of ice." [58] It seems possible that the antichrist could proclaim himself as the world's savior after shooting down some incoming comets. This idea is consistent with the antichrist being a false messiah, and leading the world to believe that they finally can dwell in *safety* from cosmic impacts. This all seems to connect to *I Thessalonians 5:3* which says:

> For when they shall say, **Peace** and *safety*; then sudden destruction cometh upon them, as travail upon a woman with child; and they shall not escape.
>
> *I Thessalonians 5:3*

The *"safety"* referred to in *I Thessalonians 5:3* could imply "safety" from comets that threaten the Earth. Safety from comets is an obvious need our scientists are just beginning to recognize, and one which *Daniel 8:10* says the antichrist is to fulfill, at least at first. The antichrist will fail because the comets that bring the Seven Trumpets and Vials are not shot down. This series of seven large impact events will bring the inescapable and "sudden destruction" of *I Thessalonians*

5:3 above. The inhabitants of the Earth will be destroyed exactly as this verse says.*

Daniel 8:10 can now be recognized as a prophecy for the future that was sealed until the "time of the end," a time when knowledge and resultant technology have been sufficiently increased (*Daniel 12:4* and *9*). In this verse, we again see the scientific insights of the Bible. Who in Daniel's time or any other time up until the current space age could have thought that anyone would be able to one day reach into heaven and "cast down" or shoot down incoming comets?

In the preceding pages, twelve points of current scientific information about comets have been presented which had been recorded within the pages of the Bible, long, long ago. How could the Bible have gotten so many scientific points correct? Has any astronomer offered a scientific treatise on comets that says all of this with a date earlier than 1986 or even 2008? The Bible has clearly established priority on a number of scientifically observed and proven facts about comets. These facts are not a random collection of scientific data about comets. Nor can the meaning of the many Bible verses illustrated be deemed coincidental. They are all interrelated and provide a comprehensive textbook on the subject of comets. When this scientifically accurate information about comets in the

---

* It would appear that the tribulation would coincide with the rule of the antichrist, where the destruction brought by cometary impacts would apply increasing pressure on man akin to labor pains, as called for by the increasingly severe comet bombardments of the Seven Trumpets and the Seven Vials. Note that *I Thessalonians 5:3* as quoted above compares the "sudden destruction" to come to the "travail upon a woman with child." This brings to mind *Isaiah 13:3–13* when destruction from comets coming in to strike the Earth is spoken about and verse 8 says that men "shall be in pain as a woman that travaileth."

Bible is evaluated along with the history of comet caused catastrophic events of the Old Testament, and the comet caused catastrophes prophesied for the future, the Bible's consistency on the subject of comets is clear. How could so many different authors of the books of the Bible written over a 1,500 year time span include so much detail specific scientific information about comets that was impossible for them to understand at the time they wrote?

We have seen that the greatest selling book of all time contains important and vital information for today's archeologists, historians, geologists, and astronomers. We can acknowledge the many points of scientific information about comets that were recorded first in the Bible. We can acknowledge the correlation of this information with the Bible's accounts of cosmic caused catastrophe in the past and its prophecies about comets causing catastrophes in the future. What is our response to the reality that the Bible contains undeniably complex scientific information about our world and galaxy? What can we conclude from this information that speaks to evidence for the existence of the God of the Bible? What does all this say about the possibility of the prophecies for the future coming to pass? How will you respond to the messages the God of the Bible has left for all of mankind?

APPENDIX A

Sumerian/Babylonian Cometary Gods

As explained in Chapter 3, the literature of ancient Near Eastern culture repeatedly describes the destruction caused by their cometary sky gods. These gods hurled flaming fire, stones and giant hailstones from heaven in what was called a "great storm of heaven" or an "evil storm."[1] A "great storm of heaven" or "evil storm" was far more destructive than a weather storm. A great storm of heaven involved cosmic impact. An impact produced both a thermal wave and a blast wave. Thus, when an ancient account tells of the "powerful winds," "scorching heat," "ground shaking" and "dust" that accompanied a "great storm of heaven" we know that an impact took place. For example, in *"Lamentation Over the Destruction of Ur"* we are told that "In *front of the storm* fires burned . . . to the battling storms was formed the *scorching heat* (lines 188 and 189) . . . (it) attacks the land and devours it" (a crater?) and "makes the land tremble and quake" and "the city they (the sky gods An and Enlil) make into ruins, and "dust was piled high." [2] In a related work "*Lamentation Over the Destruction of Sumer and Ur*," the Sumerians' fear of their sky gods and the destructive cometary storms sent by these gods is plainly stated. In lines 57 to 84 we read:

> Who can oppose the commands of An and Enlil? An *frightened* the very dwelling of Sumer, the people were *afraid*. Enlil blew an *evil storm* . . . Ningirsu (originally *Imdugud*, representing the phenomena of a dark cloud with *fiery sling stones*) wasted Sumer like milk poured to the dogs . . . The lands were confused in their *fear*. The god of that city turned away, its shepherd vanished. The people in *fear* breathed only with difficulty (possibly also indicating dust in the air) . . . The dark time was **roasted by hailstones and flames** . . . The heavens were darkened, they were covered by a shadow . . . ***dust*** passed over the mountains . . . the people were *afraid*. [3]

The ancient Sumerian hymn, "*Hymnal Prayer of Enheduanna: The Adoration of Inanna in Ur*," speaks of both the sky god Ishkur, the "Bull of Heaven," and the sky goddess Inanna, the "Queen of Heaven," (*Jeremiah 7:18* and *Jeremiah 44:17-19* and *25*) saying that it is they "who rain flaming fire over the land."[4] The Babylonian "*Epic of Gilgamesh*" also speaks of Ishkur saying, "With the ***snort*** of the Bull of Heaven, ***pits*** (***small impact craters***) were opened. Into them fell one hundred young men of Uruk (Erech)."[5] The Sumerian hymn called "*Ishkur and the Destruction of the Rebellious Land*" explains how Ishkur (Adad in Akkadian) snorts open these pits when Enil asks Ishkur, the "fiery Bull of Heaven," to come to the rebellious land and "take **small stones,** who is like you when approaching it (the rebellious land)! Take **large stones**, who is like you when approaching it! Rain down on it your **small stones**, your **large stones.** Destroy the rebellious land to your right, subdue it to your left." [6] The ancient literature also tells of Ishkur coming down from "heaven's zenith" with a "thick tail," whose "radiance has covered the land like a garment," and whose "bellow" caused the *land to shake*. [7] Considering these attributes in combination with the destructive fire, shaking, and the bombardment of stones which Ishkur also brings, makes it clear that Ishkur is cometary! Ishkur cannot be merely a weather god as generations of scholars unfamiliar with the nature and reality of cosmic bombardment have believed. (See lengthy Chapter 3 footnote quoting from *The Origin of Comets* by Bailey, Clube and Napier.) *(See Illustrations B & C)*

The previously quoted Sumerian work called "*Lamentation Over the Destruction of Sumer and Ur*" tells about cometary gods who devastated the land. The text says there was burning brought by "**hailstones and flames**" (recall the 'hail and fire' of the Seventh plague of the *Exodus – Exodus 9:24, Psalm 105:32,* and also *Isaiah 30:30*), as a harrow (comet fragment) coming from above struck the city; and heaven and earth trembled, dust passed over the mountains, the sky darkened, large trees were uprooted, the forest growth ripped out, blood as mouths and heads were crushed, and "**large stones one after another, fell with great thuds**." [8] It seems that particularly large cometary fragments or "stones" were referred to in the text as "weapons," where a "weapon makes all cower before it." [9] Individual fragments are also referred to as "spears" and "barbed arrows." In

Isaiah 13:5 the Bible refers to comets as coming "from the end of heaven, even the Lord, and the **weapons** of His indignation to destroy the whole land." In a number of places the Bible also refers to a comet or meteorite as a "glittering spear," or "sword," or "arrow."*

An example of how cometary phenomena were attributed to heavenly gods in human form can be found in the sky god Ninurta, also known as Imdugud. To the Sumerians, the winged god Ninurta or Imdugud originally had a none human form because he dwelled in a dark cloud. This led to him being represented as a black eagle with outstretched wings floating in the sky like a cloud. Later he came to be represented by a "flashing thunderbird roaring on the horizon" and "since the roar of thunder could rightly issue only from a lion's mouth, the bird was early given a lion's head." Later "seals show the bird god growing a human lower body." [10] The lion headed bird, or winged bird lion was also called Anzu, and it was regularly depicted in Sumerian, Akkadian, Babylonian, and Assyrian art. Then he was represented as half bird and half man and finally as fully human, though still winged.

Professor Thorkild Jacobsen in *The Treasures of Darkness - A History of Mesopotamian Religion* explains that in the Sumerian language, "Ninurta's older partially human form **Imdugud**, means "**sling stone**," and **Im-dugud** is associated with the phenomena of a dark cloud of "*flashing sling stones.*" [11] A dark cloud of "flashing sling stones" relates to a cloud of cometary debris. Like a truck occasionally spewing exhaust as it travels, a comet can shed a series of dense debris clouds as it travels along its orbit, and begins to fragment and disintegrate after entering the inner solar system. This

---

*   *Habakkuk 3:11-12, Deuteronomy 32:41, Psalm 18:14-15 (II Samuel 22:15-16), 144:4-6, Isaiah 34:5.*

cloud of debris can consist of massive chunks of ice, boulder, rocks and dust. In one work Ninurta is referred to as a "cloud of death," a very appropriate term for a dense and dark cloud of comet debris that can rain down massive hailstones and meteorites and bring deadly nuclear bomb like destruction to an area. [12] In a section pertaining to the Flood, The Babylonian "*Epic of Gilgamesh*" tells of Ninurta's presence as a "black cloud [that] rose up from the horizon" before comet activity began. Then it says "torches" set the land ablaze, and the "land was shattered like a pot." [13] Recall from Chapter 2 that the 1908 Tunguska airburst involved a comet fragment exploding and a black cloud then appearing (also see Chapter 11, item #2).

In the "*Lamentation Over the Destruction of Sumer and Ur*," the storm that caused the fall of the area is likened to being hit by a harrow coming from the sky. A "harrow" is a farming instrument used to break up the soil that is characterized by multiple teeth or disks. Line 80-B says "The storm was "a harrow coming from above, (where) the city was struck as by a hoe (or pickax)." [14] In "*Lamentation Over the Destruction of Ur*" a very similar statement is made. Line 272 in this text says, "My fields (are) verily like fields **torn up by the pickax** . . . ." [15] In the "*Hymn to Ninurta a God of Wrath*" lines 9-10 say, "My king, toothed pickaxe (harrow) that **uproots** the evil land, **Arrow that breaks up** the rebellious land, Lord Ninurta, toothed pickaxe (harrow) that uproots the evil land." [16]

As Ninurta is cometary, what is the reason for the comparisons to farm cultivating tools? Obviously a comet fragment coming in would break up the soil, but something more specific is also being said. We are told that these impacts left elongated craters and row like elongated craters that were parallel one to another like the crop rows created by the cultivating tools used by the Sumerians as they farmed the fertile plain of the Euphrates River.

We can now recognize in these ancient accounts a description of the land being broken up in apparent long rows, an accurate description of the effects of a low angle grazing impact from a comet or asteroid. The imagery of cultivation used to describe the destructive power of the cometary god explains the seemingly contradictory second aspect of Ninurta as a god of farming.[17] While most impact craters are typically round and deep, impact craters that are elongated or elliptical and shallow can indeed occur. Laboratory experiments with high velocity projectiles show that almost any angle of impact will produce a round impact crater. However, these experiments also show that a very low angle grazing impact within 15 degrees of horizontal will produce an elliptical or elongated impact crater rather than a round crater. Like a rock thrown to skip over the water, projectiles making such low angle approaches can make several touch downs before they totally break up. A comet could produce two or more shallow craters along its flight path, and even parallel craters like the crop rows if it split into two or more pieces. Elongated or elliptical craters that characterize grazing impacts have been observed on the Moon, Mars and Venus.[18]

In an article entitled "A Day in Hell" astronomer Peter Schultz of Brown University and J. Kelly Beatty of *Sky & Telescope* magazine reconstruct what would happen if a 150 meter in diameter cosmic object coming in at a low angle struck a broad plain in the Pampas of Argentina. They tell how the incoming object would have looked like a growing fireball. And they tell how the horizon and the sky above it would have grown steadily brighter, and appear to be burning, as "the asteroid (or comet fragment) approached at seventy times the speed of sound . . . Dazzling enough by day, the incoming sun like ball would have been unimaginable terrifying at night. The last seconds ... must have created the image of a looming fire spurting god."[19] Describing their reconstruction, Schultz and Beatty write, "the low angle of impact created a nearly instantaneous **mountain of fire** that engulfed a tract of grasslands ten kilometers wide and fifty kilometers downrange within seconds ... Hot fragments from the just born craters were drawn into hurricane force winds ... Vaporized carbonates and other debris drawn from the Pampas into the fireball produced **lethal clouds of carbon monoxide**, while ionization of the atmosphere engulfed the region with **noxious** NO and NO2 gases."[20] Thus, people in the region of a low angle impact could be killed as a result of the gases produced.

The scientific reconstruction of this type of impact event sounds similar to descriptions of disastrous events in the ancient Near Eastern literature.[21] For example, compare this to some lines from the Sumerian work called "*Lamentation Over the Destruction of Ur*," which tells about a "great storm of heaven," an "all destroying storm" that devastated the city around 1950 BC.[22] In the context of this apparent cometary storm, it seems some people were gassed so that, "Although they were not drinkers of strong drink, they drooped neck over shoulder ... standing near the weapons ... Mothers and fathers who did not leave their houses were overcome by fire; the young lying on their mother's laps" (lines 224-228).[23] (Also see the section on "The Blast That Killed 185,000 Assyrians" in Chapter 5).

In the ancient literature when the cometary god Ninurta is described in human form as driving a chariot of seven hitched winds across heaven, it brings to mind the Greek myth of Phaethon who tried to drive the chariot of the sun across heaven, which brings to mind a comet that has a multiple component tail. (Photographs have shown that in 1989 Comet Borsen-Metcalf had as many as six distinct parts to its very long tail.) While crossing the sky, Ninurta is referred to as "the light of the gods," whose "brilliance covered the mountains." Ninurta subdues by throwing "thunderbolts," and his "lightning flashes were arrows." [24] In *The Treasures of Darkness - A History of Mesopotamian Religion* by Dr. Thorkild Jacobsen of the Oriental Institute of the University of Chicago, Ninurta is described "as a warrior driving his thundering chariot (drawn by seven winds) across the skies, throwing his large and small **sling stones**." (*Zechariah 9:15* says that "The Lord of Hosts shall . . . subdue with **sling stones . . . .**")[25] Jacobsen notes that in one Sumerian myth Ninurta is subdued by "being cast into the pit," which seems to describe a comet fragment crossing the sky, and being subdued after it impacts the Earth and leaves an impact crater or pit. [26] Note that both Ninurta and Ishkur are credited with raining stones from heaven, and bringing "radiance" or "brilliance" covering the land and mountains. Their accounts make it clear that they are cometary. In ancient Near Eastern art, Ishkur was often depicted as a winged bull and Ninurta as a winged lion. Note that the *Book of Daniel* (*Daniel 7:4*) tells of a lion with eagle's wings that appears in one of Daniel's vision as a symbol of the Babylonian Empire.

APPENDIX B

Russia Is Not Magog

Therefore, thou son of man, prophesy against Gog, and say, Thus saith the Lord God; Behold, I am against thee, O Gog, the chief prince of Meshech and Tubal: and I will turn thee back, and leave but the sixth part of thee, and will cause thee to come up from the north parts, and will bring thee upon the mountains of Israel: And I will smite thy bow out of thy left hand, and will cause thine arrows to fall out of thy right hand.

Ezekiel 39:1-3

Ezekiel tells us that Gog, [referring to Magog], the nation that will lead all of the other powers of darkness against Israel, will come out of the north. Biblical scholars have been saying for generations that Gog must be Russia. What other powerful nation is to the north of Israel? None. But it didn't seem to make sense before the Russian revolution, when Russia was a Christian country. Now it does, now that Russia has become communistic and atheistic, now that Russia has set itself against God. Now it fits the description of Gog perfectly."[1]

> Future US President
> Ronald Reagan, 1971

The Native Americans of North America are called Indians, but what do these people have to do with the people of the country India? Absolutely nothing. Hundreds of years ago Christopher Columbus was looking for a water passageway to the country of India as a means of obtaining gold and valuable spices. When he arrived in the islands of the Caribbean, Columbus, seeing dark skinned people, called them Indians and declared his passage a success. However, the people we call Indians were not related to the people of India, and the Caribbean Islands are a very long way from India.

There has been a similar misconception in many Bible prophecy books over the last 100 years. As discussed in Chapter 9, the belief that Russia is the country from the North who is to attack Israel during the end times is a mistake. Although evidence that contradicts this conclusion has become common knowledge within the academic arena, this misconception is still prevalent in end times Bible prophecy books. This misconception short circuits the proper understanding of end times Bible prophecy. Just as we all know that

RUSSIA IS NOT MAGOG

the "Indians" are not citizens of the country India, we will see that Russia is not the country spoken of in *Ezekiel 38-39*.

Championed by the authors of many Bible prophecy books, the erroneous interpretation of *Ezekiel 38-39* has remained the current and popular belief among many evangelical Christians who associate the Russians with Gog and Magog. This interpretation says that sometime before the great Battle of Armageddon, there will be a Russian invasion and war with Israel. This end times Russian invasion theory is based on faulty eighteenth and nineteenth century references used to determine the identities of the ancient nations of Asia Minor (modern Turkey) which are discussed in *Ezekiel 38:2-6*. It has been mistakenly believed that the ancient nation of "Magog" in western Asia Minor was the ancient "tribal name" of the "Scythians," a group of Iranian speaking nomadic tribes from Central Asia north of Iran that traveled across the Russian steppes, and came to live in the territory north of the Black Sea. Then, in turn, and again in error, it has been believed that the ancient Scythians were the progenitors of the modern Russians. *(See Illustration E)*

Within academia today, there are no professional archeologists or historians who associate Magog with the Scythians or the Scythians with the Russians. The May 2000 issue of *National Geographic*, reflecting modern scholarship, explains that a group of **Scandinavian Viking traders called the Rus** began the Russian state around 800 AD. It says, "The Slavs and Finns there (near today's St. Petersburg) called them **Rus**, after the Finnic term for Swedes, **Ruotsi** . . . These **Rus** eventually **founded the first Russian state** centered in Kiev in today's Ukraine, and **gave their name to Russia,** a cultural inconvenience the Soviet historians were compelled to dispute for decades."[2] The name "*Rus*" or "*Rhus*" appears in the writings of Bishop Troyes in 839 AD, and according to the 12th century document known as the *Primary Chronicle* the land around Kiev was named "**Rus**" and the inhabitants called "**Russes**" in 852 AD.³ *National Geographic* goes on to tell how *Novgorod* (east of St. Petersburg), was an early Russian capital founded by the **Rus**, and how in the city of Kiev, the **Rus** Prince Vladimir converted to Christianity in 988.⁴ Also, as explained in Chapter 9, the renowned Arab Chronicler Ibn Fadlan tells how in 921 AD he met with a group of blonde haired people of Swedish origin called the **Rus**. He told how he met them on the upper Volga River in western Russia as they came up the river in their ships to trade with the Bulgars.⁵ Even the author's daughter's high school history textbook details how Russian origins are Scandinavian.⁶ Simply put, the Scythians were gone for at least 500 years before the Russians came to be a people. There is neither a genetic nor a cultural connection between the Scythians and the Russians. *(See Illustration G)*

Where did the idea begin that Magog begat the Scythians? It began with one ***very*** questionable sentence written by the First Century AD Jewish historian, Josephus in his book *The Antiquity of the Jews*. The sentence says the Greeks called the "Magogites" the "Scythians" despite the fact that Greeks **did not** call the Magogites the Scythians.* Not only is this statement in conflict with the far more detailed writings of other ancient historians (Greek and none Greek) and with ancient Assyrian texts that provide firsthand information about both the people of Magog and Scythians, but this sentence by Josephus, which identifies Magog with the Scythians, is found in a

---

*   "Magog founded those that from him were named Magogites, but who are by the Greeks are called Scythians," from Book 1, Chapter 6, line 123 of *The Antiquity of the Jews* in *The Works of Josephus* translated by William Whitson, Hendrickson Publishers, Peabody, Mass., 1987, p. 36.

passage where Josephus makes other errors in the identification of other ancient nations of Asia Minor.*

Although Josephus uses the ancient Greek historian Herodotus' book *The History* as a reference on a number of occasions,[7] in this passage about Magog and the Scythians, Josephus is not only in conflict with Herodotus, but he even conflicts with other statements he made about who he says the Greeks identified as the Scythians. In one line Josephus says the Greeks identified the Scythians with Magog, but then a few lines later Josephus using the Hebrew word for the Scythians, "*Ashchenaz*," says the Greeks identified "*Ashchenaz*" with Rhegium, which was a port city of Southern Italy (*Acts 28:13*) known for its rock of Scylla. The names "Scythians" and "Ashchenaz" are just English and Hebrew transliterations (i.e. they spell a word in the alphabet of another language) of the Greek name, *Skythes* for the same nomadic people. To those familiar with Herodotus' book *The History*, it is obvious that the Greeks didn't make either of the identifications Josephus attributes to them.

One can only speculate how and why these obviously wrong statements are in Josephus' book. Were these inaccurate statements the products of Josephus' imagination, or were they the result of alterations to the few copies of Josephus' work that existed in the first

---

*   Josephus' books provide first-hand information about the Jews of the First Century AD but when it comes to earlier periods, such as the Seventh and Eighth Century BC, Josephus usually himself relies on earlier ancient historians. Scholars today, on a case by case basis urge caution in regard to Josephus' historical reliability. Those who **blindly** accept Josephus' apparent association of Magog with the Scythians tend to elevate Josephus' words to the level of Scripture. They do not seem to be aware that Josephus' history of the Jews, *The Antiquity of the Jews* is in direct conflict with Bible scriptures in a number of instances. For example, see J. Miller and J. Hayes, *A History of Ancient Israel and Judah*, Westminster Press, Philadelphia, 1986, pp. 316, 334, 408, and 470.

century? Maybe Josephus was using the word "Magog" as a generic term for any fierce people coming down from the north. Whatever the reason, the statements don't reflect that the ancient Asia Minor nation, referred to as "Ashkenaz" or "Ashchenaz" in the Hebrew of the Old Testament (*Genesis 10:3* and *Jeremiah 51:27*), is in fact the same nation or people the Greeks called the "Scythians" (*Colossians 3:11*)?[*]

As for identifying Magog with the Scythians, who the Hebrew knew as the Ashkenaz, it is important to note that the Table of Nations found in *Genesis 10:2-3* lists Magog and the Ashkenaz as quite separate entities. Interestingly, the Assyrian word for the Scythians, *Ishkuza* or *Ashguza*, is similar to the Hebrew word for the Scythians – Ashkenaz. [8] Those who have relied on the erroneous statements found in Josephus about Magog and the Scythians must recognize the simple fact that the Scythians and the Ashkenaz are not separate nations. In truth, they are one and the same.

Further, the erroneous statements by Josephus about the Greeks identifying the Magogites with the Scythians, and the Greeks identifying Ashchenaz with Rhegiun are preceded by yet another erroneous and anachronous statement about the relationship between two other ancient nations of Asia Minor. In this statement

---

[*] Confusion has also arisen among Bible prophecy interpreters because they do not realize that the term "*Ashkenaz*," in the Old Testament denotes the "*Scythians*." Oddly enough, in the ninth-century the term "*Ashkenaz*" also came to be used to refer to medieval German Jewry. Later in time "*Ashkenaz*" came to denote all Jews whose culture originated and developed in Germany, France, and East Europe, as distinct from "*Sephardim*," whose culture originated and developed in Spain and the Mediterranean. For example, see Philip Sigal, *Judaism – The Evolution of a Faith*, Eerdman, Grand Rapids, Michigan, 1988, p. 298. And *Ashkenaz - The German Jewish Heritage*, Gertrude Hirschler editor, KTAV Publishing, Hoboken, 1989.

Josephus writes that the Greeks associated the Gomerites (*Ezekiel 38:5*) with the Galatians, when in fact the Gomerites (Cimmerians) and the Galatians had absolutely nothing to do with each other. The Greeks of Josephus' time simply did not call Gomer the Galatians. The Gomerites (Cimmerians) were invaders from an area around the northeast shore of the Black Sea. They displaced the ancient nation of Meshech in central Asia Minor and occupied this area from around 800 BC – 630 BC before pulling back. The Galatians (Gauls and not the Gauls of Roman France) were migrants from Central Europe, who came to Central Asia Minor as a result of a population explosion during the third century BC. Eventually they won over the local tribes and gained status as the Roman province of Galatia in 25 BC. While the Gomerites and the Galatians both occupied Central Asia Minor, they did so at different times and bear no relationship with each other. Despite what Josephus wrote, Gomer did not found the Galatians, and Magog did not found the Scythians.

One cannot help wondering how and why Josephus made three such totally erroneous statements in a single passage, if in fact he did. Ecclesiastical acceptance of the erroneous connection between Magog and the Scythians traces back to Saint Jerome of the late fourth century AD. It is interesting to note that several other critical statements found in Josephus' book *The Antiquity of the Jews* which was written by Josephus in Greek, and later copied in Greek by Christian copyists, are not found in an Arabic copy of his book. This indicates that the few Greek copies of his book, which were available early on and were in turn copied, were altered in some places to suit prevailing Christian beliefs. [9]

Finally, this deception about the Russian invasion of Israel was first popularized in the 19th century. A book written by a pastor at the time seems to contain a purposeful misquote of a statement about a city in Magog made by the highly regarded first century AD Roman historian Pliny. In his multi-volume work called *The Natural History*, Pliny identifies the ancient Lydian city of Hierapolis as being part of Magog. Built atop the cliffs, the city of Hierapolis controlled the water supply of the Lydian city of Laodicea six miles to the south. The city of Hierapolis was built by the Lydians and always under Lydian control until the Lydian Kingdom became part of the Persian Empire in 546 BC and then part of the Greek and Roman Empires that followed. Like Pliny, today's archeologists also identify Magog with the ancient nation of Lydia, which was in western Asia Minor. Indeed, written records from the Assyrian royal court show that the Assyrians used the name Magog as an *eponym* for the nation of Lydia. In his book Pliny writes, "Bambyx the other name of which is Hierapolis, but by the Syrians called Magog."* In other words, Pliny identifies the famed Lydian city of Hierapolis as being part of Magog. In this we see that Magog is another name for the ancient nation of Lydia. However, in the book *The Destiny of Nations* by nineteenth century Pastor John Cummings, one of those credited with being the author and popularizer of the Bible interpretation that calls for Russia invading Israel, we read, "Pliny says 'Hierapolis taken by the Scythians was afterwards called Magog.'" [10] We can see that Pliny's statement has been misquoted or misunderstood in a way that makes it seem that Magog was related to the Scythians

---

* "Bambyx (referring to the western most part of their empire at the time) the other name of which is **Hierapolis** (Holy City) but by the Syrians called **Magog**." *The Natural History of Pliny, Vol. I*, translated by John Bostock, and H.T. Riley, Henry Bohn Publishers, London, 1855 p. 439. This translation of Pliny is consistent with that found in *Pliny Natural History, Vol. II* translated by H. Rackham, Harvard University Press, Cambridge, 1942, p. 283.

rather than the Lydians. Unfortunately, this same misquote of Pliny has been carried forward in time and appears in Hal Lindsey's book *The Late Great Planet Earth.* [11] *(See Illustration E)*

It is this error which incorrectly relates Magog to the Scythians, who were erroneously believed to be progenitors of Russia that led President Ronald Reagan and others to refer to the Soviet Union as an "evil empire" and "the focus of evil in the modern world."[12] At a dinner with California legislators in 1971 (before he became president), prophecy student Ronald Reagan concisely summed up his view of Russia's end time role:

> "*Ezekiel* tells us that Gog, [referring to Magog], the nation that will lead all of the other powers of darkness against Israel, will come out of the north. Biblical scholars have been saying for generations that Gog must be Russia. What other powerful nation is to the north of Israel? None. But it didn't seem to make sense before the Russian revolution, when Russia was a Christian country. Now it does, now that Russia has become communistic and atheistic, now that Russia has set itself against God. Now it fits the description of Gog perfectly." [13]

The identification of Magog with Lydia and not Scythia is consistent with the ancient texts of the Assyrians, who at various times were either the allies or the foes of the Scythians yet had peaceful dealings with the Lydians. The identification of Magog as Lydian, and not Scythian is also consistent with the writings of the ancient historian Herodotus (ca 490-424 BC), a Greek who was born in, lived in, and traveled throughout all of ancient Asia Minor. Herodotus' famed book, *The History,* which gives a detailed history of all the ancient nations of Asia Minor, was written about 100 years after Ezekiel's writings. [14] Herodotus, who is considered the Father of History, wrote detailed information about the three different Scythian tribes and the ten different tribes that neighbored them, including their ever changing and opportunistic alliances. Herodotus also wrote in detail about the Lydian Royal Dynasty. It is clear that Magog (Lydia) and the Scythians were not related. He says that Magog (Lydia), as led by the historical figure Gog (Gyges to the Greek, Gugu to the Assyrians) and the Scythians, were in fact enemies! In Gyges of Lydia we have the leader the Assyrians called "Gugu, King of Ludu," and "Gugu of Magugu," the Bible's Gog of Magog. [15] *(See Illustration E)*

In *Foes From the Northern Frontier* Dr. Edwin Yamauchi, a professor of history at Miami University in Ohio writes that Herodotus' account of the Scythians is our chief literary source about the Scythians. He says that some Bible scholars "seem to be unaware of the numerous archeological confirmations of Herodotus' reports in general, and of his Scythian account in particular." [16] In his book, which contains a great deal of information on the Scythians, Yamauchi makes it clear that modern Russia's origins are not Scythian, and like other historians he explains that modern Russia's origins are Scandinavian (the **Rus**).[17] Also, Yamauchi emphatically explains that **none** of the ancient nations referred to in *Ezekiel* 38-39 can possibly be related to modern Russia. Further, Yamauchi says:

> "Even if one were to transliterate the Hebrew *rosh* as a proper name (as does the NAS) rather than translate it

as 'chief' (as does the KJV, NIV, and Hebrew *Tanakh*), it can have nothing to do with modern 'Russia.' This would be a **gross anachronism** for the modern name is based upon the name **Rus**, which was brought into the region of Kiev, north of the Black Sea, by the Vikings only in the Middle Ages."[18]

This Bible interpretation calling for a Russian invasion of Israel was framed at a time when little was known about Magog, the Scythians, or Russian origins. Many who have held to this interpretation have also tried to understand who the other countries listed in *Ezekiel 38-39* were (Meshech, Tubal, Gomer, and Togarmah). Some have associated Magog and/or *Rosh* with Russia, some also erroneously associate Meshech with the Russian city of Moscow, and Tubal with the Siberian city of Tobolsk. Promoted by the 1909 and 1917 editions of the *Scofield Reference Bible*, these erroneous associations are representative of a type of arm chair "name game" archeology that was quite popular during the eighteenth and nineteenth century. Based on the inappropriate transliteration of the Hebrew word *rosh*, which most translations of the Old Testament (KJV, NIV, and Hebrew *Tanakh*) properly translate as the word "chief", the NAS translation mistakenly refers to a country called **Rosh**. Historian Paul Boyer notes that "Scores, and probably hundreds of post war prophecy writers made the Rosh = Russia connection, usually citing Gesenius, Bishop Lowth, Scofield, or all three."[19] For those who insist on the erroneous transliteration of the Hebrew word for "chief" that yields a nation called **Rosh,** ancient Asia Minor had several different localities that could answer to this name and fulfill the obvious Asia Minor context of *Ezekiel 38:2, 3,* and *6* without invoking Russia. It should also be noted again that unlike Magog, Meshech, Tubal, Togarmah and Gomer of *Ezekiel 38-39*, which are **all** mentioned in the Table of Nations in *Genesis 10:2-3*, a nation called **Rosh** is notably absent.

For decades archeologists and historians have known that neither Magog, nor the Scythians, nor any of the ancient nations referred to in *Ezekiel 38-39*, could possibly have had anything to do with the origins of the modern Russians. For example, Daniel I. Block, a professor of Old Testament at Wheaton College in Wheaton, Illinois in *The Book of Ezekiel* says,

> "The popular identification of Tubal with Tubolsk in Russia (H. Lindsey, *The Late Great Planet Earth* 1970, p.53) is **ludicrous** . . . The popular identification of Meshech with Moscow (of Lindsey, *The Late Great Planet Earth* p.53) is **absurd**."[20]

The first Russians were people of a different racial stock, linguistic stock, cultural stock, and time than the Scythians. Based on excavations and historical records, archeologists and historians are absolutely sure that the progenitors of the modern Russians were the Scandinavians who border the Russians on the northwest. We know that Russia did not develop population centers until around 800 AD. By that time, the Scythians had died out, and the ancient nations of Asia Minor referred to in *Ezekiel 38-39*, (Magog, Meshech, Tubal, Gomer, and Togarmah) all had ceased to exist. It is not possible that any of these nations could be related to the origins of modern Russia. We now have primary evidence that makes this incontestable. Assyrian cuneiform court records cover the time period of Ezekiel and make reference to the nations mentioned in *Ezekiel 38-39*.

It is perplexing that even as some Bible dictionaries reflect changes in their identification of the ancient nations of *Ezekiel 38-39* based on these Assyrian records, many Bible prophecy books still persist in the incorrect theory that Russia is Magog. In many Bible prophecy books the erroneous theory that Magog and the Scythians are the founders of modern Russia is simply taken for granted, and no references for these connections are given at all. Even more disconcerting are statements made with no connection to reality. Thomas McCall and Zola Levitt in *The Coming Russian Invasion of Israel* write, "secular history books trace the fierce Scythian people . . . as forerunners of modern Russia." Hal Lindsey in *The Late Great Planet Earth* writes, "Any good history book of ancient times traces the Scythians to be a principle part of the people who make up modern Russia." [21] Few within the Christian community seem to be aware of the black eye this kind of "research and analysis" has produced in the secular academic world. Historian Paul Boyer of the University of Wisconsin writes, "So hackneyed had this scenario (Magog and Scythia as Russian) become by the 1980's that its proponents hardly bothered with the geographic and linguistic evidence marshaled by earlier writers." [22]

In *The Final Curtain* Pastor and author Chuck Smith handles the lack of accepted evidence for Magog and the Scythians connecting to modern Russia in another way. He moves Magog from Asia Minor to Southern Russia. Without giving any references, Chuck Smith writes, "Magog throughout history, has been known as the vast area north of the Caucasus Mountains. Today, it is known as Russia."[23] This is simply not true; historians and archeologists have never placed Magog north of the Caucasus Mountains. Again, there are no modern history books, research papers or studies which connect either Magog or the Scythians with the modern day Russians.

Ironically, the correct answer to the origins of modern Russia was available as early as the 1800's. As mentioned, John Cummings, a Doctor of Divinity, was one of the 19th century proponents of the Russian invasion of Israel interpretation of *Ezekiel 38-39*. Cummings was the author of *The Destiny of Nations – As Indicated in Bible Prophecy* published in 1864, a book Hal Lindsey in *The Late Great Planet Earth* quoted from a number of times.[24] While Cummings quotes and agrees with Bishop Lowth (1710) who said that the modern Russians were derived from the Scythians, and Reverend Hollis Reade who said the modern Russians have their origins in Magog and the Scythians, he also presents another view. Cummings wrote, "Gibbon (Edward Gibbon, *The Decline and Fall of the Roman Empire*, 1776) however, thinks the name ('Russia') is of Scandinavian origin, and describes the **Ruses** derived from **Ruts**, the Finnish name of Sweden."[25] Gibbon gives the very same origin of the name and people of modern Russia, the Viking tribe from Sweden, the Scandinavian **Ruses** (**Rus**), that archeologists and historians know to be true today. So, unbeknownst to Cummings, his book actually gives the truth about modern Russian origins. *(See Illustration G)*

As stated earlier, over the last couple of decades various Bible Dictionaries no longer associate Gog and Magog with Russia. Dr. Merril F. Unger, who wrote the preface to the first edition of *The Coming Russian Invasion of Israel* published in 1974 and lent his name to *The Unger Bible Dictionary* published in 1960, had believed the discussion of Gog and Magog identified a Russian led invasion of Israel during the end times. However, *The New Unger Bible Dictionary* published in 1988 (edited by R. K. Harrison) no longer associates Gog and Magog with Russia. Instead it ***correctly*** associates Gog and

Magog with the ancient western Asia Minor nation of **Lydia**. [26] *The New Unger Bible Dictionary* says, "Gog is described by the prophet (Ezekiel) as belonging to the land of Magog, the situation of which is defined by its proximity to the isles of the Aegean. It is clear that **Lydia** is meant and that by Magog we must understand the land of Gog." [27] The capital of Lydia was Sardis.

The New Bible Dictionary, Second Edition published in 1982 says, "The only reasonable identification of Gog is with Gyges, the King of **Lydia** (c. 660 B. C.) – Assyrian Gugu; Magog could be Assyrian *ma(t)gugu*, "land of Gog." [28] *Harper's Bible Dictionary* published in 1985 also identifies Gog with the historical Gyges of Lydia, and identifies "Magog" as a phrase in the Assyrian language which means 'land of Gog.'" [29] A closer look at Gog and the nation of Lydia, as referred to in other places in the Bible, makes it quite clear that *Ezekiel 38-39* referred to the historical king called Gog and the land of Gog [Magog], that is, the ancient nation also known as Lydia.) *Encyclopedia Judaica* also identifies Gog with Gyges, the historical King of Lydia, and Magog, his country, with Lydia when it says: "Since in the list of the sons of Noah (*Genesis 10:2*), Magog is mentioned, the most reasonable identification put forward is with Gyges, also known as Gugu, King of Lydia; and Magog, was his country." [30]

Archeologists know that within the Assyrian cuneiform records which have been translated, there are references to all of the ancient nations listed in *Ezekiel 38-39*. Bible scholars now use these same Assyrian texts and related Babylonian texts to provide independent verification and edification of the Bible's historical accounts from about 805 BC to 530 BC.[31] For example, the Bible (*Ezra 4:15, 19* and *5:17-6:7*) tells how the Jews of the fifth century

BC (538 BC – 457 BC) overcame opposition by the local Persian governor to the rebuilding of the Temple in Jerusalem by referring to these **same** Assyrian cuneiform court records. These Assyrian records show that Magog, Meshech, Tubal, and Togarmah (*Ezekiel 38:5*) were all nations of ancient Asia Minor that dealt with the Assyrian court. They stretched across Asia Minor. Gomer (*Ezekiel 38:5* – Gimmiraia = Cimmerians) actually invaded Asia Minor by coming down from an area around the northeast shore of the Black Sea.

These nations all co-existed in Asia Minor (Modern Turkey) at a time when Magog (also known as Ludu or Lydia in both the Bible and the Assyrian texts) was led by a militant leader called Gog (685-652 BC), about 100 years before Ezekiel wrote. Gog is the Hebrew spelling of the name of a militant leader from western Asia Minor known as Gyges of Lydia to the Greeks. This same leader and country was known as Gugu of Ludu to the Assyrians. [32] The Assyrians made great use of *eponyms* (i.e. words or names derived from the name of a person) and even published eponym lists. In the Assyrian language "the land of Gugu" is rendered as *Ma-gugu,* just a "the land of Zamua" is rendered as *Ma-zamua*.[*] The Hebrew spelling of *Magugu* is "Magog," and thus "Magog" simply means "the land of Gog."

The ruler referred to in this "land of" eponym format was usually the very first ruler from this land to become known to the Assyrian court. Assyrian court records indeed show that Gyges of Lydia (Gugu of Ludu) was the first ruler from Lydia they ever met, since all dealings with western Asia Minor up to then were mediated

---

[*]   In Akkadian the word *mat* means "land." From *Mat-guggu* we get *ma-guggu* which means "the land of guggu."

by Meshech in central Asia Minor.³³ However, after Meshech fell to Gomer, during the reign of Gyges, direct contact between Lydia and Assyria began, and Gyges became the first Lydian ruler known to the Assyrian Court. Hence, Lydia would have also been known as the "land of Gyges" and from this we get Ma-gog.*

When *Ezekiel 38:2* refers to Gog from the land of Magog, as the chief prince of Meshech and Tubal, the Scripture is referring to a specific geographic area – Asia Minor – and to a specific time period and ruler – when Magog, Meshech, and Tubal were all co-existent. The Assyrian records do speak of such a time period and person, and we now know that this Gog, the King of Lydia (Gugu of Ludu = Gyges of Lydia), led the defensive efforts of Magog (Lydia), Meshech, and Tubal against invading Gomer (the Cimmerians). This was a very narrow time period because Meshech fell to Gomer (the Cimmerians) and was taken over during the reign of Gog.†

From these same cuneiform records, a corroborating inscription found in Magog (Lydia), and from accounts by ancient historians, we also know that this King Gog sent troops to Egypt to serve as mercenaries sometime after 664 BC. He helped Psammetichus I (the son of Necho I of *II Chronicles 35:20*) suppress native rivals, and secure his kingship over all of Egypt. ³⁴ The fact that Gog sent troops to help Psammetichus and serve as his personal bodyguard explains the seeming incongruity of a nation of Asia Minor being counted among the north African nations of Ethiopia and Libya as allies of the Egyptians. It also explains why Magog (Lydia) is named in prophecies about the fall of Egypt during the end times. <u>Magog </u>or Lydia is counted among those "in league" with and

---

* By 540 BC Lydia was incorporated into the Persian empire.

† Magog, Meshech, Tubal, and Gomer are listed together in *Genesis 10:2*.

who "uphold Egypt" in *Ezekiel 30:4-6* and in *Jeremiah 46:8-10*. In *Ezekiel 38-39* we once again see Lydia (Magog), Ethiopia, and Libya as allies (*Ezekiel 38:5*). It is the Lydian army's expertise with the bow (*Jeremiah 46:9-10* and possibly Lud in *Isaiah 66:19*) that is addressed in *Ezekiel 39:3*, when it says, "I will smite the bow out of thy left hand, and cause thine arrows to fall out of thy right hand." Magog's or Lydia's alliance with Egypt and use of the bow are among the added verifiable insights of the Bible that emerge when knowledge from archeology and history is used as an aid in the interpretation of the Bible. Again, the archeological record shows that Magog, Meshech, Tubal, Gomer, and Togarmah all ended long before population centers began to develop in Russia. Thus, archeologists know that **none** of the nations referred to *Ezekiel 38-39* had anything to do with the origins of modern Russia.

Since the modern day political group or country that is supposed to invade and go to war with Israel is not Russia, what information does the historical record and Bible provide to help identify the mystery leader and his nation that the Bible calls Gog and Magog? The historical King Gog and the nations of Magog, Meschek, and Tubal over which Gog was a "prince" as referred to in *Ezekiel 38-39* served as "historical archetypes" of the antichrist and the multi-national confederacy over which the antichrist is to rule during the tribulation period. The antichrist and his empire will participate in the great end times Battle of Armageddon. The Bible uses the historical Gog as an archetype or model of the antichrist to come in the exact same ways that Bible uses the historical King David as a type or model of Jesus the Christ's Second Coming (*Ezekiel 34:23* and *Ezekiel 37:22-24*).* More is revealed by examining the Sumerian

---

*   *Ezekiel 38:17* gives added insight about Gog being a "type" of the

language used by the Assyrians as a sacred language, much like Latin is used today by the Roman Catholic Church. In Sumerian the word *gug* (Gog) means "darkness." In the chief prince Gog from Magog in *Ezekiel 38:2*, we have a reference to the chief prince of darkness from the land of darkness, another type of reference to the antichrist and the lands of his kingdom.* Chapter 9 reveals the connections between Gog, Magog, and the battle spoken of in *Ezekiel 38-39* to the end times battle spoken of in the *Book of Revelation*.

Beyond the misunderstanding about the Russian invasion of Israel, the entire nature of the Battle of Armageddon has been misunderstood. This misunderstanding includes information about why this battle is to take place and about who will be battling whom. The current understanding of specific events to take place during the battle and even where the battle is to take place must be reevaluated. See Chapter 9 for more details.

---

antichrist when it asks, "Art thou he of whom I have spoken in old times by my servants the prophets of Israel . . .? And it shall come to pass at the same time when Gog shall come against the land of Israel, saith the Lord God, that my fury shall come up in my face." Space does not permit but there are indeed passages in the Old Testament where the antichrist and the Lord saving Israel are spoken about. (There are no passages where Russia is spoken about.)

*   In the chief prince Gog from Magog leading, Meshech, Tubal and Gomer, we have in ancient terms "the chief prince of darkness from the land of darkness" leading men that "caused their terror in the land of the living" (*Ezekiel 32:26*) and "creatures from hell" (Assyrian term for Gomer).

"Gog and Magog" are also referred in *Revelation 20:8* as opposing the saints of Christ (Messiah) at the end of the Millennium and here "Gog and Magog" (Gog and the land of Gog) in their most basic meaning, are taken to be types or references to any "antichrist" and "the land of this antichrist"; just as *I John 2:18, 22, 4:3*, and *II John 7* all make it clear that "there are many antichrists."

NOTES

NOTE ONE (For context, see Chapter 2) There is a limit to the amount of shaking that fault-generated earthquakes and volcanic eruptions can produce. The issue of a giant fault generated earthquake gained public attention when the *NBC* Network aired a miniseries called "*10.5*" in May of 2004. The size of the earthquake and the damage done by the earthquake in the miniseries is the stuff of urban legend. In terms of math and geology, geophysicists say while a 12-14 magnitude or more ***impact generated*** earthquake is possible, a magnitude 10.5 ***fault generated*** earthquake is **not** possible. Indeed, seismologists and governmental authorities were quick to warn the public that the whole mini-series was fictional from beginning to end. The producers of the miniseries admitted that they played loose with the facts and had not consulted with any scientists. Dr. Lucy Jones, the U. S. Geological Survey scientist in charge of Southern California said:

> The prediction is blatantly inconsistent with everything we know about earthquakes . . . It is complete science fantasy. There is nothing in it that's connected to reality.[*]

---

[*]   "Quake experts find fault with miniseries," *Arizona Daily Star, Associated Press,* April 11, 2004, p. A6. Also "Scientists, Government Decry *NBC* Miniseries '10.5,'" April 30, 2004, *Reuters News Media* on http://www.au.new.yahoo.com.

The largest earthquake ever recorded was 9.5 in magnitude and occurred on May 22, 1960 in Chile on a fault almost 1,000 miles long. Geophysicists explain that a magnitude 10.5 fault generated earthquake is not possible because "The magnitude of an earthquake is related to the length of the fault on which it occurs—the longer the fault, the larger the earthquake. In order to have a magnitude 10.5 earthquake you would have to have a fault that circles the Earth – no such fault exists".* Obviously, if you can't have a 10.5 fault generated earthquake, you can't have a 14.5 fault generated earthquake, an earthquake powerful enough to cause the shaking and movement called for in Revelation.† In other words, earthquakes caused by faulting and volcanic eruptions simply can not cause every mountain and island to be moved out of their place.‡

A popular theory of the nuclear interpretation school of thought is that a worldwide nuclear war could set off "all the earthquake faults previously undiscovered by man. . . in an

* http://www.Earthquakecountryinfo/10.5.
† e.g. *Revelation 16:18-20*.
‡ The college text Perry H. Rahn, *Engineering Geology: An Environmental Approach (Second edition)*. Upper Saddle River, New Jersey: Prentice Hall, 1996, p. 440 gives a nomogram (diagram) that relates the magnitude of an Earthquake to the amplitude of seismogram displacement it produces at a particular distance from its epicenter. The nomogram shows that a magnitude 2.5 earthquake produces one millimeter of displacement at a distance of 40 kilometers. Going up 4 magnitudes, the diagram shows that a magnitude 6.5 earthquake produces 100 millimeters of displacement at a distance of 400 kilometers. Projecting up 4 magnitudes, a magnitude 10.5 earthquake would produce what would be the equivalent of 10 meters (10,000 millimeters) of displacement at a distance of 4,000 kilometers, if such a huge seismogram measurement were possible. Again projecting up 4 magnitudes gives some perspective to the devastating world-wide shaking and ground movement an impact generated magnitude 14.5 earthquake would produce.

interrelated shaking of our planet."* Unfortunately, this theory of nuclear war as the cause of the massive earthquakes described in Revelation is in error. There is no geological or geophysical basis for this nuclear theory. Nuclear bombs do not produce big enough seismic waves and ground movement to cause significant and far-reaching earthquakes, let alone worldwide earthquakes. Neither can nuclear bombs somehow "set off" massive earthquakes along all the previously undiscovered faults of the world for an "interrelated shaking of our planet." The energy effects of nuclear explosions have been extensively studied and this idea that nuclear warfare can in one way or another bring about global shaking is pure geo-fantasy. There have been many studies on underground nuclear tests, even some conducted in fault zones. It is a fact that nuclear explosions do not cause big and far reaching earthquakes; nuclear explosions do not even "set off" earthquakes along nearby faults, let alone along distant faults. Nuclear explosions do not set off earthquakes because most of the "energy from a nuclear blast dissipates quickly along the Earth's surface," and does not travel downward enough to reach the centers of faults.†

In the aforementioned fictional TV miniseries "*10.5*," a nuclear explosion is used to stop a giant earthquake. This is pure nonsense, because the points of origins of earthquakes are typically several miles to hundreds of miles underground, depths the energy released by nuclear bombs cannot reach! In response to the mini-series, Rick Wilson, an engineering geologist with the California Geological

---

*   Hal Lindsey, *There's a New World Coming: An In-Depth Analysis of the Book of Revelation*, (Eugene, Oregon: Harvest House Publishers, 1984), p. 96.

†   http://www.sklponline.co.uk/Earthquakes.

Survey said, "there is no evidence that a nuclear explosion could help cause or prevent earthquakes of a significant magnitude."[*]

A special statement about nuclear explosions and earthquakes on the U.S. Geological Survey's web site says, "Scientists agree that even large nuclear explosions have little effect on seismicity outside the area of the blast itself."[†] For example, on November 6, 1971 a five megaton underground nuclear test (code name: Cannikin) was conducted on Amchitka in the Aleutian Islands. While the blast produced a sizable wave in the test area, "it did not trigger any earthquakes in the seismically active Aleutian Islands."[‡]

Nuclear explosions, nuclear triggered earthquakes, and/or fault generated earthquakes cannot fulfill the premise of many popular prophecy books that they will one way or another cause "the whole world to be literally shaken apart."[§] Neither can they produce enough shaking to cause every mountain and island to be moved out of their places. However, what is impossible with nuclear power is terrifyingly possible with cosmic impact. *(See Illustration I)*

NOTE TWO (For context, see Chapter 2) Linguistically speaking, every translation from the original language is an interpretation, and the better the translator understands the subject matter, the better the translation. An interpretation should also be compatible with the language usage and culture of that particular time period. When the document in question is the Bible, then the interpretation

[*] Rick Wilson, "A Geo-logical Movie Review," *California Department of Conservation – Earthquakes*, http://www.consrv.ca.gov.
[†] http://www.Earthquake.usgs.gov/bytopic/megagk
[‡] http://www.Earthquake.usgs.govfaq. Also see Bruce A. Bolt, *Nuclear, Explosions and Earthquake, the Parted Veil*, (San Francisco, CA: W. H. Freeman & Co., 1976).
[§] Lindsey, 98.

needs to also be consistent and compatible with the other Scriptures and teachings of the Bible. Needless to say, errors will arise in Bible interpretation due to the use of an inaccurate translation of a word or passage.

With these guidelines for interpretation in mind, it is important to know that the Greek word *aggelos* (#32 in *Strong's Concordance*) that is translated as "angels" in the Sixth Trumpet (*Revelation 9:14-15*) can also be translated as "messengers," and that in the ancient world, comets were often referred to as messengers. In the Old Testament the Hebrew word *malawk* (#4397 in *Strong's Concordance*) is translated "angel" or "angels" approximately 110 times and translated as "messenger" or "messengers" approximately 100 times. Yet in the New Testament, the Greek word *aggelos* (#32 in *Strong's Concordance*) is predominately translated "angel" or "angels," and only translated "messenger" or "messengers" seven times.[*]

Context is the key in determining whether *malawk* or *aggelos* should be translated as "messenger" or as "angel." For example, *Psalm 78:47-49* tells how the God of the Bible in his wrath cast "hail" ('great chunks of ice' in the Amplified Bible) and "hot thunderbolts" during the Exodus by sending "a band of destroying angels"(NAS). In the Hebrew *Tanakh* (the Hebrew version of the Old Testament) this last phrase reads, "a band of deadly messengers." Since the "hail" and " and "hot thunderbolts" referred to in this passage are indications that cometary activity was involved, translating *malawk* as "messengers" (an ancient reference to comets) is the more accurate translation.

---

[*] *The Encyclopedia Judaica* – Corrected Edition (Keter Publishing House, Jerusalem Israel, Vol. 2, p. 958) notes that after the Greek word *aggelos* for "messenger" or "angel" was translated in the Latin Bible as "angel and then passed into other European languages it acquired the almost exclusive meaning of "angel."

In other words, *Psalm 78:47-49* tells how "deadly messengers" (cometary material not heavenly beings) brought "hail" and "hot thunderbolts."

In an effort to fully understand the Sixth Trumpet, it should also be noted that prepositions can have a wide range of meanings, and translators make a decision on which meaning to assign based on what they think is being said. Therefore, in the context of the first four Trumpets clearly describing events related to cometary impact (hail and fire hitting the Earth, a great burning mountain falling to the sea, a great star hitting rivers, and the sky being smitten), a consistent pattern is set for the causes of *Revelation's* catastrophes. Since the Sixth Trumpet describes the heads, tails, comas, outgassing vents, and destructive power of comets, it is easy to recognize that the Sixth Trumpet in *Revelation 9:14* begins by speaking about the loosing of four messengers or comets that are bound for the Euphrates River.

Prophecy books that describe the loosing of four demonic angels that are **bound in** the Euphrates River are presenting a translation and explanation that is inconsistent with the context of everything described previously and everything described thereafter.* Just as we would not put the head and neck of a giraffe on the body of a lion, a translation that describes demonic angels bound in the Euphrates does not fit with the rest of the passage, which gives a scientifically detailed description of active comets coming in to strike the Earth.

NOTE THREE (For context, see Chapter 3) Few Bible authors of today are aware that the religion practiced by the Babylonians

---

* The Greek word *epi* (#1909 in *Strong's Concordance*) which some have translated as "in" or "at" in *Revelation 9:14* is a preposition which can also be used to convey direction and be translated as "toward" or "for."

was almost totally derived from the Sumerians. As a result, interpretations of scripture when it relates to the Babylonian religion have been inaccurate. Generally, these errors can be traced back to a popular 19th century book, *The Two Babylons* by Alexander Hislop. His book was written decades before archeologists learned that the Sumerian people had actually existed, and decades before Sumerian, Akkadian, Babylonian, and Assyrian literature would be translated revealing the true religious beliefs and practices of these cultures. When faced with the historical record, it is clear that *The Two Babylons* is a work born of myth and misdirected religious and Masonic zeal rather than facts.

Astonishingly, many of the references cited in *The Two Babylons* do not actually exist, and other references are out of context. The book was written using sources that were clearly unaware of the Sumerians, and oblivious of the nature of the Babylonian religion and the pantheon of Babylonian gods. While scholars have come late to the knowledge of the Sumerians (whose existence was lost to the world for over two thousand years), it is clear that the scripture reflects knowledge of them. For example, scripture refers to the Sumerian cities of Erech (*Genesis 10:10*) and Ur (*Genesis 11:28, 31, 15:7,* and *Nehemiah 9:7*), the land of Shinar (*Genesis 10:10, 11:2, 14:1, 9, Isaiah 11:11, Daniel 1:2,* and *Zechariah 5:11*) which scholars now identify with Sumer, and the deified Sumerian shepherd-king of Erech called Dumuzi who the Babylonians called Tammuz (*Ezekiel 8:14).* Ancient mythographers writing in Sumerian and Akkadian have celebrated Dumuzi or Tammuz in literary works as the consort of the Sumerian goddess Inanna, (the "Queen of Heaven" – *Jeremiah 7:18* and *Jeremiah 44:17-19* and *25*) who the Babylonians called Ishtar. These ancient literary works reveal that Hislop's ideas about Tammuz and the origins and nature of Babylon religion are simply not true. Interpretation of Bible scripture should not be based on 17th, 18th, and 19th century assessments of ancient cultures, especially when more recent archeological discoveries reveal something entirely different. The information from ancient records must be allowed to speak and students of the Bible should not ignore what these ancient records say just to hold on to interpretations about the past that support long-held ideas. *(See Illustrations B & C)*

A standard college reference such as *Ancient Near Eastern Texts – Relating to the Old Testament – Third Edition*, Edited by James B. Pritchard, (Princeton University Press, 1969) relates what these ancient cultures believed according to their own records. The classic scholarly work, *The Sumerians: Their History, Cultures and Character* by Samuel Noah Kramer (University of Chicago Press, 1963) is an excellent book for anyone seeking to understand the religion the Babylonians adopted from the Sumerians. Also see *The Treasures of Darkness - A History of Mesopotamian Religion* by Thorkild Jacobsen, (Yale University Press, 1976). A standard Christian reference work, *The New Bible Dictionary – Second Edition* (Tyndale Publishing 1988), provides a simple overview of the Sumerians. Looking into any recent encyclopedia or high school history text will make it clear that the Sumerians preceded the Babylonians.

NOTE FOUR (For context, see Chapter 4) Not only are the two phrases of *Genesis 7:11* consistent with what scientists know today, there are other scriptures in the Bible that reiterate the specific scientific information given in these two phrases. Just as *Genesis 7:11* accurately states that comets were the sources of the oceans' waters, the Bible refers to water-bearing comets as the "bottles of heaven" or as the "water jars of the heavens" in *Job 38:37 KJV* and *NIV*. In *Job 38:29-30* and *Psalm 147:16-17*, the "water bowls or basins of heaven", obviously refer to comets when the Hebrew word *kef-ore* (#3713 in *Strong's Concordance*), which is traditionally translated as "hoarfrost" is more accurately translated as "water bowls" or "water basins" based on the context.* And, as *Genesis 7:11* accurately tells of comets residing or lying in wait in heaven, the passage is consistent with other Bible passages that refer to the spherical Oort Cloud of comets that encloses our solar system. The Oort Cloud of comets is referred to "ye waters that be above the heavens" in *Psalm 148:4*, and "the waters above (and under) the firmament (heaven)" in *Genesis 1:6-8*.†

Job 38:22-23 NIV describes the Oort Cloud of comets and related Kuiper Belt of comets as "the **storehouses** of the snow... the **storehouses** of hail (**snow and hail represent frozen waters as in comets**), which I reserve for times of trouble, for days of war and battle." In addition to the Biblical references to comets as "waters" and the Oort Cloud as the "waters above the heavens," 'water' appears in the Bible in a surprising way. The Hebrew word for "heaven," the word *shaw-mayim* (#8064 in *Strong's Concordance*) is in itself an adaptation of the Hebrew word for "water," the word *mayim* (#4325 in *Strong's Concordance*). (See Illustration A)

*   For example, the Hebrew word, *kef-ore*, #3713 in *Strong's Concordance*, is translated as "basin" or "basins" in *I Chronicles 28:17, Ezra 1:10*, and *8:27*.

†   *Genesis 1:6-8* tells of the separation of the waters of the Oort Cloud of comets by the "firmament" [KJV] or "expanse" [NIV], which the Bible also calls "heaven," but to us is the solar system; this should not be confused with *Genesis 1:9-10* which tells of the separation of the waters **within** the solar system [heaven], so that dry ground called "land" or "Earth" appeared, and the gathered together waters were called seas.

NOTE FIVE (For context, see Chapter 6) The things that should be considered when using the Bible's own point of view to evaluate an interpretation of a Bible prophecy include:

1) <u>It takes two or three witnesses to establish a point</u> (*Deuteronomy 19:15, Genesis 41:32, II Corinthians 13:1,* and *Matthew 18:16*). For example, to be confident in the interpretation of a catastrophic event prophesied to occur in the Seven Trumpets and the Seven Vials, this same event should be referred to in at least one other place in scripture. In this way scripture will interpret itself through related scriptures (*II Timothy 3:16*).

2) <u>That scripture is internally consistent</u> (*I Corinthians 14:33* and *Mark 3:25*). For example, the end times catastrophic events called for in the Trumpets and Vials should be consistent with the end times catastrophic events called for in the Old Testament and/or in other parts of the New Testament. These interpretations should also reflect the Bible's use of certain key words in a consistent manner. Once the Bible establishes certain meanings for words by repeated usage through the course of scripture, the interpretation of end times prophecy should not suddenly attach entirely new meanings to these words if the established meanings clearly work. For example, Biblical precedent for the meanings of the words "star," "messenger," and "horsemen" should be used in interpreting the Trumpets and Vials; interpretations that attach entirely new meanings for these words should be suspect.

3) <u>Mankind reaping what they sow</u> (*Galatians 6:7, Hosea 8:7,* and *Job 4:8*). The Bible is consistent within itself so the interpretation of the catastrophic events prophesied to occur in the Trumpets and Vials should reflect nations reaping the kinds of things their people have at one time or another sown. *Hosea 8:7* specifically says,

"For they have sown the wind, and they shall reap the whirlwind." Here, it is noteworthy to recognize that comets can literally bring "whirlwinds," and that comets were perceived, depicted, and referred to as "whirlwinds" in the ancient world.

4) Human behavior and events are interrelated (*Leviticus 18:24-28, 20:22, Numbers 35:33,* and *Psalm 106:38*). In addition to saying that man reaps what he sows, the Bible says that the land itself can become defiled and spew out the nations that inhabit it. To be consistent with these Biblical laws, the catastrophic events prophesied to occur should involve nations that have particularly defiled their land.

5) There are patterns within scripture (*I Corinthians 10:6, 11, Matthew 11:13-14, 17:11-12,* and *Hosea 12:10*). The Bible says that some of the people, places, and events it refers to constitute patterns and types for other people, places, and events. In the interpretation of the events prophesied to occur in the Trumpets and Vials there should be patterns or relationships with some of the same people, places, and events already spoken about in the Bible. For example, the place of a great battle victory for Israel against a multinational enemy force prophesied to occur in the Trumpets and Vials should have some relationship to where an earlier battle victory for Israel against a multinational enemy force occurred.

6) That which has been prophesied to occur must have already in essence been (*Ecclesiastes 1:9* and *3:15*). While the previous Biblical criteria refers to patterns stemming from people, places, and events referred to in the Bible, this item refers to patterns stemming from people, places, and events over all of time (not just patterns arising during the Biblical period), and events the Bible prophesies to occur. The secular world refers to this principle as history repeating itself. The interpretation of the events prophesied to occur in the Trumpets and

Vials should involve Biblical precedent and historical precedent for the types of events prophesied to occur, including the geological and astronomical events. For example, the great battle victory prophesied to occur in the Trumpets and Vials also involves catastrophic events occurring so there should be a historical precedent for a great battle victory associated with catastrophic events, ideally the same type of catastrophic events. We will see that a comet will once again be the God of the Bible's weapon of choice for bringing about victory in battle and for "bringing about" catastrophe.

7) <u>The God of the Bible does not change</u> (*Malachi 3:6* and *Hebrews 13:8*). The Bible says that God does not change. This is consistent with the previous two rules that tell of patterns and types being behind the fulfillment of prophecy. The God of the Bible 'not changing' means that, if the God of the Bible responded to certain circumstances in a certain way in the past, then God will probably respond to similar circumstances in a similar way in the future. For example, in the Old Testament the God of the Bible responded to the rise of the Akkadian Empire's multinational empire by destroying it (see Chapter 8 "The Truth about What Really Happened at the Tower of Babylon") as told in *Genesis 10:10* and *11:1-11*. The God of the Bible's response to the rise of the multinational empire prophesied to arise during the end times should be the same, since the Bible says that God changes not.

8) <u>All wonders and catastrophic events must be in accord with the laws of heaven and Earth</u> (*Jeremiah 31:35-37* and *33:25*). The ordinances of heaven and Earth referred to in the Bible are the same as what we now call the laws of nature and the laws of physics. Since the God of the Bible says that these "laws" cannot be broken, any interpretation of Bible prophecy must not involve circumstances

that break these "natural" laws. When "wonders" are spoken about in Bible prophecy, these "wonders" must involve natural phenomena of some sort, albeit rare natural phenomena.

9) <u>The God of the Bible passes judgment on people based on their own words</u> (*Luke 19:22, Matthew 12:37,* and *Numbers 14:2, 28-29*). In the Bible the God of the Bible has sent catastrophe according to the things that man himself has said he feared. As the Trumpets and Vials prophesies a series of catastrophes to be brought by comets, it is important to note that since knowledge has recently increased, scientists now believe that cosmic bombardment is the number one threat to humanity.

10) <u>The God of the Bible gives his people notice prior to certain actions</u> (*Amos 3:7, Genesis 18:17, I Thessalonians 5:20,* and *Matthew 24:25*). The Bible says that the God of the Bible gave notice to his people through his prophets prior to certain actions. The Trumpets and Vials of the *Book of Revelation* constitute final prophetic notice of the catastrophes to come during the end time, catastrophes that were first prophesied in the Old Testament.

11) <u>There is a Biblical standard for the accuracy of prophecy</u> (*Deuteronomy 18:21-22* and *I Thessalonians 5:20-21 NIV*). The Biblical standard for prophecy truly sent by the God of the Bible is 100 percent accurate. The proper interpretation of the catastrophic events prophesied to occur in the Trumpets and Vials should involve Biblically-sound and scientifically-sound interpretations for each and every one of the seven separate rounds of cosmic impact prophesied to occur.

NOTE SIX (For context, see Chapter 7) In trying to identify the type of bacteria described in the Fifth Trumpet, there are three possible candidates from presently known types of bacteria: streptococcal bacteria, spiral bacteria, and tubercle bacillus.

Streptococcal bacteria is a type of bacteria that is spherical in form. Streptococcal bacteria can cause the type of very painful sores called for by the Fifth Trumpet and Fifth Vial. The physical description of the "locusts" in the Fifth Trumpet begins by saying that the "shapes of the locusts were like horses . . . ." Most *streptococci* form chains, that can take a sinuous shape similar to that of a "seahorse" (a small marine fish that swims in an upright position that is characterized by horse-like head, a body covered by bony plates, and a prehensile [serpent-like] tail) and may, in effect, resemble the "horse," called for by the Fifth Trumpet.*

Spiral bacteria (e.g. a *spirochete*) is a type of bacteria that is spiral in form. The spirochete or spiral bacteria, which can have one or more corkscrew-like twists and curved ends is another type of bacteria with a shape that can be said to resemble a seahorse, and thus a horse.†

Tubercle bacillus is a type of bacteria that is rod-like in form. One description of the micro-organism that causes tuberculosis, the tubercle bacillus, says "under the eye of a microscope the tubercle bacillus . . . looks like a maneless sea-horse."‡

---

*   Gerald J. Tortora, Berdell Funke, Christine Case, *Microbiology: An Introduction (Third Edition)*, (Redwood City, California: Benjamin Cummings Publishing Co., 1989), pp. 273 and 75.

†   Gerald J. Tortora, Berdell Funke, Christine Case, pp. 75-76 and 643. For example, the spiral bacteria, *Leptospira interrogans* [a *spirochete*] has a characteristic shape that resembles that of a sea horse.

‡   Nikiforuk, Andea. *The Fourth Horseman: A Short History of Epidemics, Plagues, Famines, and Other Scourges,* (New York: M. Evans & Co., 1991), p. 3.

The Fifth Trumpet gives more information about the "shape" of the locusts (bacteria) when it says that the shapes of the locusts were not only like "horses," but like *"**horses prepared for battle**"* (*Revelation 9:7*), that is, like horses wearing headdresses and padding of some sort. This imagery is not unusual since putting headdresses on horses to prepare them to go into battle was a common practice during ancient times. The Fifth Trumpet emphasizes this headdress on a horses' head by saying "and on their heads they wore something like crowns of gold" which seems to indicate a circular and spiked headdress. A circular and spiked headdress on one end of a bacterial form could be one way of describing cilia, or fimbria, relatively short structural components of bacteria that can appear as "tufts" projecting from one end (or both ends) of a bacterial cell wall. (Recalling the seahorse analogy made earlier, the description of horses with crowns made in *Revelation 9:7*, again brings the sea horse to mind, since sea horses typically have crown-like projections on their heads.)

The Fifth Trumpet goes on describing these "locusts" (bacteria) saying, "their faces were like the faces of men" (*Revelation 9:7*).[*] This may refer to the circular eye-like nucleus within the cell of a bacteria that can give it the general appearance of a human face in profile.

Then, the Fifth Trumpet says the locusts or bacteria "had hair as the hair of women" (*Revelation 9:8*), indicating long flowing hair. This may refer to the "flagella" some bacteria have. "Flagella" are "long **hair-like** appendages which project from the body (cell wall) of the bacterium, and with a lashing motion propel it through any liquid medium in which it may be."[†] Several flagellum could project

---

[*]   *Revelation 9:7 NIV* "their faces resembled human faces."
[†]   Joseph Laffan Morse, editor, *The New Funk and Wagnall's Encyclopedia,* (New York: Unicorn Publishers, Inc., 1950), p. 892.

from only one end of a bacterium (e.g. lophotrichous) making the Fifth Trumpet's words that "they had hair as the hair of a woman" (*Revelation 9:8*) a fitting description. The description of locusts with faces like that of men and hair like that of women gives us bacteria with an eye-like nucleus and hair-like flagellum. Next, the Fifth Trumpet in describing those locusts (bacteria) says, "their teeth were as the teeth of lions" (*Revelation 9:8*) indicating the presence possibly of a few relatively thicker fimbriae, or relatively shorter and thicker flagellum (cilia) than previously described. (Or the relationship between pili and fimbriae could be involved here, since fimbriae are shorter and finer than pili.)

Then, the Fifth Trumpet in describing these locust (bacteria) says, "they had breastplates as it were breastplates of iron" (*Revelation 9:9*) indicating some relationship between these locust (bacteria) and iron. Since the rocky material of asteroids and comets is typically rich in iron, the term "breastplates of iron" could refer to the iron rich dust particles that would arise after impact to carry these "locusts" (bacteria), as if they were on a plate, wherever the wind blows. (Also note that comet's coma, the envelope of gas and dust that surrounds an active comet's nucleus is rich in various metals including iron.) After impact, the iron-rich rocky matrix that the "locusts" (bacteria) were embedded in would result in dust rich in iron particles. The "locusts" (bacteria) could be carried by these iron rich dust particles which are described in *Revelation 9* as "breastplates of iron." If the bacteria involved formed thick-walled and extremely heat resistant endospores, they could very well survive the tremendous heat of impact.

When it is recognized that the "locusts" spoken of in *Revelation 9* represent "bacteria," this reference to bacteria borne by iron-rich dust particles ("breastplates of iron") becomes very important because some types of bacteria (chemolithic) get their energy by oxidizing inorganic substances such as iron and sulfur.* In the section on bacteria, *Microsoft's Encarta 98 Encyclopedia* says, "Many bacteria have very complex metabolic systems. Some can live on ***iron*** and other metal deposits." Thus, the Fifth Trumpet's reference to "iron" is consistent with a description of a particular type of bacteria. Cometary dust would be rich in sulfur as well as iron. Certain types of bacteria can derive energy by oxidizing iron, others by oxidizing sulfur, and some from both iron and sulfur. Interestingly, bacteria that oxidize sulfur can deposit "sulfur granules in the cell, where they serve as an energy reserve," and bacteria that can oxidize iron can deposit iron granules in their cells. This gives a second possible interpretation to the phrase "breastplates of iron" given in *Revelation 9:9*.

In addition to the "breastplates of iron" as a reference to bacteria carried by iron rich dust particles, it could also be a reference to the iron oxide (Fe3O2-magnetite) crystals the bacteria may contain within their cell walls. The Hebrew word for "breastplate" (which stands behind the Greek word for "breastplate" [*thorax* - #2382 in *Strong's Concordance*] as used in *Revelation 9:9*) is *kho-shen* (#2833 in *Strong's Concordance*) which means to "contain" as in the sense of a pocket. In saying that the locust (bacteria) had "breastplates of iron," *Revelation 9:9* could mean that these bacteria "contain iron" within their cell walls. Interestingly, NASA is studying one of the bacterial candidates found in a meteorite from Mars that is associated with iron oxide (magnetite) crystals.† When *Revelation 9:9* says that

*   Gerald J. Tortora, Berdell Funke, Christine Case, p. 92.
†   "Studies of meteorite boost life on Mars theory," *L.A. Times, Washington Post News Series*, in *Tucson Citizen*, March 14, 1997, p. 11A.

the locusts (bacteria) had "breastplates of iron," this phrase could convey two separate pieces of information; one, that these bacteria are carried by dust particles that are rich in iron, and two, that these bacteria "contain iron" within their cell walls.

After saying these "locusts" had "breastplates of iron," the Fifth Trumpet continues the "going to battle" imagery adding "the sound of their wings was as the sound of chariots of many horses running to battle" (*Revelation 9:9*). Locusts can make a sound like the sound crickets make by rubbing parts of their body together (a process called stridulation), but the sound referred to here is the sound a great swarm of locusts flying through the air. Some people who have seen and heard a great swarm of locusts approaching have described the sound of the approaching swarm to be like the terrifying sounds of an ancient army approaching. Conceivably the "breastplates of iron," the iron particles contained in the dust raised by a comet impact could rub together when driven by the wind and make a sound like the clattering of the chariots wheels and the thundering of horses galloping across the plain like called for in the Fifth Trumpet. In describing these "locusts" (bacteria), the Fifth Trumpet also says, "they had tails like unto scorpions" (*Revelation 9:10*). This analogy tells us these "locusts" (bacteria) have curved and segmented tails like the tail of a scorpion that is curved and segmented. A curved tail describes bacteria that have a curved shape at one or both ends and a segmented tail describes bacteria that appear to be made of separate parts joined together. There are indeed bacteria that appear to be **both** curved and segmented. A segmented appearance is characteristic of bacteria whose cells form chains such as *streptococcus bacteria*, where each cell in the chain appears as a segment of the chain.

Bacteria with a curved shape at one or both ends is consistent with *Revelation 9's* earlier description of these bacteria that said the "shapes of the locusts were like horses" (*Revelation 9:7*) since the shape of a seahorse is curved. So bacteria with "shapes like horses" (*Revelation 9:7*) and "tails like scorpions" (*Revelation 9:10*) indicates bacteria having curved heads and curved tails or bacteria that are curved at both ends. Bacteria that are curved at both ends supports the earlier description of bacteria with shapes similar to sea horses with their heads curved on their bodies and their tails curving at the other end. There are indeed bacteria that appear to be curved at **both** ends and with an overall shape similar to that of a sea horse. Some of the *streptococcal* bacteria mentioned earlier can form chains that take a shape similar to that of a sea horse. **Spiral bacteria** such as *Leptospira interrogans* (a *spirochete*) can also be shaped like a sea horse, and "the *tubercle bacillus* (the micro-organism that causes tuberculosis) . . . looks like a maneless sea-horse."[*] Streptococcal bacteria, spirochetes, and the tubercle bacillus are all examples of bacteria that can appear to be curved at both ends like a seahorse. Actually, among already known bacteria there are many types that can have both a seahorse like sinuous shape and a segmented body, and thus meet the specifications of both *Revelation 9:7* and *Revelation 9:10*.[†]

The Fifth Trumpet also says these "locusts" (bacteria) will "torment" men for five months (*Revelation 9:5*). The Fifth Vial explains this torment will be the result of the very painful sores (*Revelation 16:10-11*) caused by these locusts or bacteria.[‡] Those tormented by

---

[*]   Nikiforuk, Andrea, 3.
[†]   "The Rise of Life on Earth," Richard Monastersky, *National Geographic*, Vol. 193, No 3, March 1998, pp. 74-75.
[‡]   Recall, that the impact event of the First Trumpet and First Vial

the painful sores will have come into direct contact with the dust that carries these bacteria or possibly infected by others already affected by these bacteria. The Bible, indicating that these "locusts," (that is, bacteria) will cause painful sores, speaks of them causing some type of skin disease. Today it is well known that bacteria can cause a variety of painful skin diseases. *Streptococcus pyogenes* can cause impetigo, a skin infection characterized by pustules, and erysipelis, a skin infection characterized by areas of reddened and swollen skin.*
"*Staphylococci* are responsible for many serious infections that are characterized by boils or abscesses."† These skin infections can sometimes enter the bloodstream and cause serious toxemic diseases such as "scolded skin syndrome" where the "skin of the palms and soles peel off in sheets when it is touched." ‡

The only difficult spot in interpreting the discourse of the Fifth Trumpet is the question of whether comets can really contain bacteria. The "locusts" referred to in the Fifth Trumpet are shown to perform in ways that are uncharacteristic of "locusts." The Fifth Trumpet says these locusts: 1) do not eat grass, green things nor any trees; 2) have shapes like horses with crowns on their heads; 3) have curved and segmented tails like scorpions; 4) have faces with

---

tells how "there fell a noisome and grievous sore upon men" [*Revelation 16:2*]. As discussed, the sores of the First Trumpet and First Vial seem to come from noxious substances being released into the atmosphere after impact; while the sores of the Fifth Trumpet and Fifth Vial seen to come from bacteria being released into the atmosphere; but the sores of the First Trumpet and First Vial could also come from bacteria.)

* Gerald J. Tortora, Berdell Funke, Christine Case, *Microbiology*, pp. 506-507.

† P. Rivers, R. Evert, & S. Eickhous, *Biology of Plants*, (New York: Worth Publishing, 1981), p. 198.

‡ Gerald J. Tortora, Berdell Funke, Christine Case, *Microbiology*, 1989, pp. 506-507.

eyes like men; 5) have long and flowing hair like women; 6) have teeth like lions; 7) have breastplates that are associated with and/or contain iron; 8) make a sound like chariots; 9) cause disease in the form of painful sores for up to five months (*Revelation 16:10-11*); 10) have movements that are controlled by a comet both in regard to their travel from heaven to earth, and in regard to their being carried around the Earth by the cometary dust that arises from the comet's impact crater and is blown by the wind.

Are the "locusts" of the Fifth Trumpet bacteria? Do the "locusts" of the Fifth Trumpet symbolically represent bacteria associated with a comet? In interpreting the "locusts" of the Fifth Trumpet to be bacteria, it seems compelling that the ten different descriptive aspects of the "locusts" as given in the Fifth Trumpet and listed above actually give scientifically sound information about bacteria. This multi-faceted scientifically accurate information about bacteria, given long before bacteria were even discovered, seems to go beyond chance and coincidence, especially when the *context* is considered. The Fifth Trumpet is about a cometary impact, and the "locusts" (bacteria) spoken about come out of the smoke and dust that arises after the impact.

Astronomer John S. Lewis specifically identified the Fifth Trumpet of the ninth chapter of the *Book of Revelation* as pertaining to "impact phenomena." Astronomers Clube and Napier saw the events of the eight and ninth chapters of the *Book of Revelation* as a "series of impacts," and even remarked "on the clarity of the astronomical associations." While Clube and Napier did not specifically identify the "locusts" of the Fifth Trumpet, they did list "locusts emerging from the abyss" as one of six "relevant items" from the *Book Revelation's* chapters that "are impact-inspired."[*] If the "locusts" of the Fifth Trumpet are not "bacteria," then what are they? (Prions, a precursor to bacteria and viruses?) In consideration of the ten descriptive points given in the Fifth Trumpet that have a scientifically sound hypothesis, what else could these impact-inspired "locusts emerging from the abyss" be? If the "locusts" of the Fifth Trumpet are in fact "bacteria," how did the Bible come to give a scientifically sound description of bacteria long before bacteria were discovered and long before scientists began to make a connection between comets and bacteria?

---

[*] Victor Clube and Bill Napier, *The Cosmic Serpent*, (London: Faber and Faber, 1982), pp. 216 & 217.

CITATIONS

CHAPTER ONE - ARE THERE CORRECT ANSWERS

1 Nancy Gibbs, "Apocalypse Now," *Time Magazine*, 1 July 2002:42.
2 Kenneth L. Woodward, "The Way the World Ends," *Newsweek Magazine*, 1 Nov. 1999: 68.
3 Paul Boyer, *When Time Shall Be No More: Prophecy Belief in Modern American Culture* (Cambridge, Mass: Harvard University Press, 1992) 144.
4 Daniel David Luckenbill, *Ancient Records of Assyria and Babylonia, Vol. 1 and Vol. II* (New York: University of Chicago Press, 1926, Greenwood Press Reprint, 1968).
5 Dewey M. Beegle, *Prophecy and Prediction*, (Ann Arbor, MI: Pryor Pettengill, 1978) 2.
6 Boyer, 131 and 143.
7 Michael D. Lemonick, "Is the Bible Fact or Fiction?" *Time Magazine*, 18 Dec. 1995: 64.
8 Mark A. Noll, *The Scandal of the Evangelical Mind*, (Grand Rapids, MI: William B. Eerdmans, 1994) 186-187.
9 Noll, 183.
10 "The Bible in Science," *New York Observer*, 26 March 1863: 98- 99. As quoted in Noll, 183-184.
11 Sandra Blakeslee, "Ancient Crash, Epic Wave – Did Catastrophe fall from above in 2807 BC?" *The New York Times*, 14 Nov. 2006, with 16

Nov. 2006 correction: 1-4.
12   Sandra Blakeslee, 4.
13   Anna Salleh, "Mega-tsunamis more common than we think," Australian Broadcasting Corp. *Science Online,* 16 Nov. 2006. <http://www.abc.net.au/news/newsitems/200611/51790224.htm>. Dallas Abbott, Edward Bryant and others, "Impact Craters As Sources of Megatsunami Generated Chevron Dunes," *Geological Society of America Abstracts*, paper No. 119-20, Vol. 38, No. 7, Philadelphia Annual Meeting, 22-25 Oct 2006: 229.
14   As above. Dallas Abbott, Edward Bryant and others, "Report of International Tsunami Expedition to Madagascar," 28 Aug. – 12 Sept. 2006, *<http://www.ldeo.columbia.edu/users/menke/slides/madagascar06/report.pd>*.
15   Dallas Abbott, W. W. Masse, D. Breger, L. Burckle, "Burckle abyssal impact crater," *Black Rock Forest 2005 Research Symposium*, 20 June 2005, <http://www.blackrockforest.org/research/symposium2005.htm>. Dallas Abbott, Edward Bryant and others, "Impact Craters" and "Report of International Tsunami...."
16   Sandra Blakeslee, 1 and 4.
17   Sandra Blakeslee, 1 and 4.
18   Simon Rocker, "Catch a Falling Star", *Jewish Chronicle*, London, 6, March 1998.
19   Victor Clube & Bill Napier, *The Cosmic Winter*, (Cambridge, MA :Basil Blackwell, Inc., 1990) 23 & 244.
20   Robert Matthews, "Meteor Clue to end of Middle East Civilization," 16 April 2002, *<http://www.news.telegraph.co.uk.>*
21   Dr. Sharad Master, "A Possible Holocene Impact Structure in the Al' Amarah Marshes Near the Tigris-Euphrates Confluence, Southern, Iraq," *Meteorites and Planetary Science*, Vol. 36, Supplement, A 124 (2001). Also "Umm al Binni Lake, a possible Holocene impact structure in the marshes of Southern Iraq. Geological evidence for its age, and implications for Bronze-age Mesopotamia," presented in *Environmental Catastrophes and Recoveries in the Holocene, Dept. of Geography and Earth Sciences, Brunel University*, Uxbridge, UK, 29 Aug. – 2 Sept. 2002, <http://www.atlas-conferences.com/cgi-bin/abstract/caig-15>.
22   Dr. Sharad Master, as above.

23 Robert Roy Britt, "Comets, Meteors and Myth: New Evidence for Toppled Civilizations and Biblical Tales," <http://www.space.com>, 13 Nov. 2001. "Collapse of Early Bronze Age Civilization: Has the Smoking Gun Been Found?" CC Net Special – *The Cambridge Conference*, Liverpool Moores University, 16 April 2002 <http://www.abob.libs.uga.edu/bobk/ccc/cc110501.html.>

24 Beegle, 61.

CHAPTER TWO - STARS THAT FALL FROM HEAVEN

1 Duncan Steel, *Rogue Asteroids and Doomsday Comets: The Search for the Million Megaton Menace That Threatens Life on Earth*, (New York: John Wiley and Sons, 1995), 1.
2 Marcelo Gleiser, *The Prophet and the Astronomer: A Scientific Journey to the End of Time*, (New York: W. W. Norton & Co., 2001), 14 & 20-21.
3 John S. Lewis, *Rain of Iron and Ice: The Very Real Threat of Comet and Asteroid Bombardment*, (Reading, MA: Addison-Wesley - Helix Books, 1996), 11-13.
4 Victor Clube & Bill Napier, *The Cosmic Serpent: A Catastrophic View of Earth History*, (London: Faber and Faber, 1982), 213 & 217.
5 Victor Clube & Bill Napier, 212.
6 Victor Clube & Bill Napier, *The Cosmic Winter*, (Cambridge, MA: Basil Blackwell Inc., 1990), 194-195.
7 Gerrit L. Verschuur, *Impact: The Threat of Comets and Asteroids*, (New York: Oxford University Press, 1996), 103-104.
8 Gary DeMar, *End Times Fiction: A Biblical Consideration of the Left Behind Theology*, (Nashville, TN: Thomas Nelson, 2001), 102.
9 Virgil. Aeneid, Vol. II, Book II, H. Rushton Fairclough, translator. London: Leob Classical Library, 1978.
10 Samuel Noah Kramer translator, "Hymnal Prayer of Enheduanna: The Adoration of Inanna in Ur," line 13, in *Ancient Near Eastern Texts - Relating to the Old Testament*, James Pritchard, editor (Princeton, NJ: Princeton University Press, 1969), 580.
11 A. K. Grayson translator, "The Epic of Gilgamesh," Tablet VI, lines 122-126, in *Ancient Near Eastern Texts*…505.

12 Piotr Michalowski, *The Lamentation Over the Destruction of Sumer and Ur*, (Winona Lake, IN: Eisenbrauns Publisher, 1989), 41, 47, 61 & 79. Samuel Noah Kramer, translator "Lamentation Over the Destruction of Sumer and Ur," in *Ancient Near Eastern Texts...* 613.
13 Victor Clube & Bill Napier, *The Cosmic Winter*, 272.
14 John & Mary Gribbin, *Fire on Earth*, (New York: St. Martin's Griffin, 1996), 32-33 and 133. "A Single Intruder, A Chain of Craters?" *National Geographic*, Geographical Section, Oct. 1998.
15 Marcelo Gleiser, 100.
16 John S. Lewis, 149.
17 Duncan Steel, 257. Gary D. Goodwin, "Machholz 2," <http://www.tmgnow.com/repository/cometary/>.
18 John S. Lewis, 149 and John & Mary Gribbin, 132.
19 Robert Roy Britt, "A Comet's Life: Icy Adventures from Birth to Death," <http://www.space.com/scienceastronomy/solarsystem/comet_linear>, 17 May 2001.
20 Paul Boyer, *When Time Shall Be No More: Prophecy Belief in Modern American Culture* (Cambridge, Mass: Harvard University Press, 1992) 128.
21 "The Atomic Bombing of Hiroshima and Nagasaki" by *Manhattan Engineering District*, "Avalon Project," <http://www.yale.edu/lawweb/avalon/abomb/mpmenu.htm> Chapter 15 Ground shock.
22 *Voyage Through the Universe - Comets, Asteroids and Meteorites, Editors* (Alexandria, Virginia: Time-Life Books, 1990), 128.
23 "Meteor: Fire in the Sky," The History Channel, 2005.
24 Marcelo Gleiser, 95.
25 Editors of Time-Life Books, 121 and 127.
26 Duncan Steel, 1.
27 Editors of Time-Life Books, 121 & 127.
28 "Earth called Sitting Duck for Asteroids," Associated Press, *Tucson Citizen*, 4 July 1991: 17A.
29 Editors of Time-Life Books, 128.
30 Editors of Time-Life Books, 128.
31 John S. Lewis, 11.
32 "Halley's Gas and Dust Jets," *Sky and Telescope*, Nov. 1988: 456.
33 Tim LaHaye & Jerry B. Jenkins, *Are We Living in the End Times?* (Wheaton, IL: Tyndale House, 1999) 5. Tim LaHaye, *No Fear of the*

CITATIONS 513

Storm, (Sisters, OR: Multnomah, 1992), 240.

34 Nigel Calder, *Comets: Speculation and Discovery*, (New York: Dover Publishing, 1994), 3.

35 John S. Lewis, G. Watkins, H. Hartman, & R. G. Prinn, "Chemical consequences of major impact events on Earth," *Geological Society of America*, Special Paper 190, 1982. Victor Clube & Bill Napier, *The Cosmic Winter*, 291.

36 John S. Lewis, G. Watkins, H. Hartman, & R. G. Prinn, as above. John S. Lewis, *Rain of Iron and Ice*, 110.

37 John S. Lewis, 110-111. Editors of Time-Life Books, 37. "The Tunguska Explosion." <http://www.madladdesigns,co.uk/unexplained/enigmas/tunguska.htm.>.

38 Same as above.

39 "The Tunguska Event – 100 years Later," Tony Phillips editor, *NASA Science News*, 30 June 2008, http://www.science.nasa.gov/rss.xml! Duncan Steel, "Tunguska at 100," *Nature*, 26 June 2008: 1157-1159.

40 Victor Clube & Bill Napier, *The Cosmic Winter*, 270-271. John S. Lewis, 54. John C. Brandt, & Robert D. Chapman, *Rendezvous in Space: The Science of Comets*, (New York: W. H. Freeman & Co., 1992), 185-186. Roy A. Gallant, "The Sky Has Split Apart," <http://www.galisteo.com/tunguska/does/p4>. William K. Hartman, "1908 Siberian Explosion" <http://www.psi.edu/projects/siberia/>.

41 John S. Lewis, 63 & 179.

42 Victor Clube & Bill Napier, *The Cosmic Winter*, 159. "An Unseen Meteor Storm," News Notes, *Sky & Telescope*, May 1990: 476.

43 John S. Lewis, 148.

44 L. Mansinha, & E. Smylie, "The Rotation of the Earth," *Scientific American*, Dec. 1971: 80-89. "Earthquake and the Earth's Wobble," Science, 13 Sept. 1967, 1127.

45 John S. Lewis, 89 & 146. Victor Clube & Bill Napier, *The Cosmic Winter*, 259-260.

46 Editors of Time-Life Books, 121.

47 "Moon Blast! New insights into the birth of our satellite," *Time*, 27 August 2001: 56.

48 Marcelo Gleiser, 98 & 106.

49 Clark Chapman & David Morrison, "The Next Doomsday Impact," (Planetary Science Institute – Tucson, NASA - Ames Research Center),

Astronomy, Nov. 1989: 8. Clark Chapman & David Morrison, *Cosmic Catastrophes*, (New York: Plenum Publishing, 1989), 275.
50 Gerrit Verschuur, "The End of Civilization?" *Astronomy*, Sept. 1991: 54.
51 M. E. Bailey, S.V.M. Clube & W. M. Napier, *The Origin of Comets*, (New York: Pergamon Press, 1990), 400.
52 Donald K. Yeomans, *Comets: A Chronological History*, (New York: John Wiley & Sons, 1991), 334.

CHAPTER THREE – THE LORD OF HOSTS

1 Samuel Noah Kramer, *The Sumerians: Their History, Culture and Character*, (Chicago, Illinois: University of Chicago Press, 1963), 6 and 19.
2 Tim LaHaye, *Revelation Unveiled*, (Grand Rapids, Michigan: Zondervan Publishing, 1999), 266-271. Harry A. Ironside, *Lectures on the Book of Revelation*, 12th edition, (Neptune, New Jersey: Loizeaux Brothers, 1942), 287-291. Clarence Larkin, *Dispensational Truth*, (Philadelphia, Pennsylvania: Clarence Larkin Estate, 1920), 140. Hal Lindsey, *The Late Great Planet Earth*, (Grand Rapids, Michigan: Zondervan Publishing, 1970), 105-107 & 111-113. Chuck Smith, *Dateline Earth*, (Old Tappan, New Jersey: Chosen Books – Fleming H. Revell Co., 1989), 151-152. Rev. Alexander Hislop, *The Two Babylons: The Papal Worship of Nimrod and His Wife*, Second Edition, (Neptune, New Jersey: Loizeaux Brothers, 1959), 21-31.
3 Rev. Alexander Hislop, 21-31.
4 Diane Wolkstein & Samuel Noah Kramer, "Sumerian History, Cultures and Literature," Samuel Noah Kramer, translator in *Inanna - Queen of Heaven and Earth*, (New York: Harper & Row, 1983), 122.
5 Samuel Noah Kramer, 288-289.
6 Thorkild Jacobsen, *The Treasures of Darkness - A History of Mesopotamian Religion*, (New Haven: Yale University Press, 1976), 234.
7 Samuel Noah Kramer translator, "Self-Laudatory Hymn of Inanna and Her Omnipotence," and "Hymnal Prayer of Enheduanna: The Adoration of Inanna in Ur," in *Ancient Near Eastern Texts Relating to the Old Testament*, James Pritchard, editor, (Princeton, NJ: Princeton

University Press, 1969), 578-581.
8 Samuel Noah Kramer, *Sumerian Mythology*, (Philadelphia, Pennsylvania: University of Pennsylvania Press, 1972), 76 -82.
9 Diane Wolkstein & Samuel Noah Kramer, "Her Stories and Hymns from Sumer," 95- 97, 103, 105-106.
10 Thorkild Jacobsen, 234-235.
11 Victor Clube & Bill Napier, *The Cosmic Winter*, (Cambridge, MA: Basil Blackwell Inc., 1990), 150-151, 153, 166-167, and 244.
12 Victor Clube & Bill Napier, *The Cosmic Serpent*, (London: Faber and Faber, 1982), 283.
13 Samuel Noah Kramer, translator, "Self-Laudatory . . ." 579.
14   Samuel Noah Kramer, translator, "Hymnal Prayer . . ." lines 12-13, 580.
15 Samuel Noah Kramer, translator, "Self-Laudatory . . ." and "Hymnal Prayer . . .", 578 579.
16 Samuel Noah Kramer, translator, "Self-Laudatory . . ." and "Hymnal Prayer . . .", 578-580.
17 Marcelo Gleiser, *The Prophet and the Astronomer: A Scientific Journey to the End of Time*, (New York: W. W. Norton & Co., 2001), 67.
18 Marcelo Gleiser, 25.

CHAPTER FOUR – OLD TESTAMENT CATASTROPHES
Part 1: Noah's Flood and Sodom and Gomorrah

1 Chaim Potok, *In The Beginning*, (New York: Alfred A. Knopf, 1975), 400.
2 Victor Clube & Bill Napier, *The Cosmic Winter*, (Cambridge, MA: Basil Blackwell Inc., 1990), 154 & 192.
3 Gerrit L. Verschuur, *Impact: The Threat of Comets and Asteroids*, (New York: Oxford University Press, 1996), 106.
4 Victor Clube & Bill Napier, 192-195, 244 and 284.
5 Bruce Feiler, *Walking the Bible- A Journey By Land Through the Five Books of Moses*, (New York: William Morrow/Harper Collins, 2001), 183.
6 Trevor Palmer, "Catastrophes: The Diluvial Evidence," Paper presented at the *Society of Interdisciplinary Studies*, (Easthamstead Park: Silver

Jubilee Conference, Sept. 19, 1999), 1.
7   Simon Rocker, "Catch a Falling Star," *Jewish Chronicle*, London, 6 March 1998.
8   Victor Clube & Bill Napier, *The Cosmic Winter*, 23 & 244.
9   Duncan Steel, *Rogue Asteroids and Doomsday Comets: The Search for the Million Megaton Menace That Threatens Life on Earth*, (New York: John Wiley and Sons, 1995), 16.
10 Gerrit Verschuur, 103.
11 *Voyage Through the Universe - Comets, Asteroids and Meteorites*, Editors (Alexandria, Virginia: Time-Life Books, 1990), 128.
12 Davis A. Young, "Theology and Natural Science," *Reformed Journal*, May 1988: 14-16, as quoted in Mark A. Noll, *The Scandal of the Evangelical Mind*, (Grand Rapids, Michigan: William Eerdmans Publisher, 1995), 231-232.
13 Davis A. Young, 14-16.
14 Davis A. Young, 14-16.
15 "Are We Drinking Comet Water?" *Astrobiology Magazine*, NASA, 24 March 2006, astrobio.net/news. Phil Beradelli, "Main Belt Comets May Have Been Source Of The Earth's Water", 23 March 2006, http://www.spacedaily.com/reports. "New Class Of Comets May Be The Source Of The Earth's Water," 23 March, 2001, http://www.science.nasa.gov/headlinesy2001. Dr. Louis A. Frank, *The Big Splash*, (Secaucas, NJ: Carol Publishing, 1980), 42-45 & 48. Damond Benningfield, "Where Do Comets Come From?" *Astronomy*, September 1990, 30. John C. Brandt & Robert D. Chapman, *Rendezvous In Space: The Science of Comets*, (New York: W.H. Freeman & Company, 1992), 145 &193.
16 Ker Than, "Comet Swarm Delivered the Earth's Oceans?" *National Geographic News*, 5 Aug 2009, http://www.nationalgeographic.com/news/.
17 Ker Than, as above.
18 Ker Than, as above.
19 Ker Than, as above.
20 Dr. Louis A. Frank, 48.
21 Marsha Walton, "Trembler big enough to 'vibrate the whole planet," CNN ONLINE, http://www.story.muddy.wave.jpg, 5/19/05.
22 In the KJV the Hebrew word kes-eel) (#3685 in *Strong's Concordance*) is translated "Orion" three times, but in Job 38:31 it can also be translated

as "brutish." Job 38:31 could read, "Canst thou ... loose the bands of the brutish ones (#3685 in *Strong's Concordance*)?" The phrase speaks of comets being loosed.

23 Douglas Myles, *The Great Waves*, (New York, NY: McGraw-Hill Book Co., 1985), 39-42.
24 Douglas Myles, 42.
25 Douglas Myles, 58.
26 E. A. Speiser translator, "The Epic of Gilgamesh," in *Ancient Near Eastern Texts - Relating to the Old Testament*, James Pritchard, editor (Princeton, NJ: Princeton University Press, 1969), 93-94.
27 *The Epic of Gilgamesh*, (Hamondsworth: Penguin Books, 1960), 128. Alexander Heidel, *The Gilgmagesh Epic and Old Testament Parallels*, (Chicago: University of Chicago Press, 1949), 84-86.
28 Dr. Sharad Master, "A Possible Holocene Impact Structure In the Al'Amarah Marshes, Near the Tigris-Euphrates Confluence, Southern Iraq," *Meteorites and Planetary Science*, Vol. 36, Supplement A 124, 2001. (Cambridge Conference).
29 Samuel Noah Kramer, *The Sumerians: Their History, Culture and Character*, (Chicago, Illinois: University of Chicago Press, 1963), 328.
30 Victor Clube & Bill Napier, *The Cosmic Winter*, 166.
31 John S. Lewis, *Rain of Iron and Ice: The Very Real Threat of Comet and Asteroid Bombardment*, (Reading, MA: Addison-Wesley - Helix Books, 1996), 186.
32 John S. Lewis, 152.
33 Sandra Blakeslee, "Ancient Crash, Epic Wave – Did Catastrophe fall from above in 2807 BC?" The *New York Times*, 14 Nov. 2006, with 16 Nov. 2006 correction: 1-4.
34 Sandra Blakeslee, F1 and F4.
35 Sandra Blakeslee, F4. Dallas Abbott, Edward Bryant and others, "Impact Craters As Sources of Megatsunami Generated Chevron Dunes," *Geological Society of America Abstracts*, paper No. 119-20, Philadelphia Annual Meeting, Vol. 38, No. 7, 22-25 Oct 2006: 229.
36 Sandra Blakeslee, "Mega-tsunamis," *New York Times*: 16 November 2006. Sandra Blakeslee, "Ancient Crash ...", 4.

37 Dallas Abbott, Edward Bryant and others, "Report of International Tsunami Expedition to Madagascar," 28 Aug. – 12 Sept. 2006, <http://www.ldeo.columbia.edu/users/menke/slides/madagascar06/report pd>.
38 Dallas Abbott, Edward Bryant and others, "Impact Craters." Anna Salleh, "Mega-tsunamis more common than we think," *Australian Broadcasting Corp. Science Online*, 16 Nov. 2006.
39 Dallas Abbott, Edward Bryant and others, as above.
40 Dallas Abbott, W. W. Masse, D. Breger, L. Burckle, "Burckle abyssal impact crater," *Black Rock Forest 2005 Research Symposium*, 20 June 2005, <http://www.blackrockforest.org/research/symposium2005.htm>. Editors of Time-Life Books, 127.
41 Sandra Blakeslee, 4. Dallas Abbott, and others, "Burckle Abyssal Impact Crater."
42 Sandra Blakeslee, 4. Dallas Abbott, and others, "Burckle Abyssal Impact Crater."
43 Sandra Blakeslee, 4. Dallas Abbott, and others, "Burckle Abyssal Impact Crater."
44 Dallas Abbott, and others, "Report of International Tsunami…" 4.
45 Sandra Blakeslee, 4.
46 Viacheslav Gusiakov, "Report of International Tsunami Expedition to Madagascar" 28 Aug–12 Sept 2006, *Novosibirsk Tsunami Laboratory*, Russia 630090, http://www.tsun.sscc.ru/proj.htm.
47 Sandra Blakeslee, 4.
48 Jerry L. Coffman and Carl A. Von-Hake editors, *Earthquake History of the United States*, Revised edition, US Dept. of Commerce, 1970: 109.
49 Dallas Abbott, and others, "Burckle Abyssal Impact Crater."
50 Sandra Blakeslee, F4.
51 E. A. Speiser translator, 94, lines 108-110.
52 John S. Lewis, 156.
53 Lauren Ristvet and Harvey Weiss, "The Habur Region in the Late Third and Early Second Millenium BC," in *The History of Archeology of Syria*, Vol. 1, Fig. 3:6, Winfried Orthmann, ed., Saabrucken: Saabrucken Verlog, 2005.
54 Harvey Weiss, "Late Third Millennium Abrupt Climate Change and Social Collapse in West Asia and Egypt," in *Third Millennium BC*

Climate Change and Old World Collapse; N. Dulfes, G. Kukla, H. Weiss (eds.), (Heidelberg/Berlin: Springer Verlag, 1997), 718-720.

55 Harvey Weiss, 718-720. Gerry Lemcke, and Michael Sturm, "Oxygen (18) Isotope and Trace Element Measurements as Proxy for the Reconstruction of Climate Change at Lake Van (Turkey): Preliminary Results," both in *Third Millennium BC* … 653-678.

56 Benny J. Peiser, Trevor Palmer, and Mark E. Bailey editors, "Natural Catastrophes During Bronze Age Civilizations: Archeological, geological astronomical and cultural perspectives" *British Archeological Reports* – 5728, Oxford: Archaeopress, 1998.

57 S. R. O'Brien, P. A. Mayewski et.al., "Complexity of Holocene Climates as Reconstructed from a Greenland Ice Core," *Science*, Vol. 270, Issue 5244, 22 Dec 1995, 1962-1964.

58 Lars G. Franzen, "Cosmic Activity as detected from raised bog stratigraphies in Northern Europe and Siberia," abstract from *Environmental Catastrophes and Recoveries in the Holocene Conference* 8/29-9/02/02, Suzanne Leroy and Iain Stewart Organizers, Department of Geography and Earth Sciences, Brunel University Uxbridge, United Kingdom, received 5/13/02.

59 Richard A. Kerr, "Sea-Floor Dust Shows Drought Felled Akkadian Empire," *Science*, Vol. 279, No. 5349, 16 Jan 1998, 325-326.

60 Victor Clube & Bill Napier, *The Cosmic Winter*, 166.

61 Lamb, H F et.al., *Nature*, 373 134 (1995).

62 Baillie, M.G.L. and M.A.R. Munro, "Irish Tree Rings, Santorini and Volcanic dust Veils," *Nature*, (24 March 1988), 332-345.

63 Zielinski, G.A., et.al., "Record of Volcanism since 7000 BC from GISPZ Greenland ice core and implication for the volcano-climate system," *Nature* 264 (1994), 948. D.A. Meese, et al., "The Accumulation record from the GISPZ core as an indicator of climate change throughout the Holocene," *Science*, 266 (1994), 1680.

64 Trevor Palmer, *Perilous Planet Earth: Catastrophes and Catastrophism Through the Ages*, (Cambridge: Cambridge University Press, 2003), 119.

65 Trevor Palmer, 119.

66 Dr. Sharad Master, "Umm al Binni Lake, a possible Holocene impact structure in the marshes of Southern Iraq. Geological evidence for its age, and implications for Bronze-age Mesopotamia," presented in

Environmental Catastrophes and Recoveries in the Holocene, Dept. of Geography and Earth Sciences, Brunel University, Uxbridge, UK, 29 Aug. – 2 Sept 2002, <http://www.atlasconferences. com/cgi-bin/abstract/caig-15>.

67 William Ryan and Walter Pitman, *Noah's Flood: The New Scientific Discoveries About the Event that Changed History*, (New York: Simon & Schuster, 1998).
68 Robert Cooke, "Noah's Flood," *Newsday*: 14 Jan 2003. "Noah's Flood Hypothesis May Not Hold Water," Rensselaer Polytechnic Institute Magazine, Sept 2002. Ali E. Aksu, et.al., "Persistent Holocene Outflow from the Black Sea to the Eastern Mediterranean Contradicts Noah's Flood Hypothesis," *GSA Today* (Geological Society of America), May 2002: 4-10. Also see Marine Geology, 15 Oct 2002.
69 Victor Clube & Bill Napier, *The Cosmic Winter*, 166.
70 John S. Lewis, 35.
71 John S. Lewis, 28 & 166.
72 *Virgil*. Aeneid, Vol. II, Book II and VII-XII. The Minor Poems. H. Rushton Fairclough, translator. London: Leob Classical Library, 1978.
73 Wil Milan, "The 1833 Leonid Meteor showers: A Frightening Flurry," *SPACE.com*, 14 Nov 2001. Victor Clube & Bill Napier, *The Cosmic Winter*, 286.
74 R. Sanderson, "The night it rained fire," *Griffith Observer*, Nov 1984: 2, quoted in Victor Clube & Bill Napier, *The Cosmic Winter*, 286.
75 John S. Lewis, 170.
76 John S. Lewis, 12-13.
77 Samuel Noah Kramer translator, "Hymnal Prayer of Enheduanna: The Adoration of Inanna in Ur," line 13, and "Lamentation Over the Destruction of Sumer and Ur," in *Ancient Near Eastern Texts…* 580 & 613. Piotr Michalowski, *The Lamentation Over the Destruction of Sumer and Ur*, (Winona Lake: Eisenbrauns Publisher, 1989), 41, 47, 61 & 79.
78 Josephus, *The Wars of the Jews*, Book 4/Chapter 8 lines 483-485, in *Josephus*, complete and Unabridged, (Peabody, Mass: Hendrickson Publishers, 1987), 687.
79 Timo Niroma, "Sodom and Gomorrah," Helsinki, Finland, http://www.eunet.fi./pp/tilmari/tilmari2.htm. Lemcke and Sturm, 653-678.

CHAPTER FIVE – OLD TESAMENT CATASTROPHES
Part 2: The Exodus, Joshua's Great Victory,
Debra and Barak's Victory and
the Blast that Killed 185,000 Assyrians

1   Victor Clube & Bill Napier, *The Cosmic Serpent: A Catastrophic View of Earth History*, (London: Faber and Faber, 1982), 217 & 222. The title of the book is based on the serpent-like configurations comet smoke trails can produce in the sky.
2   Victor Clube & Bill Napier, 220.
3   Victor Clube & Bill Napier, *The Cosmic Winter*, (Cambridge, MA :Basil Blackwell, Inc., 1990), 192-195.
4   Victor Clube & Bill Napier, *The Cosmic Winter*, 284.
5   Michael Lemonick, "Score One for the Bible - Fresh Clues support the story of Joshua at the walls of Jericho," *Time*, 5 March 1990, 59.
6   Bryant Wood, "Did the Israelites Conquer Jericho? A New Look at the Archeological Evidence," *Biblical Archeological Review*, March/April 1990, 53.
7   John S. Lewis, *Rain of Iron and Ice: The Very Real Threat of Comet and Asteroid Bombardment*, (Reading, MA: Addison-Wesley - Helix Books, 1996), 110.
8   Norman Wessels & Janet Hopon, *Biology*, (New York: Random House, Inc., B.I. 1988), 489. Gerard Tortora, Berdell Funke and Christine Case, *Microbiology - Third Edition*, (New York: Benjamic/Cummings Publishing Co., 1989), 293 & 296.
9   John Wilson translator, "The Admonitions of Ipuwer," in *Ancient Near Eastern Texts - Relating to the Old Testament*, James Pritchard, editor (Princeton, NJ: Princeton University Press, 1969), 441.
10 John Gribbon, *This Shaking Earth*, (New York: G. P. Putnam's, 1978), 15. James Penick Jr., *The New Madrid Earthquakes of 1811-1812*, (Columbia: University of Missouri, 1976), 62.
11 *Voyage Through the Universe - Comets, Asteroids and Meteorites*, Editors (Alexandria, Virginia: Time-Life Books, 1990), 19-20.
12 Nahum Glatzer translator, "A Treatise on the Life on Moses," *Philo*, Book 1, Section XXXII in *The Essential Philo*, (New York: Schocken Books, 1971), 218-219. F. H. Colson translator, *Philo - Volume VI*,

(Cambridge: Harvard University Press, 1935, reprinted in The Loeb Classic Library, 1984), 367.
13. John Noble Wilford, "Red Sea parting may have been wind caused, scientists say," *New York Times* copyright in the *Arizona Daily Star*, 15 March 1992: 11A.
14. As above.
15. As above.
16. As above.
17. John Garstang, *The Foundations of Bible History*, (New York: Richard R. Smith Inc., 1931), 137.
18. Victor Clube & Bill Napier, *The Cosmic Serpent*, 220.
19. Victor Clube & Bill Napier, *The Cosmic Serpent*, 154 & 28. Duncan Steel, *Rogue Asteroids and Doomsday Comets: The Search for the Million Megaton Menace That Threatens Life on Earth*, (New York: John Wiley and Sons, 1995), 133-136 & 149-150. Gerrit L. Verschuur, *Impact: The Threat of Comets and Asteroids*, (New York: Oxford University Press, 1996), 134-135.
20. Duncan Steel, 149-154. Both dates are plus or minus 500 years.
21. John & Mary Gribbin, *Fire on Earth*, (New York: St. Martin's Griffin, 1996), 136-140.
22. Duncan Steel, 152.
23. Victor Clube & Bill Napier, *The Cosmic Serpent*, 217, 238, & 255-256. Victor Clube & Bill Napier, *The Cosmic Winter*, 40, 55 & 185-195. A. H. Gardiner translator, *Admonitions of an Egyptian Sage from a hieratic papyrus in Leiden*, 1909.
24. John Wilson translator, "The Admonitions of Ipuwer," in *Ancient Near Eastern Texts…*, 441.
25. H. Rackham translator, *Natural History*, Pliny Loeb Classical Library, (Cambridge: Harvard University Press, 1938), Book 2, Section 91.
26. Victor Clube & Bill Napier, *The Cosmic Serpent*, 221.
27. Victor Clube & Bill Napier, *The Cosmic Serpent*, 222.
28. Victor Clube & Bill Napier, *The Cosmic Winter*, 270-271.
29. Marie-Agnes Courty, "Causes and Effects of the 2350 BC Middle East Anomaly Evidenced by Micro-debris Fallout, Surface Combustion and Soil Explosion." Benny J. Peiser, "Comparative Stratigraphy of Late Holocene Sediments and Destruction Layers Around the World: Geological, Climatological and Archaeological Evidence and

CITATIONS

Methodological Problems," both in *Natural Catastrophes during Bronze Age Civilization: Archeological, geological, astronomical and cultural perspectives,* edited by Benny J. Peiser, Trevor Palmer and Mark E. Bailey, Oxford England: Archaeopress Gordon House, 1998. Also see http://www.knowledge.co.uk/sis/abstract/peiser.htm; and http://www.knowledge.co.uk/sis/abstract/courty.htm.

30 John S. Lewis, 168.
31 A. K. Grayson translator, "The Epic of Gilgamesh," Tablet VI, lines 122-126, in *Ancient Near Eastern Texts…*, 505. Piotr Michalowski, *The Lamentation Over the Destruction of Sumer and Ur,* (Winona Lake, IN: Eisenbrauns Publisher, 1989), 41, 47, 61 & 79. Also see Samuel Noah Kramer translator, "Lamentation Over the Destruction of Sumer and Ur," in *Ancient Near Eastern Texts…*, 613.
32 Victor Clube & Bill Napier, *The Cosmic Winter,* 231.
33 "Moon Blast! New Insight into the Birth of our Satellite," *Time*, August 27, 2001, 56. Ben Harder, "Was Moon Born From Planet's Crash into Earth," *National Geographic News,* 20 Aug 2001, http://www.nationalgeographic.com/news/2001/08/0820–moonimpact.html. Robert Roy Britt, "24 Hours of Chaos: The Day the Moon was Made," 15 Aug 2001, http://www.space.com/solarsystem /oon_making_0108-1.html.
34 Victor Clube & Bill Napier, *The Cosmic Winter,* 159 & 161-162. "An Unseen Meteor Storm," News Notes, *Sky & Telescope*, May 1990, 476.
35 Mansinha, L. and Smylie, "The Rotation of the Earth," *Scientific American*, Dec 1971, 80-89. "Earthquake and the Earth's Wobble," *Science*, 13 Sept 1967, 1127.
36 Marsha Walton, "Trembler big enough to 'vibrate the whole planet,'" *CNN ONLINE*, http://www.story.muddy.wave.jpg, 5/19/05. Lely T. Djuhari, "Asian Quakes' Tsunami Kill More Than 7,000," Jakarta, Indonesia AP, 26 Dec 2004.
37 "Scientists: Quakes may have made Earth Wobble," *CNN Online*, http://www.12/29/quake.wobble.reut/index.html, 12.29.04.
38 John S. Lewis, 168-169.
39 Victor Clube & Bill Napier, *The Cosmic Winter,* 39 & 40.
40 Victor Clube & Bill Napier, *The Cosmic Serpent*, 271.
39 Nigel Calder, *Comets: Speculation and Discovery*, (New York: Dover Publishing, 1994), 92-93.

40 Alan Mac Robert, "The Messier Crater Pair," *Sky & Telescope*, April 1992, 424-425. Peter Schultz & J. Kelly Beatty, "Teardrops on the Pampas," *Sky and Telescope*, April 1991, 391-392. Ben Harder, "What Caused Argentina's Mystery Craters?," *National Geographic News*, 9 May 2002, http://www.nationalgeographic.com/news.

41 Samuel Noah Kramer translator, "A Sumerian Lamentation: Lamentation Over the Destruction of Ur," in *Ancient Near Eastern Texts...*, 458-463.

42 Samuel Noah Kramer 459.

CHAPTER SIX - THE FIRST FOUR TRUMPETS AND VIALS
Of the Book of Revelation

1 John S. Lewis, *Rain of Iron and Ice: The Very Real Threat of Comet and Asteroid Bombardment*, (Reading, MA: Addison-Wesley - Helix Books, 1996), 12.

2 S. V. M. Clube, "The Dynamics of Armageddon," (Dept. of Astrophysics, University of Oxford), *Speculations of Science and Technology*, Vol. 11, No. 4, 1989, 263.

3 Victor Clube and Bill Napier, *The Cosmic Winter*, (Cambridge, MA :Basil Blackwell, Inc., 1990), 272.

4 John & Mary Gribbin, *Fire on Earth*, (New York: St. Martin's Griffin, 1996), 32-33 & 133. "A Single Intruder, A Chain of Craters?" *National Geographic*, Geographical Section, Oct. 1998.

5 Victor Clube & Bill Napier, *The Cosmic Winter*, 272.

6 Victor Clube & Bill Napier, *The Cosmic Serpent: A CatastrophiView of Earth History*, (London: Faber and Faber, 1982), 212

7 Victor Clube & Bill Napier, *The Cosmic Serpent*, 213.

8 Victor Clube & Bill Napier, *The Cosmic Winter*, 194-195.

9 Gerrit L. Verschuur, *Impact: The Threat of Comets and Asteroids*, (New York: Oxford University Press, 1996), 103-104. In regard to *Revelation 6:13* Verschuur unfortunately passed over the added astronomical information in the next verse, *Revelation 6:14*; and in very similar passages in *Isaiah 34:4* and *Matthew 24:29*.

10 John S. Lewis, 12.

11 As noted earlier, during Biblical times the word "blood" was sometimes

used as a metaphor for "death." See Chapter Five and the discussion of how Exodus 7:20-21 says "the waters that were in the river (Nile) were turned to blood (death). And the fish that were in the river died."

12 A. K. Grayson translator, "The Epic of Gilgamesh," Tablet VI, lines 122-126, in *Ancient Near Eastern Texts - Relating to the Old Testament*, James Pritchard, editor (Princeton, NJ: Princeton University Press, 1969), 505. Piotr Michalowski, *The Lamentation Over the Destruction of Sumer and Ur,* (Winona Lake, IN: Eisenbrauns Publisher, 1989), 41, 47, 61 & 79. Samuel Noah Kramer, translator "Lamentation Over the Destruction of Sumer and Ur," in *Ancient Near Eastern Texts…*, 613.
13 "Were the Dinosaurs Roasted to Death?" *Newsweek*, 14 March 1988: 66. Victor Clube & Bill Napier, *The Cosmic Winter,* 227. John & Mary Gribbin, 35.
14 Clark Chapman & David Morrison, *Cosmic Catastrophes*, (New York: Plenum Publishing, 1989), 93.
15 John & Mary Gribbin, 51.
16 John & Mary Gribbin, 45.
17 "New Clovis – Age Comet Impact Theory," University of Oregon, Asteroid and Comet Mission News, Science and Technology, *Newswise*, 21 May 2007, http://www.newswise.com/p/articles/view/ 530208.
18 "New Clovis – Age Comet Impact Theory," Iron and Ice, 23 May 2007, http://www.spacedailly.com/reports/New_Clovis_Age. "Comet Cools Clovis," *Astrobio.net*, from a National Science Foundation News release, NASA, 23 Aug 2007.
19 Michael Abrams, "Stone-Age Asteroid May Have Wiped Out Life in America," 14 Jan 2008, http://www.discovermagazine.com/2008/jan/stone-age. Heather Pringle, "Did a Comet Wipe out Prehistoric Americans?" New Scientist.com news service, 22 May 2007.
20 The word "blood" is being used as a metaphor for "death" in this passage.
21 Clark Chapman & David Morrison, 93.
22 *Voyage Through the Universe - Comets, Asteroids and Meteorites,* Editors (Alexandria, Virginia: Time-Life Books, 1990), 121-133.
23 Editors of Time-Life, 125-127.
24 Editors of Time-Life, 128.
25 Editors of Time-Life, 128.
26 Gerrit Verschuur, 122.

CITATIONS

27 A Single Intruder, A Chain of Craters?" *National Geographic,* Geographical Section, October 1998.
28 Donald K. Yeomans, *Comets: A Chronological History,* (New York: John Wiley & Sons, 1991), 261.
29 Nigel Calder, *Comets: Speculation and Discovery,* (New York: Dover Publishing, 1994), 25-26 & 92-93.
30 Nigel Calder, 93. John C. Brandt & Robert D. Chapman, *Rendezvous in Space: The Science of Comets,* (New York: W. H. Freeman & Co., 1992), 107 & 111.
31 Victor Clube & Bill Napier, *The Cosmic Winter,* 291.
32 John S. Lewis, G. Watkins, H. Hartman, & R. G. Prinn, "Chemical consequences of major impact events on Earth," *Geological Society of America,* Special Paper 190, 1982. John S. Lewis, 110.
33 Encarta 98 - under Lake Baikal.
34 Encarta 98 - under Great Lakes.
35 Encarta 98 – Great Lakes.
36 Jack J. Szpytman, *The Origin and Formation of the Great Lakes Region- the Global Deluge and Ice Age,* Grosse Point Woods, Michigan, 1980, copy #98 http://www. universal-flood.net/.
37 John & Mary Gribbin, 44 & 46. Nigel Calder, *Comets: Speculation and Discovery,* (New York: Dover Publishing, 1994), 125.
38 Victor Clube & Bill Napier, *The Cosmic Winter,* 157.
39 William J. Broad, "Meteoroids Hit Atmosphere in Atomic–Size Blasts," 25 Jan 1994, *New York Times,* NY Times.com. E. Tagliaferri, A. Erlich, et al, "Detection of Meteoroid Impacts by Optical Sensors in Earth Orbit," in *Hazards Due to Comets and Asteroids,* T. Gehrels ed., University of Arizona Press, Tucson, AZ 1995, Section II. Also see *CommSpacTrans* Sec3.4. html-nasa.gov.
40 As above.
41 As above.
42 As above.
43 As above.
44 As above.
45 John & Mary Gribbon, 45. Victor Clube & Bill Napier, *The Cosmic Winter,* 158.
46 Editors of Time-Life Books, 121-133.
47 John S. Lewis, 12.

48 "As in the days when you came out of Egypt, I will show them my wonders" (Micah 7:15 NIV).

CHAPTER SEVEN - THE THREE WOES
The Fifth, Sixth and Seventh Trumpets And Vials

1 Victor Clube and Bill Napier, *The Cosmic Winter,* (Cambridge, MA: Basil Blackwell, Inc., 1990), 272.
2 John S. Lewis, *Rain of Iron and Ice: The Very Real Threat of Comet and Asteroid Bombardment*, (Reading, MA: Addison-Wesley - Helix Books, 1996), 12.
3 Victor Clube & Bill Napier, *The Cosmic Serpent: A Catastrophic View of Earth History,* (London: Faber and Faber, 1982),156.
4 *The Works of Josephus*, translated by William Whiston, (Peabody, Massachusetts: Hendrickson Publishing, 1987), Book 10, Chapter 1, line 2: 267 & 891.
5 Samuel Noah Kramer translator, "The Curse of Agade," in *Ancient Near Eastern Texts - Relating to the Old Testament*, James Pritchard, editor (Princeton, NJ: Princeton University Press, 1969), 646-652. Harvey Weiss, "Late Third Millennium Abrupt Climate Change and Social Collapse in West Asia and Egypt," in *Third Millennium BC Climate Change and Old World Collapse*, Dalfes, Kukla, Weiss (eds.), Springer Verlag, Heidelberg/Berlin, 1997, 719. Robert Roy Britt, "Comets, Meteors and Myth: New Evidence for Toppled Civilizations and Biblical Tales," <http://www.space.com>, 13 Nov. 2001. "Collapse of Early Bronze Age Civilization: Has the Smoking Gun Been Found?" *CC Net Special – The Cambridge Conference,* Liverpool Moores University, 16 April 2002 <http://www.abob.libs.uga.edu/bobk/ccc/cc110501.html.>
6 Ann Gibbons, "How the Akkadians' Empire Was Hung Out to Dry," *Science*, Vol. 261, 20 Aug 1993, 985.
7 H. Weiss, M. A. Courty, W. Wetterstrom, & others, "The Genesis and Collapse of Third Millenium North Mesopotamia Civilization," *Science*, Vol. 261, 20 Aug 1993, 996.
8 Samuel Noah Kramer, "The Curse of Agade …" 649, lines 120, 171, 174.
9 Robert Roy Britt, "Comets, Meteors and Myth: New Evidence for

Toppled Civilizations and Biblical Tales," <http://www.space.com>, 13 Nov. 2001.

10 Harvey Weiss, "Late Third Millennium Abrupt Climate Change and Social Collapse in West Asia and Egypt," in *Third Millennium BC Climate Change and Old World Collapse,* Dalfes, Kukla, Weiss (eds.), Springer Verlag, Heidelberg/Berlin, 1997.

11 Victor Clube & Bill Napier, *The Cosmic Winter,* 10, 141, & 153.

12 Anne Denogean, "UA helps study big Moon of Jupiter," *Tucson Citizen,* 16 July 1998, 1C & 4C.

13 Jack J. Szpytman, *The Origin and Formation of the Great Lakes Region- the Global Deluge and Ice Age,* Grosse Point Woods, Michigan, 1980, copy #98 http://www. universal-flood.net/.

14 Gerrit L. Verschuur, *Impact: The Threat of Comets and Asteroids,* (New York: Oxford University Press, 1996), 183 &189.

15 John & Mary Gribbin, *Fire on Earth,* (New York: St. Martin's Griffin, 1996), 132. Duncan Steel, *Rogue Asteroids and Doomsday Comets: The Search for the Million Megaton Menace That Threatens Life on Earth,* (New York: John Wiley and Sons, 1995), 257.

16 John S. Lewis, 11.

17 M. E. Bailey, S. V. M. Clube, W. M. Napier, *The Origins of Comets,* (Oxford: Pergamon Press, 1990), 4 & 430. Victor Clube & Bill Napier, *The Cosmic Winter,* 140.

18 Donald K. Yeomans, *Comets: A Chronological History,* (New York: John Wiley & Sons, 1991), 253.

19 "Halley's Gas and Dust Jets," *Sky and Telescope,* Nov 1988: 456.

20 *Voyage Through the Universe - Comets, Asteroids and Meteorites,* Editors (Alexandria, Virginia: Time-Life Books, 1990), 31. Charles Morris, "Comets Encke and Levy: Something Old and Something New," *Astronomy,* Sept 1990, 67.

21 John S. Lewis, 183-205.

22 Harry Thomas Frank editor, *Atlas of the Bible Lands,* (Maplewood, NJ: Hammond Inc., 1984), B-20.

23 Jerry L. Coffman & Carl von Hake, editors, *Earthquake History of the United States,* Revised edition, U.S. Dept. of Commerce, No. 41-1, 1977, 43-44. James Penick Jr., *The New Madrid Earthquakes of 1811-1812,* (Columbia: University of Missouri, 1976), 62.

24 Duncan Steel, 1.

25 "Kaboom! Ancient impacts scarred the Moon to its core, may have created 'man in the moon,'" 2/10/06, http://www.theallineed.com/astronomy/06022201.htm. Article is also in *Ohio State Research*, Ohio State University, Columbus, Ohio, 2/8/06, www.osuiedu/units/research and also in *Science Daily*, 10 Feb. 2006, http://www.sciencedaily.com/releases/2006 /02/060210091105.htm.
26 "The Impact Through the Moon," *Softpedia News*-Science, http://www.news. softpedia.com/news/.
27 As above, "Kaboom! Ancient impacts scarred the Moon to its core . . . .," 2/10/06.
Science Daily, 10 Feb 2006, as above.
28 Laramie V. Potts and Ralph R. B. Von Frese, "Impact-induced mass flow effects on lunar shape and the elevation dependence of nearside maria with longtitude," *Physics of the Earth and Planetary Interiors*, Vol. 153, Nov 2005, 165-174.
29 Sridhar Nadamuni, "For 'Man in the Moon,' a bump in the head," *The Boston Globe*, 13 Feb 2006, http://www/boston.com/news/ globehealth_science/articles/2006/02.
30 "Big Bang In Antarctica – Killer Crater Found Under Ice," Ohio State Research, Columbus, Ohio, 06/01/06. Also in "Possible Extinction Crater Found Under Antarctica," Staff Writers, http://www.spacedaily.com/reports 6/03/06.
31 "Antarctic Smackdown," Poh Si Teng, *The Columbus Dispatch*, Columbus, Ohio, 6/27/06.
32 Larry O'Hanlon, "Siberian Traps," http://www.dsc.discovery.com/convergence/ supervolcano/others/ others_07.html.
33 "Magma's Makeup Yields New Clues to Catastrophic Eruptions," from *University of Rochester Research*, 24 July 1998, http://wwv.sciencedaily.com/releases /1998/07/980724080426.htm. Also see "Massive Volcanic Eruptions in Siberia Linked to Largest Extinction," *Science*, 7/12/91, http://www.rochester.edu/pr/ releases/ear/basu2.htm.
34 "Antarctic Smackdown," Pho Si Teng, *The Columbus Dispatch*, Columbus, Ohio, 06/27/06.
35 "Caloris Basin," *Wikipedia*, 8 Oct 2006. W. S. Keifer & B. C. Murray, "The Formation of Mercury's smooth plains," *Icarus*, International Journal of Solar System Studies, Cornell University, Ithaca, NY, 1987, v. 72, 477-491.

36 James Marusek, "What Would Happen if a Massive Oort Cloud Comet Strikes Earth," *Cambridge Conference*, ccNet, Issue 51/2003, and http://www.personals. galaxyinternet.net/tunga/Venus.htm 6/13/2003. Sean C. Solomon, "The Resurfacing controversy for Venus: An overview and mechanistic perspective," in MIT *Tectonic History of the Terrestrial Planets*, 03, 1993, http://www.adsabs.Harvard.edu/ abs/1993thtp. reptu...s. *Wikipedia*, "Geology of Venus," http://www.en.wikipedia.org/wiki/venus. R. G. Strom and others, "The global resurfacing of Venus," *Journal of Geophysical Research*, Vol. 99, 10, 899-10, 926, 1995. David H. Grinspoon, "Venus Unveiled: A Great Volcanic Flood Must Have Resurfaced Earth's Sister World Some 600 million years Ago," *Astronomy*, May 1957. Ellen Foxxe editor, *Size, Composition, and Surface Features of the Planets Orbiting the Sun*, the Roses Publishing Group, NY, NY, 2005, 47-54.
37 *Voyage Through the Universe - Comets, Asteroids and Meteorites*, Editors (Alexandria, Virginia: Time-Life Books, 1990), 125.
38 "Ancient impact may explain Mars mystery," *CNN.com*/2008/Tech/space/, 25 June 2008. Alicia Chang, "Scientists Think Big Object Whacked Mars," *AOL.com/*Science News, 25 June 2008. J.R. Minkel, "Giant Asteroid Flattened Half of Mars, Studies Suggest Hemisphere-size Crater," *Scientific American*, 25 June 2008, http://www.sciam.com/articles.cfm?d=giant-asteroid-flattened & prin .... Alan Fischer, "Study: Mars Smacked by huge asteroid," *Tucson Citizen*, 26 June 2008, 1.
39 As above.
40 Jeffrey C. Andrews-Hanna & Maria T. Zuber, et al, "The Borealis basin and the origin of the martian crustal dichotomy," (Massachusetts Institute of Technology), *Nature*, 26 June 2008, 1212. F. Nimmo & S. D. Hart, et al, "Implications of an Impact Origin for the Martian hemispheric dichotomy," *Nature*, 26 June 2008, 1220-1223. Walter S. Kiefer, "Forming the martian great divide," *Nature*, 26 June 2008, 1191-1192.
41 Margarita M. Marinova & Oded Aharonson, et al, "Mega-Impact formation of the Mars hemispheric dichotomy," *Nature*, 26 June 2008, 1216.
42 "Ancient impact may explain Mars mystery," *CNN.com*/2008/Tech/space/, 25 June 2008. Alicia Chang, "Scientists Think Big Object

Whacked Mars," *AOL.com/Science News*, 25 June 2008.
43 Margarita M. Marinova & Oded Aharonson, et al, 1217-1218.
44 Margarita M. Marinova & Oded Aharonson, et al, 1218.
45 Margarita M. Marinova & Oded Aharonson, et al, 1218.
46 John & Mary Gribbin, 68-69.
47 "Moon Blast! New Insight into the Birth of our Satellite," *Time*, 27 Aug 2001, 56. Ben Harder, "Was Moon Born From Planet's Crash into Earth," *National Geographic News*, 20 Aug 2001, http://www.nationalgeographic.com/news/2001/08/0820-moonimpact.html. Robert Roy Britt, "24 Hours of Chaos: The Day the Moon was Made," 15 Aug 2001, http://www.*space.com*/solarsystem/ oon_making_0108-1.html.
48 As above.
49 Larry Copenhauer, "UA gets $1.2m. to aid in asteroid mission," *Tucson Citizen*, 31 Oct 2006, 4a.
50 Robert Roy Britt, "Earth Deemed Older," *Space.com*, 28 Aug 2002, http://www.space.com/scienceastronomy/older_earth-020828html
51 Both the Greek and Hebrew words for "heaven" (Greek #3772 and Hebrew #8064 in *Strong's Concordance*) can be translated as "air" or "sky." Revelation 21:1 makes references to "a new heaven and a new Earth." Isaiah 65:17 says, "For behold, I create new heavens and a new Earth . . . ." Isaiah 66:22 makes references to "the new heavens and the new Earth" God is to make. Also note how II Peter 3:12 says that "the heavens (air or atmosphere) being on fire shall be dissolved and the elements shall melt with fervent heat." Relating the atmosphere being on fire to the elements melting is insightful, since the elements of nitrogen (77%) and oxygen (21%) make up 98% of the atmosphere.

CHAPTER EIGHT - THE TALE OF THE THE TOWER OF BABYLON

1 Gerrit L. Verschuur, *Impact: The Threat of Comets and Asteroids*, (New York: Oxford University Press, 1996), 97.
2 Gerrit L. Verschuur, 101.
3 Samuel Noah Kramer translator, "The Curse of Agade," in *Ancient Near Eastern Texts - Relating to the Old Testament*, James Pritchard, editor

(Princeton, NJ: Princeton University Press, 1969), 646.
4   H. Weiss, M. A. Courty, W. Wetterstrom, & others, "The Genesis and Collapse of Third Millenium North Mesopotamia Civilization," *Science*, Vol. 261, 20 Aug 1993, 996.
5   Chaim Potok, *In The Beginning*, (New York: Alfred A. Knopf, 1975), 400.
6   T. Caldwelll, J. Oswalt & J. Sheehan, *An Akkadian Grammar*, (Marquette University Press, 1974), 11, 4, & 12 in Part II.
7   Charles F. Pfeifer & Everett F. Harrison editors, *The Wycliffe Bible Commentary*, (Chicago: Moody Bible Institute, Moody Press, 1962), 16.
8   *The Baker Encyclopedia of the Bible*, (Grand Rapids, Michigan: Baker Book House, 1988), 243.
9   A. Leo Oppenheim translator, The "Sargon Chronicle," Texts from the Beginnings to the First Dynasty of Babylon – Historiographic Documents, in *Ancient Near Eastern Texts…*, 266.
10 Samuel Noah Kramer, *The Sumerians: Their History, Culture and Character*, (Chicago, Illinois: University of Chicago Press, 1963), 6.
11 Samuel Noah Kramer, *The Sumerians*, 42-43. T. A. Caldwell, J. N. Oswalt, & J. F. Sheehan, *An Akkadian Grammar*, (Milwaukee, Wisconsin: Marquette University Press, 1975), 1.
12 Samuel Noah Kramer, *The Sumerians*, 288-289.
13 "Nimrod" *New Bible Dictionary* Second ed., (Wheaton, Illinois: Tyndale House Publisher, 1982), 836.
14 Samuel Noah Kramer, "The Curse of Agade… 646.
15 Gerrit Verschuur, 101.
16 Samuel Noah Kramer, "The Curse of Agade… 648.
17 Samuel Noah Kramer, "The Curse of Agade… 646-652. Harvey Weiss, "Late Third Millennium Abrupt Climate Change and Social Collapse in West Asia and Egypt," in *Third Millennium BC Climate Change and Old World Collapse*, Dalfes, Kukla, Weiss (eds.), Springer Verlag, Heidelberg/Berlin, 1997, 719.

CHAPTER NINE- ARMAGEDDON, THE SEALS, AND A ROCK

1. Priit J. Veslind, "In Search of Vikings," *National Geographic*, Vol. 197, No. 5, May 2000, 18.
2. Otto Friedrich, "The Soviet Empire – A Land Great and Rich in Search of Order," *Time*, 12 March 1990, 46. Andrew Sherratt editor, *The Cambridge Encyclopedia of Archaeology*, (New York: Crown Publishing, 1980), 317. Edwin Yamauchi, *Foes From the Northern Frontier*, (Grand Rapids, Michigan: Baker Books, 1982), 20. "Historical Map of the Soviet Union," *National Geographic*, March 1990. "In Search of Vikings," *National Geographic*, May 2000, 18-19.
3. "In Search of Vikings," 18-19.
4. *Viking Answer Lady*, "Risala: Ibn Fadlan's Account of the Rus," http://www.vikinganswerlady.com. Based on the translation of a composite of surviving manuscript versions, passage 80 and 82, part 2.
5. *Viking Answer Lady*, http://www.vikinganswerlady.com. For the full text and commentary of Ibn Fadlan's account of the Rus see: H. M. Smyser, "Ibn Fadlan's Account of the Rus with some Commentary," Franciplegius: *Medieval and Linguistic Studies – Honor of Francis Peabody Margoun, Jr.*, eds. Jess B. Bessinger Jr. & Robert P. Creed, editors, (New York: New York University Press, 1965), 92-119. James E. Montgomery, "Ibn Fadlan and the Rusiyyah," *Journal of Arabia & Islamic studies*, Vol. 3 (2000), 1-15. ISSN:0806-198x.
6. Paul Boyer, *When Time Shall Be No More: Prophecy Belief in Modern American Culture* (Cambridge, Mass: Harvard University Press, 1992) 162.
7. Paul Boyer, 162.
8. Edwin Yamauchi, 20-21.
9. Edwin Yamauchi, 20-21.
10. Paul Boyer, 160.
11. Daniel I. Block, *The Book of Ezekiel: Chapters 25-48*, New International Commentary on the Old Testament series, (Grand Rapids, Michigan: Eerdmans, 1998), 434.
12. Daniel David Luckenbill, *Ancient Records of Assyria and Babylonia*, Vol. 1 and Vol. II (New York: University of Chicago Press, 1926, Greenwood

Press Reprint, 1968).
13　Daniel David Luckenbill, Vol. I, p. 144; & Vol. II, p. 95. Walther Zimmerli, *Ezekiel 2*, (Philadelphia, Pennsylvania: Fortress Press, 1983), 66. Regarding Togarmah trading horses (Ezekiel 27:14) see above Vol. II, 297, 325, & 352.
14　Kristian Kristiansen, *Europe Before History*, (Cambridge: Cambridge University Press, 1998), 194.
15　Charles F. Pfeifer & Howard F. Voss, *The Wycliffe Historical Geography of Bible Lands*, (Chicago, Illinois: Moody Press, 1967), 329.
16　By 540 BC Lydia was incorporated into the Persian empire.
17　Daniel David Luckenbill, Vol. II, section 784, 296 & section 904, 351.
18　Grahame Clark, *World Prehistory*, (Cambridge: Cambridge University Press, 1969), 167.
19　Charles Pfeifer and Howard Voss, 498-499. David Greene translator, *The History*, Herodotus, (Chicago, Illinois: University of Chicago Press, 1987), 1.94, 78. Andrew Sherratt editor, 230.
20　Plutarch, *Life of Romulus*, 75 AD, part II. Published by printed for Jacob Tonson, 1711. Princeton, New Jersey: Princeton University, digitized July 14, 2008.
21　Mary T. Boatwright, Richard, J.A. Talbert, & Daniel J. Gargola, *The Romans: from Village to Empire*, (New York: Oxford University Press, 2004), 39, Box 2.1.
22　Nicholas Wade, "DNA Boosts Herodotus' Account of Etruscans as Migrants to Italy," *New York Times.com*. 3 April 2007.
23　Nicholas Wade.
24　"DNA Shows Etruscans came from Anatolia," *Turkish Daily News*, 9 Feb 2007, feature story.
25　David Greene translator, 1.94, 78.
26　Dr. Ajmone–Marsan and others, "The Proceedings of the Royal Society" – online report, February 2007, as quoted in *The New York Times*, 3 April 2007, 3.
27　Dr. Ajmone–Marsan and others.
28　Francis Brown, *The New Brown-Driver-Briggs-Gesenius Hebrew and English Lexicon*, (Peabody, Mass: Hendrickson Publishers, 1979), 356.
29　Aryeh E. Shimron & Barbu Lang, "Preliminary Geological Map of the SE Hermon Range," *New Geological Data and K-Ar Geochronology of the Magmatic Rocks on the Southeastern Flanks of Mt. Hermon*, Geological

Survey of Israel, Report No. GSI/41/88, Jerusalem, Dec 1988, Fig. 1-2, 7.

30 The Seven Seals – *Revelation 6:1-17,* & *8:1-5.* Seven Trumpets – *Revelation 8:6-9:21, 11:15-19.* The Seven Vials – *Revelation 16:1-21.*

31 Those who think the battle of Ezekiel 38/39 is not the same as the battle of Armageddon and that the battle of Ezekiel 38/39 occurs at the ***beginning*** or ***middle*** of the tribulation period have ***not*** been able to give passages before Ezekiel where God has spoken of the battle of this day as called for by Ezekiel 39:8.

32 "Migratory birds know no boundaries", *International Center for the Study of Bird Migration* http://www.birds.org.il/show_item.asp?levelId=457. Hadoram Shirihai, *The Birds of Israel*, Poyser Publishing, 73102.

33 *Volcano*, Editors (Alexandria, Virginia: Time-Life Books, 1982), 142.

34 Editors of Time-Life, 163.

35 Editors of Time-Life, 142.

36 Editors of Time-Life, 163.

37 "Early Cretaceous Magmatism Along the SE Flank of the Hermon Range," *Israel Geologic Society, Annual Meeting, Ramot, Guidebook for Excursions*, 1989, 1-5. Figure 1 on p. 2 contains a geologic map of the area. Aryeh E. Shimron, "The Early Cretaceous 'Nimrod' Volcanic Episode on Mount Hermon," *Israel Geologic Society, Annual Meeting, En Boqeq, Abstract*, 1988, 107-108. The Mount St. Helens' crater is 2.4 miles long and 1.2 miles wide in size.

38 Aryeh E. Shimron, "The Early Cretaceous 'Nimrod' Volcanic episode on Mount Hermon," 1988 and "Early Cretaceous Magmatism Along the SE Flank of the Hermon Range," *New Geological Data and K-Ar Geochronology of the Magmatic Rocks on the SE Flank of Mount Hermon*, Geological Survey of Israel; Report No. GSI/41/88, Ministry of Energy and Infrastructure, Jerusalem, Dec 1988, 4 & 6 and *Israel Geologic Society, Annual Meeting, Ramot, Guidebook for Excursians*, 1989 1-5, & 10. Mishirav, "Boreholes for the Location of Scoria in Har Hermon" – 1987.

39 *New Geological Data and K-Ar Geochronology…*, 4.

40 Editors of Time-Life Books, 19 & 32-36.

41 *New Bible Dictionary*, (Lercester England: Inter-Varsity Press, 1982), 478.

42 John Gribbon, *This Shaking Earth*, (New York: G. P. Putnam's, 1978), 106-107.
43 Naomi Lubick, "Volcanic Accomplice," *Scientific American.com*, 17 March 2001. Dallas Abbott & Ann Isley, "Extraterrestrial Influences on Mantle Plume Activity," *Earth Science and Planetary Letters*, Vol. 205, Dec 2002, 53-62.
44 Naomi Lubick and Dallas Abbott & Ann Isley.
45 "Columbia University Research Finds Correlation Between Meteorite and Comet Impacts and an Increase in Volcanic Activity Development," *Earth Institute News*, http://wwwearthinstitute.columbia.edu/news, 17 Jan 2003. Dallas Abbot & Ann Isley, 53-62.
46 Robert Roy Britt, "How Impacts Can Trigger Volcanoes," *Space.com*, 4 Feb 2003, http://www.space.com/scienceastronomy/asteroids_volcanoes_030204-1.html.
47 Robert Roy Britt, "How Impacts Can Trigger Volcanoes."
48 Kate Ravilious, "Earth's Volcanism Linked to Meteorite Impacts?" *New Scientist*, Issue 2373, 13 Dec 2002, http://www.Newscientist.com. Adrian Jones, David Price & others, "Impact Induced melting and the development of large igneous provinces," *Earth and Planetary Science Letters*, Vol. 202, 551. Robert Roy Britt, "How Impacts Can Trigger Volcanoes."
49 Robert Roy Britt, "How Impacts Can Trigger Volcanoes."
50 *Leviticus 13:57; Deuteronomy 9:3; Isaiah 6:5-7; 66:15-16, Matthew 25:41, II Thessalonians 1:8* and *II Peter 3:10-12*.

CHAPTER TEN - THE SIGN OF THE SON OF MAN IN HEAVEN

1 Donald K. Yeomans, *Comets: A Chronological History*, (New York: John Wiley & Sons, 1991), 354.
2 Donald K. Yeomans, 331. Armand Delsemme, "Whence Come Comets?" *Sky and Telescope*, March 1989, 262-263.
3 Donald K. Yeomans, 331 and Armand Delsemme, 262-263.
4 Armand Delsemme, 263.
5 M. Davis, P. Hunt & R. A. Muller, "Extinction of Species by Periodic Comet Showers," *Nature*, 19 April 1984, 308, 715-717. D. P. Whitemire

& A. A. Jackson "Are Periodic Mass Extinctions Driven By a Distant Solar Companion?" *Nature*, 19 April 1984, 308, 713-715.
6   Isaac Asimov, "Nemesis," *American Way*, July 1984, 23-24.
7   Richard Muller, *Nemesis: The Death Star - The Story of A Scientific Revolution*, (New York: Weidenfeld and Nicholson, 1988), 92.
8   Clark Chapman & David Morrison, *Cosmic Catastrophes*, (New York: Plenum Publishing, 1989), 106.
9   "Teegarden's Star," Sol Station, http://www.solstation.com/stars/50-02530.htm.
10 "Red dwarf star releases giant burst of light," 20 May 2008, http://www.chinaview.com and http://www.news.xinhuanet.com/english/2000-05/20content_821549.htm.
11 As above.
12 "Star Search Finds Neighborly Red Dwarf," *Space.com*, 20 May 2007. B. J. Teegarden, S. H. Pravdo & others, "Discovery of a New Nearby Star," *Astrophysics*, 11 Feb 2003, arXiv.org>astro-ph. "News – Newly Discovered Star May Be Third Closet" – *NASA*, 20 May 2003, http:www.jpl.nasa.gov/news/news-print.cfm?release=2003-072
13 As above.
14 "Teegarden's Star," *Wikapedia*.
15 "News' Newly Discovered Star May Be Third-Closest," *NASA Gov News* – as above.
16 Clark Chapman & David Morrison, 106.

CHAPTER ELEVEN - WHAT THE BIBLE KNEW FIRST

1   The "heavens" is an area where both man and God have made and can make mutual observations.
2   John S. Lewis, *Rain of Iron and Ice: The Very Real Threat of Comet and Asteroid Bombardment*, (Reading, MA: Addison-Wesley - Helix Books, 1996), 13.
3   Victor Clube & Bill Napier, *The Cosmic Winter*, (Cambridge, MA :Basil Blackwell, Inc., 1990), 194-195. Victor Clube & Bill Napier, *The Cosmic Serpent: A Catastrophic View of Earth History*, (London: Faber and Faber, 1982), 212-223.
4   William J. Broad, "Meteoroids Hit Atmosphere in Atomic–Size Blasts,"

25 Jan 1994, *New York Times*, NY Times.com. E. Tagliaferri, A. Erlich, et al, "Detection of Meteoroid Impacts by Optical Sensors in Earth Orbit," in *Hazards Due to Comets and Asteroids*, T. Gehrels ed., University of Arizona Press, Tucson, 1995, Section II.

5   *Asteroid Impact*, TV documentary aired on the Learning Channel on 5/30/99.
6   *Impact Earth*, TV documentary aired on the Learning Channel on 5/30/99.
7   *Impact Earth*, 5/30/99.
8   "Tsunami – Other historic tsunami," *Wikipedia*, http://en.wikipedia.org.
9   Victor Clube & Bill Napier, *The Cosmic Winter*, 272.
10 John & Mary Gribbin, *Fire on Earth*, (New York: St. Martin's Griffin, 1996), 32-33 & 133.
11 John & Mary Gribbin, 32-33. Richard Stone, "Target Earth," *National Geographic*, August 2008, 146.
12 "Crater clue to dinosaur deaths," *Tucson Citizen*, Associate Press, December 6, 1990, 6C. Norma Cole, "Caribbean meteor wiped out dinosaurs?" *Tucson Citizen*, 18 May 1990, 1B.
13 Anne Denogean, "UA helps study big Moon of Jupiter," *Tucson Citizen*, 16 July 1998, 1C & 4C.
14 John & Mary Gribbin, 105.
15 Clark Chapman & David Morrison, *Cosmic Catastrophes*, (New York: Plenum Publishing, 1989), 67.
16 Donald K. Yeomans, *Comets: A Chronological History*, (New York: John Wiley & Sons, 1991), 243.
17   Ker Than, "Comet Swarm Delivered the Earth's Oceans?" *National Geographic News*, 5 Aug 2009, http://www.nationalgeographic.com/news/."Earth Got Water From Comets?" The Associated Press – Los Angeles, *Tucson Citizen*, 3 Feb 2006, 1B. "Are We Drinking Comet Water?" *Astrobiology Magazine*, NASA, 24 March 2006, http.//www.astrobio.net/news. Phil Beradelli, "Main Belt Comets May Have Been Source Of The Earth's Water", 23 March 2006, http://www.spacedaily.com/reports. "New Class Of Comets May Be The Source Of The Earth's Water," 23 March, 2001, http://www.science.nasa.gov/headlinesy2001. "A Taste For Comet Water," *NASA*, May 18, 2001, http://www.science.nasa.gov/headlinesy2001. Dr. Louis A. Frank, *The Big Splash*, (Secaucas,

NJ: Carol Publishing, 1980), 42-45 & 48. Damond Benningfield, "Where Do Comets Come From?", *Astronomy*, September 1990, 30. John C. Brandt, & Robert D. Chapman, *Rendezvous in Space: The Science of Comets*, (New York: W. H. Freeman & Co., 1992), 145 & 193.
18 Jeffrey Krager, "The Gentle Cosmic Rain," *Time*, June 9, 1997, 52.
19 Louis Frank, *The Big Splash*, 1990, plate 1, 2, and 3. Note that the term "cometary water cloud" is used throughout the book.
20 Anne Keller reporter, *CNN Today*, http://www.cnn.com, 29 May 1997.
21 M. E. Bailey, S.V.M. Clube & W. M. Napier, *The Origin of Comets*, (New York: Pergamon Press, 1990), 4.
22 "Celestial Celebrity," *Time*, 4 Oct 2004, 12.
23 William Hartman, "The Changing Face of Chiron," *Astronomy*, Aug 1990, 47-48.
24 William Hartman , 47-48.
25 Donald K. Yeomans, 253 & 296.
26 John Bortle, "Comet Digest," *Sky & Telescope*, Aug 1991, 215.
27 Charles Morris, "Comets Encke and Levy: Something Old and Something New," *Astronomy*, Sept 1990, 67.
28 "Mystery illness strikes after meteorite hits Peruvian village" (Update) 9/19/2007, *Agence France-Presse* URL-Physorg.com/news. Liubomar Fernandez & Patrick J. McDonnell, "Meteorite Causes a stir in Peru," *Los Angeles Times*, 9/21/07, http://www.latimes.com/news/nationworld/latinamerica/la-fg-meteor21Sep21. "Earth's Newest Impact Crater," 9/15/07, http://www.livinginperu.com/news/4719. Note that Peter Schultz a meteor crater specialist at Brown University made the size guesstimate.
29 As above – Agence France-Presse.
30 As above – Agence France-Presse.
31 www.nuggetshooter.iphost.com, November 9, 2007, which posted a report excerpted from The American Meteor Society.
32 Victor Clube & Bill Napier, 291. John S. Lewis, G. Watkins, H. Hartman, & R. G. Prinn, "Chemical consequences of major impact events on Earth," *Geological Society of America*, Special Paper 190, 1982.
33 John S. Lewis, 110-111.
34 "Germs in Space," *Sky and Telescope*, April 1991, 357. Fred Hoyle & N.

C. Wickramasinghe, *Diseases From Space*, Harper & Row, New York, 1979.

35 John Horgan, "In the Beginning," *Scientific American*, Feb 1991, 124.
36 Richard Stenger, "Scientists discovers possible microbe from space," *CNN.com*, 24 Nov 2000.
37 Richard Stenger. John Horgan, 128.
38 Richard Stenger. John Horgan, 128.
39 Richard Stenger. John Horgan, 128.
40 Richard Stenger. John Horgan, 128.
41 Adam Rogers, "Eking Out Life in the Ice," *Newsweek*, 6 July 1998, 62.
42 "New Organism raises Mars questions," *CNN.com*, 24 Feb 2005, http://www/cnn.com/2005/TECH/science/02/23/bacterian.defrost.reat/index.html.
43 Robert Roy Britt, "Frozen 32,000 Years and Now It's Alive," *Live Science.com*, 2/28/05.
44 Robert Roy Britt.
45 Robert Roy Britt.
46 Michael Kahn, "Ancient bacteria could point to life on Mars," *Reuters*, 8/27/07.
47 Mark Floyd, "Mars meteorite similar to bacteria-etched Earth rocks," Oregon State University, 3/23/06, *Phyorg.com*.
48 Ted Koppel, "Flirting with Disaster - The Dilemma of Looking for Life in Space," *Nightline, ABC News*, 16 May 1997.
49 William Broad, "There's a 'Doomsday" Rock,' But When Will it Strike?," *New York Times*, 18 June 1991, B-10.
50 William Broad, B-10.
51 William Broad, B-10.
52 "Panel skeptical an Asteroid will Hit the Earth - but just in case . . ." Associated Press (San Juan Capistrano) in the *Arizona Daily Star*, 5 July 1991, 13A.
53 David Noland, "The Asteroid Threat is Out There," *Popular Mechanics*, 11 Jan 2007.
54 David Noland.
55 Jonathan Leake, "Mission to destroy it before it get us," *The Sunday Times*, 24 Dec 2006. David Noland.
56 G. Siegfried Kutter, *The Universe and Life*, (Boston, Mass.: Jones and Bartlett Publishers, 1987), 192.

57 "Save the Earth," *Time*, 1 Feb 1993, 56.
58 Lucy McFadden, "Death from Above?" *Sky & Telescope*, January 1996, 54.

APPENDIX A - SUMERIAN/BABYLONIAN COMETARY GODS

1 Samuel Noah Kramer, translator "Lamentation Over the Destruction of Sumer and Ur," in *Ancient Near Eastern Texts- Relating to the Old Testament*, James Pritchard, editor (Princeton, NJ: Princeton University Press, 1969), lines 12 -13, 580.
2 Samuel Noah Kramer, 580.
3 Piotr Michalowski, *The Lamentation Over the Destruction of Sumer and Ur*, (Winona Lake, IN: Eisenbrauns Publisher, 1989), 41.
4 Samuel Noah Kramer, translator "Hymnal Prayer of Enheduanna: The Adoration of Inanna in Ur," *Ancient Near Eastern Texts…*, line 13, 580.
5 A. K. Grayson translator, "The Epic of Gilgamesh," Tablet VI, lines 122-126, in *Ancient Near Eastern Texts…*, 505.
6 Samuel Noah Kramer, "Ishkur and the Destruction of the Rebellious Land," *Ancient Near Eastern Texts…*, 578.
7 Samuel Noah Kramer translator, "Hymnal Prayer …," line 13, 580, and "Ishkur and the Destruction…" 578. E. A. Speiser translator, "The Epic of Gilgamesh" in *Ancient Near Eastern…*, 85.
8 Piotr Michalowski, 41, 47, 61 & 79. Samuel Noah Kramer, translator "Lamentation Over …," *Ancient Near Eastern…*, 613.
9 Piotr Michalowski, 41, 47, 61 & 79. Samuel Noah Kramer, 618.
10 Thorkild Jacobsen, *The Treasures of Darkness - A History of Mesopotamian Religion*, (New Haven: Yale University Press, 1976), 7, 17 & 128.
11 Thorkild Jacobsen, 17 & 130.
12 A. K. Grayson, "The Myth of Zu," in *Ancient Near Eastern Texts…*, 515-517.
13 E. A. Speiser, "The Epic of Gilgamesh" in *Ancient Near Eastern Texts…*, 94.
14 Michalowski, 1989, 41. Samuel Noah Kramer "Lamentation…" 613.

15 Samuel Noah Kramer translator, "A Sumerian Lamentation: Lamentation Over the Destruction of Ur," in *Ancient Near Eastern Texts...*, 463.
16 Samuel Noah Kramer translator, "Hymn to Ninurta As a God of Wrath," *Ancient Near Eastern Texts...*, 577.
17 Samuel Noah Kramer translator, "Hymn to Ninurta as God of Vegetation," *Ancient Near Eastern Texts...* 576.
18 Alan Mac Robert, "The Messier Crater Pair," *Sky & Telescope*, April 1992, 424-425. Peter Schultz & J. Kelly Beatty, "Teardrops on the Pampas," *Sky & Telescope*, April 1991, 391-392. Ben Harder, "What Caused Argentina's Mystery Craters?," *National Geographic News*, 9 May 2002, http://www. nationalgeographic .com/news.
19 Peter Schultz & J. Kelly Beatty, "Teardrops ..." 391-392.
20 Peter Schultz & J. Kelly Beatty, "Teardrops ..." 391- 392. Ben Harder, "What Caused Argentina ..."
21 The broad featureless plain of the Pampas is similar to the flat plain of the Euphrates River where Ur and other Sumerian cities were built, but the annual flooding of the Euphrates River would have quickly filled in and buried any shallow elongated or elliptical craters cut into the deep soil of the Euphrates flood plain. Because of flooding, there would be no evidence of a grazing cosmic impact left in the Euphrates flood plain today.
22 Samuel Noah Kramer translator, "A Sumerian Lamentation...," *Ancient Near Eastern Texts...*, 463.
23 Samuel Noah Kramer, 459.
24 A. K. Grayson, "The Myth of Zu," *Ancient Near Eastern Texts...*, 515.
25 A. K. Grayson, 515.
26 Jacobsen, 132-133, &135.

APPENDIX B – RUSSIA IS NOT MAGOG

1 As quoted in Paul Boyer, *When Time Shall Be No More: Prophecy Belief in Modern American Culture* (Cambridge, Mass: Harvard University Press, 1992), 162. At a dinner with California legislators in 1971 (before he became president), prophecy student Ronald Reagan concisely

summed up his view of Russia's end-time role.
2. Priit J. Vesilind, "In Search of Vikings," *National Geographic*, Vol. 197, No. 5, May 2000, 18.
3. Otto Friedrich, "The Soviet Empire – A Land Great and Rich in Search of Order," *Time*, 12 March 1990, 46. Andrew Sherratt editor, *The Cambridge Encyclopedia of Archaeology*, (New York: Crown Publishing, 1980), 317. Edwin Yamauchi, *Foes From the Northern Frontier*, (Grand Rapids, Michigan: Baker Books, 1982), 20. "Historical Map of the Soviet Union," *National Geographic*, March 1990, "In Search of Vikings," *National Geographic*, May 2000, 18-19.
4. *National Geographic*, 18-19.
5. Viking Answer Lady, http://www.vikinganswerlady.com. Based on the translation of a composite of surviving manuscript versions, passage 80 and 82 part 2. For the full text and commentary of Ibn Fadlan's account of the Rus see: H. M. Smyser, "Ibn Fadlan's Account of the Rus with some Commentary," *Franciplegius: Medieval and Linguistic Studies – Honor of Francis Peabody Margoun, Jr.*, Jess B. Bessinger Jr. & Robert P. Creed editors, (New York: New York University Press, 1965), 92-119 and James E. Montgomery, "Ibn Fadlan and the Rusiyyah," *Journal of Arabia & Islamic studies*, Vol. 3 (2000), 1-15. ISSN:0806-198x.
6. *World History: The Human Experience*, (Gerville, Ohio: McGraw-Hill Company, 1997), 260-262.
7. William Whiston translator, *The Works of Josephus*, (Peabody, Massachusetts: Hendrickson Publishing, 1987), Book 10, Chapter 1, line 2: 891.
8. *Encyclopedia Judaica*, Vol. 3, 718. *The New Bible Dictionary* second edition, (Wheaton, Illinois: Tyndale House Publishers, 1982), 1080.
9. James Charlesworth, *Jesus Within Judaism*, (New York: Doubleday, 1988), 95.
10. John Cumming, D. D. Hurst & Blackette, *The Destiny of Nations as Indicated in Prophecy*, London, 1864, 126.
11. Hal Lindsey, *The Late Great Planet Earth*, (Grand Rapids, Michigan: Zondervan Publishing, 1970), 53.
12. Paul Boyer, 162.
13. Paul Boyer, 162.
14. David Greene translator, *The History*, Herodotus, (Chicago, Illinois:

University of Chicago Press, 1987).
15 Daniel David Luckenbill, *Ancient Records of Assyria and Babylonia*, Vol. II (New PYork: University of Chicago Press, 1926, Greenwood Press Reprint, 1968). Section 784, p. 296 & section 909, 351.
16 Edwin Yamauchi, 110.
17 Edwin Yamauchi, 20-21.
18 Edwin Yamauchi, 20-21.
19 Paul Boyer, 157 & 383.
20 Daniel I. Block, *The Book of Ezekiel: Chapters 25-48*, New International Commentary on the Old Testament series, (Grand Rapids, Michigan: Eerdmans, 1998), 73.
21 Thomas McCall and Zola Levitt, *The Coming Russian Invasion of Israel, Updated*, (Chicago, Illinois: Moody Press, 1987), 29. Hal Lindsey, 53.
22 Paul Boyer, 160.
23 Chuck Smith, *The Final Curtain*, (Costa Mesa, California: The Word for Today, 1984), 30.
24 John Cummings, *The Destiny of Nations – As Indicated in Bible Prophecy*, (London: D.D Hurst & Blackette, 1864). Hal Lindsey, 52-54.
25 Cummings, 127-128.
26 Merril Unger, Preface aurthor First Edition, Thomas McCall and Zola Levitt, *The Coming Russian Invasion...*, 1987, 11. Merril Unger, "Gog," *The Unger Bible Dictionary*, Second Edition, (Chicago, Illinois: Moody Press, 1960), 490.
27 R. K. Harrison, editor, *The New Unger Bible Dictionary*, (Chicago, Illinois: Moody Press, 1988), 804.
28 J. D. Dorylan, editor, *The New Bible Dictionary* (Second Edition), (Wheaton, Illinois: Tyndale House Publishing, 1982), 432.
29 Madeleine S. Miller & J. Lane Miller, editors, *Harper's Bible Dictionary*, (New York: Harper Bros., 1985), 352.
30 "Gog and Magog," *Encyclopedia Judaica*, Vol. 7, (Jerusalem: Keter Publishing House), 691.
31 Daniel David Luckenbill, Vol. II, 290-354.
32 Daniel David Luckenbill, Vol. II, section 784, 296 & section 909, 351.
33 Daniel David Luckenbill, Vol. II, 297, 298, & 351-352.
34 J. Miller, J. H. Hayes, *A History of Ancient Israel and Judah*, (Philadelphia, Pennsylvania: Westminster Press, 1986), 370.

BIBLIOGRAPHY

Books

Bailey, M. E, S.V.M. Clube & W. M. Napier. *The Origin of Comets*. New York: Pergamon Press, 1990.

Beegle, Dewey M. & M. Pryor. *Prophecy and Prediction*. Ann Arbor, Michigan: Pettengill Publishing, 1978.

Block, Daniel I. *The Book of Ezekiel: Chapters 25-48* (New International Commentary on the Old Testament Series). Grand Rapids, Michigan: Eerdmans, 1998.

Boatwright, Mary T., Richard J. A. Talbert, & Daniel J. Gargola. *The Romans: From Village to Empire*. New York: Oxford University Press, 2004.

Bolt, Bruce A. *Nuclear, Explosions and Earthquake: The Parted Veil*. San Francisco, California: W. H. Freeman & Co., 1976.

Boyer, Paul. *When Time Shall Be No More: Prophecy Belief in Modern American Culture*. Cambridge, Massachusetts: Harvard University Press, 1992.

Brandt, John C. & Robert D. Chapman. *Rendezvous in Space: The Science of Comets*. New York: W. H. Freeman & Co., 1992.

Brown, Francis. *The New Brown-Driver-Briggs-Gesenius Hebrew-English Lexicon*. Peabody, Massachusetts: Hendrickson Publisher, 1979.

Burnham, Robert. *Burnham's Celestial Handbook*. New York: Dover Publication, 1966.

Calder, Nigel. *Comets: Speculation and Discovery*. New York: Dover Publicist, 1994.

BIBLIOGRAPHY

Caldwell, T. A., J. N. Oswalt & J. F. Sheehan. *An Akkadian Grammar Part II.* Milwaukee, Wisconsin: Marquette University Press, 1974.

Chapman, Clark & David Morrison. *Cosmic Catastrophes.* New York: Plenum Publishing, 1989.

Charlesworth, James. *Jesus Within Judaism.* New York: Doubleday, 1988.

Clark, Grahame. *World Prehistory.* Cambridge, England: Cambridge University Press, 1969.

Clube, Victor & Bill Napier. *The Cosmic Serpent: A Catastrophic View of Earth History.* London, England: Faber & Faber, 1982.

The Cosmic Winter. Cambridge, Massachusetts: Basil Blackwell, Inc., 1990.

Colson, F. H. Translation. *Philo - Volume VI.* Cambridge, Massachusetts: Harvard University Press, 1935. Reprinted in The Loeb Classic Library, 1984.

Corliss, William R. *The Unexplained: A Sourcebook of Strange Phenomena.* Glen Arm, Maryland: Sourcebook Project, 1974.

Cumming, John. *The Destiny of Nations as Indicated in Prophecy.* London: Hurst & Blackette, 1864.

DeMar, Gary. *End Times Fiction: A Biblical Consideration of the Left Behind Theology.* Nashville, Tennessee: Thomas Nelson, 2001.

Dorylan, J. D., editor. *The New Bible Dictionary – Second Edition.* Wheaton, Illinois: Tyndale House Publishers, 1982.

Eller, Vernard. *The Most Revealing Book of the Bible: Making Sense Out of Revelation.* Grand Rapids, Michigan: Eerdmans, 1974.

Elwell, Walter A., editor. *The Baker Encyclopedia of the Bible.* Grand Rapids, Michigan: Baker Book House, 1988.

Feiler, Bruce. *Walking the Bible - A Journey By Land Through the Five Books of Moses.* New York: William Morrow/Harper Collins, 2001.

Foxxe, Ellen editor. *Size, Composition, and Surface Features of the Planets Orbiting the Sun.* New York: Roses Publishing Group, 2005.

Frank, Harry Thomas editor. *Atlas of the Bible Lands.* Maplewood, New Jersey: Hammond Inc., 1984.

Frank, Dr. Louis A. *The Big Splash.* Secaucas, NJ: Carol Publishing, 1980.

Gardiner, Alan. H. *Admonitions of an Egyptian Sage from a Hieratic Papyrus in Leiden.* G. Olms Verlag, 1909.

Garstang, John. *The Foundations of Bible History.* New York: Richard R. Smith Inc., 1931.

Glatzer, Nahum. "A Treatise on the Life on Moses." Philo, Book 1, in *The Essential Philo*. New York: Schocken Books, 1971.

Gleiser, Marcelo. *The Prophet and the Astronomer: A Scientific Journey to the End of Time*. New York, NY: W. W. Norton & Co., 2001.

Grayson, A. K., translator. "The Epic of Gilgamesh," Tablet VI, lines 122-126, and "The Myth of Zu." In *Ancient Near Eastern Texts - Relating to the Old Testament*. James B. Pritchard, Ed. Princeton, New Jersey: Princeton University Press, 1969.

Gribbon, John. *This Shaking Earth*. New York: G. P. Putnam's, 1978.

Gribbin, John & Mary. *Fire on Earth*. New York: St. Martin's Griffin, 1996.

Gurzadyan, G. A. *Flare Stars*. New York: Pergamon Press, 1980.
"The Atmosphere of M Dwarks," *The M-Type Stars* - NASA (J. Johnson & F. Querci editors) 1986.

Harrison, R. K. editor. *The New Unger Bible Dictionary*. Chicago, Illinois: Moody Press, 1988.

Heidel, Alexander. *The Gilgmagesh Epic and Old Testament Parallels*. Chicago, Illinois: University of Chicago Press, 1949.

Herodotus. *The History*. David Greene, translator. Chicago, Illinois: University of Chicago Press, 1987.

Hirschler, Gertrude editor. *Ashkenaz - The German Jewish Heritage*. Hoboken, New Jersey: KTAV Publishing, 1989.

Hislop, Rev. Alexander. *The Two Babylons: The Papal Worship of Nimrod and His Wife, Second Edition*. Neptune, New Jersey: Loizeaux Brothers, 1959. Original publication date 1859.

Ironside, Harry A. *Lectures on the Book of Revelation 12th edition*. Neptune, New Jersey: Loizeaux Brothers, 1942.

Jacobsen, Thorkild. *The Treasures of Darkness - A History of Mesopotamian Religion*. New Haven, Connecticut: Yale University Press, 1976.

Josephus, Flavius & William Whiston, translator. "The Antiquity of the Jews" and "The Wars of the Jews," in *The Works of Josephus*. Peabody, Massachusetts: Hendrickson Publishers, 1987.

Karls, Farah. *World History: The Human Experience*, Gerville, Ohio: Glencoe/McGraw-Hill Company, 1997.

Kelly, Allan & Frank Dachille. *Target: Earth The Role of Large Meteors in Earth Science*. Carlsbad, 1953, as quoted in *Impact*, Gerrit I. Verschuur, Oxford England: Oxford University Press, 1996.

Kramer, Samuel Noah, translator. "Hymn to Ninurta as God of Vegetation."

"Hymn to Ninurta As a God of Wrath." "Hymnal Prayer of Enheduanna: The Adoration of Inanna in Ur." "Ishkur and the Destruction of the Rebellious Land." "Lamentation Over the Destruction of Sumer and Ur." "Self-Laudatory Hymn of Inanna and Her Omnipotence." "The Curse of Agade." *Ancient Near Eastern Texts - Relating to the Old Testament*. James B. Pritchard, Ed. Princeton, New Jersey: Princeton University Press, 1969.

Kramer, Samuel Noah & Diane Wolkstein. "Sumerian History, Cultures and Literature." Samuel Noah Kramer, translator. In *Inanna - Queen of Heaven and Earth*. New York: Harper & Row, 1983.

Kramer, Samuel Noah. *Sumerian Mythology*. Philadelphia, Pennsylvania: University of Pennsylvania Press, 1972.

The Sumerians: Their History, Culture and Character. Chicago, Illinois: University of Chicago Press, 1963.

Kristiansen, Kristian. *Europe Before History*. Cambridge, England: Cambridge University Press, 1998.

Kutter, G. Siegfried. *The Universe and Life*. Boston, Massachusetts: Jones & Bartlett Publishers, 1987.

LaHaye, Tim. *No Fear of the Storm*. Sisters, Oregon: Multnomah, 1992.

Revelation Unveiled. Grand Rapids, Michigan: Zondervan Publishing, 1999.

LaHaye, Tim & Jerry B. Jenkins. *Are We Living in the End Times?* Wheaton, Illinois: Tyndale House Publishers, 1999.

Larkin, Clarence. *Dispensational Truth or God's Plan and Purpose in the Ages*. Philadelphia, Pennsylvania: Rev. Clarence Larkin Estate, 1920.

Lemcke, Gerry & Michael Sturm. "Oxygen (18) Isotope and Trace Element Measurements as Proxy for the Reconstruction of Climate Change at Lake Van (Turkey): Preliminary Results." In *Third Millennium BC Climate and Old World Collapse*. N. Dulfes, G. Kukla, H. Weiss (eds.). Heidelberg/Berlin: Springer Verlag, (1997).

Lewis, John S. *Rain of Iron and Ice: The Very Real Threat of Comet and Asteroid Bombardment*. Reading Massachusetts: Addison-Wesley Publishing (Helix Books), 1996.

Lindsey, Hal. *The Late Great Planet Earth*. Grand Rapids, Michigan: Zondervan Publishing, 1970.

There's a New World Coming: An In-Depth Analysis of the Book of Revelation. Eugene, Oregon: Harvest House Publishers, 1984.

Luckenbill, Daniel David. *Ancient Records of Assyria and Babylonia- Historical Records of Assyria, from Sargon to the end, Volumes I and. II*. Chicago, Illinois: University of Chicago Press, 1926 and Greenwood Press Reprint, 1968.

Maimonides, Moses. *The Guide for the Perplexed*. Indianapolis, Indiana: Hackett Publishing Company, 1995.

McCall, Thomas & Zola Levitt. *The Coming Russian Invasion of Israel, Updated*. Chicago, Illinois: Moody Press, 1987.

Michalowski, Piotr. *The Lamentation Over the Destruction of Sumer and Ur*. Winona Lake, Indiana: Eisenbrauns Publisher, 1989.

Miller, J. & J. Hayes. *A History of Ancient Israel and Judah*. Philadelphia, Pennsylvania: Westminster Press, 1986.

Miller, Madeleine S. & J. Lane Miller, editors. *Harper's Bible Dictionary*. New York: Harper Bros., 1985.

Mitton, Simon. *The Cambridge Encyclopaedia of Astronomy*. New York: Crown Publishers, 1978.

Morse, Joseph Laffan editor. *The New Funk and Wagnall's Encyclopedia*. New York: Unicorn Publishers, Inc., 1950.

Muller, Richard. *Nemesis: The Death Star - The Story of A Scientific Revolution*. New York: Weidenfeld & Nicholson, 1988.

Myles, Douglas. *The Great Waves*. New York, NY: McGraw-Hill Book Co., 1985.

Nikiforuk, Andea. *The Fourth Horseman: A Short History of Epidemics, Plagues, Famines, and Other Scourges*. New York: M. Evans & Co., 1991.

Noll, Dr. Mark A. *The Scandal of the Evangelical Mind*. Grand Rapids, Michigan: William B. Eerdmans Publishing, 1994.

Oppenheim, A. Leo, translator. "The Sargon Chronicle" in *Ancient Near Eastern Texts - Relating to the Old Testament*. James Pritchard, ed. 1969.

Palmer, Trevor. *Perilous Planet Earth: Catastrophes and Catastrophism Through the Ages*. Cambridge: Cambridge University Press, 2003.

Peabody, Jr. Francis, Jess B. Bessinger Jr. & Robert P. Creed, editors. *Franciplegius: Medieval and Linguistic Studies – Honor of Francis Peabody Margoun, Jr*. New York: New York University Press, 1965.

Penick Jr., James. *The New Madrid Earthquakes of 1811-12*. Columbia, Missouri: University of Missouri Press, 1988.

Petit, Michel. *Variable Stars*. New York: John Wiley & Sons, 1982.

Pfeifer, Charles F. & Everett F. Harrison, editors. *The Wycliffe Bible Commentary*. Chicago, Illinois: Moody Press, 1962.

Pfeifer, Charles F. & Howard F. Voss. *The Wycliffe Historical Geography of Bible Lands*. Chicago, Illinois: Moody Press, 1967.

Pliny. *The Natural History of Pliny Vol. I*. John Bostock & H. T. Riley, translators. London England: Henry Bohn Publishers, 1855.

Pliny. *Natural History/Pliny Book 2*. H. Rackham, translator. Cambridge, Massachusetts: Harvard University Press, 1938.

Pliny. *Pliny: Natural History Vol. II*. H. Rackham, translator. Cambridge, Massachusetts: Harvard University Press, 1942.

Plutarch, *Life of Romulus*, 75 AD, part II. Published by printed for Jacob Tonson, 1711. Princeton, New Jersey: Princeton University, digitized July 14, 2008.

Potok, Chaim. *In The Beginning*. New York: Alfred A. Knopf, 1975.

Pritchard, James editor. "Historical Documents – Cyrus" in *Ancient Near Eastern Texts – Relating to the Old Testament*. Princeton, New Jersey: Princeton University Press, 1969.

Rahn, Perry H. *Engineering Geology: An Environmental Approach (Second edition)*. Upper Saddle River, New Jersey: Prentice Hall, 1996.

Ristvet, Lauren & Harvey Weiss. "The Habur Region in the Late Third and Early Second Millenium BC," and, Fig. 3, p. 6, in *The History of Archeology of Syria, Vol. 1*, Winfried Orthmann, ed., Saabrucken: Saabrucken Verlag, 2005.

Rivers P., R. Evert, & S. Eickhous, *Biology of Plants*, New York: Worth Publishing, 1981.

Roth, Cecil editor. *Encyclopaedia Judaica, Vol. 5*. "Covenant." Vol. 7, "Gog and Magog." Jerusalem, Israel: Keter Publishing House.

Ryan, William & Walter Pitman. *Noah's Flood: The New Scientific Discoveries About the Event that Changed History*. New York: Simon & Schuster, 1998.

Sanders, N. K., translator. *The Epic of Gilgamesh*. Harmondsworth: Penguin Books, 1960.

Sherratt, Andrew, editor. *The Cambridge Encyclopedia of Archaeology*. New York: Crown Publisher/Cambridge University Press, 1980.

Shirihai, Hadoram. *The Birds of Israel*. Princeton, New Jersey: Princeton

University Press, 1996.

Sigal, Philip. *Judaism – The Evolution of a Faith*. Grand Rapids, Michigan: Eerdman, 1988.

Simon, Maurice, translator. *Babylonian Talmud*, Sedar Zerafim, Tractate Berakoth Chapter IX Folio 59a. Edited by I. Epstein. London: Soncino Press, 1948.

Smith, Chuck. *Dateline Earth*. Old Tappan, New Jersey: Chosen Books – Fleming H. Revell Co., 1989.

The Final Curtain. Costa Mesa, California: The Word for Today, 1984.

Smyser, H. M. "Ibn Fadlan's Account of the Rus with some Commentary," in *Franciplegius: Medieval and Linguistic Studies – Honor of Francis Peabody Margoun, Jr.*, Jess B. Bessinger Jr. & Robert P. Creed editors. New York: New York University Press, 1965.

Speiser, E. A., translator. "The Epic of Gilgamesh" in *Ancient Near Eastern Texts - Relating to the Old Testament*. James B. Pritchard, Ed. Princeton, New Jersey: Princeton University Press, 1969.

Steel, Duncan Ph. D. *Rogue Asteroids and Doomsday Comets: The Search for the Million Megaton Menace That Threatens Life on Earth*. New York: John Wiley & sons, 1995.

Tortora, Gerard, Berdell Funke & Christine Case. *Microbiology - Third Edition*. New York: Benjamic/Cummings Publishing Co., 1989.

Unger, Merril. "Gog" in *The Unger Bible Dictionary, Second Edition*. Chicago, Illinois: Moody Press, 1960.

Unger, Merril. Preface to the First Edition of *The Coming Russian Invasion of Israel,* Updated. Thomas McCall & Zola Levitt. Chicago, Illinois: Moody Press, 1987.

Verschuur, Gerrit L. *Impact: The Threat of Comets and Asteroids*. New York: Oxford University Press, 1996.

Virgil. *Aeneid, Vol. II, Books VII-X11. The Minor Poems*. H. Rushton Fairclough, translator. London: Leob Classical Library, 1978.

Voyage Through the Universe - Comets, Asteroids and Meteorites. Time-Life Book Editors. Alexandria, Virginia: Time-Life, 1990.

Weiss, Harvey. "Late Third Millennium Abrupt Climate Change and Social Collapse in West Asia and Egypt." In *Third Millennium BC Climate and Old World Collapse*. N. Dulfes, G. Kukla, H. Weiss (eds.). Heidelberg/Berlin: Springer Verlag, 1997.

Wessels, Norman & Janet Hopon. *Biology*. New York: Random House, Inc., 1988.
Wilson, John, translator. "The Admonitions of Ipuwer," in *Ancient Near Eastern Texts - Relating to the Old Testament - Third Edition*, James Pritchard, editor. Princeton, New Jersey: Princeton University Press, 1969.
Wiseman, Dr. D. J. "Nimrod" and "Babel." *New Bible Dictionary Second Edition*. J. D. Douglas editor. Leicester, England: Inter-Varsity Press and Wheaton, Illinois: Tyndale House Publisher, 1982.
Yamauchi, Edwin. *Foes from the Northern Frontier*. Grand Rapids, Michigan: Baker Book House, 1982.
Yeomans, Donald K. *Comets: A Chronological History of Observation, Science, Myth and Folklore*. New York: John Wiley & Sons, 1991.
Zimmerli, Walther. *Ezekiel 2*. Philadelphia, Pennsylvania: Fortress Press, 1983.

Periodicals

Abarca, Jaime, et al. "Increase in Sunburn and Photosensitivity Disorders." *Journal of the American Academy of Dermatology, Part 1*. (Feb 2002).
Dr. Ajmone–Marsan & others. "The Proceedings of the Royal Society." Online report, Feb. 2007, as quoted in *The New York Times*. (April 3, 2007), p. 3.
Aksu, Ali E., et.al. "Persistent Holocene Outflow from the Black Sea to the Eastern Mediterranean Contradicts Noah's Flood Hypothesis." *GSA Today* (Geological Society of America). (May 2002). Also see *Marine Geology*. (October 15, 2002).
Andrews-Hanna, Jeffrey C. "The Borealis basin and the origin of the martian crustal dichotomy." *Nature*. (Massachusetts Institute of Technology) Maria T. Zuber, et al. (June 26, 2008), p. 1212.
"A Single Intruder, A Chain of Craters?" *National Geographic*. Geographical Section. (Oct 1998).
"An Unseen Meteor Storm." News Notes, *Sky & Telescope*. (May 1990), p. 476.
Asimov, Isaac. "Nemesis." *American Way*. (July 1984), pp. 23-24.
Baillie, M.G.L. and M.A.R. Munro, "Irish Tree Rings, Santorini and Volcanic dust Veils," *Nature*, (24 March 1988), pp. 332-345.

Benningfield, Damond. "Where Do Comets Come From?" *Astronomy*. (Sept 1990), p. 30.

"Big Bang In Antarctica – Killer Crater Found Under Ice." *Ohio State Research*, Columbus, Ohio, (06/01/06).

Blakeslee, Sandra. "Ancient Crash, Epic Wave – Did Catastrophe fall from above in 2807 BC?" *The New York Times*. (Nov 14, 2006, with Nov 16, 2006 correction), pp. 1-4.

Bortle, John. "Comet Digest." *Sky & Telescope*. (Aug 1991), p. 215.

Broad, William J. "Meteoroids Hit Atmosphere in Atomic–Size Blasts." *New York Times*. NY Times.com. (Jan 25, 1994).

"New Proof of Asteroids Devastation," *New York Times*, Science Section, (Feb 18, 1997).

"There's a 'Doomsday Rock,' But When Will it Strike?" *New York Times*. (June 18, 1991), p. B-10.

"Celestial Celebrity." *Time*. (Oct 4, 2004), p. 12.

Chapman, Clark & David Morrison. "The Next Doomsday Impact." *Astronomy*. (Nov 1989), p. 8.

Clube, S. V. M. "The Dynamics of Armageddon," *Speculations of Science and Technology* Vol. 11, No. 4. (1989), p. 263.

Cole, Norma. "Caribbean meteor wiped out dinosaurs?" *Tucson Citizen*. (Tucson, Arizona), (May 18, 1990), p. 1B.

Cooke, Robert. "Noah's Flood." *Newsday*. (Jan 14, 2003).

Copenhauer, Larry. "UA gets $1.2m. to aid in asteroid mission." *Tucson Citizen*. (Tucson, Arizona), (Oct 31, 2006), p. 4a.

"Crater clue to dinosaur deaths." *Tucson Citizen* - Associate Press, (Tucson, Arizona), (Dec 6, 1990), p. 6C.

Davis, M., P. Hunt, & R. A. Muller. "Extinction of Species by Periodic Comet Showers." *Nature*. (April 19, 1984), 308, pp. 715-717.

Delsemme, Armand. "Whence Come Comets?" *Sky and Telescope*. (March 1989), pp. 262-263.

Denogean, Anne. "Solar System gets bigger, with UA Seniors' help." *Tucson Citizen* (Tucson, Arizona), (June 5, 1997), pp. 1A and 6A.

"UA helps study big Moon of Jupiter." *Tucson Citizen* (Tucson, Arizona), (July 16, 1998), pp. 16 and 46.

Djuhari, Lely T. "Asian Quakes' Tsunami Kill More Than 7,000." *AP*. Jakarta, Indonesia: (Dec 26, 2004).

"DNA Shows Etruscans came from Anatolia." *Turkish Daily News.* (Feb 9, 2007).
"Earth called Sitting Duck for Asteroids." Associated Press. *Tucson Citizen* (Tucson, Arizona), (July 4, 1991), p. 17A.
"Earth Got Water From Comets?" *Tucson Citizen* The Associated Press, (Feb 3, 2006), p. 1B.
Fischer, Alan. "Study: Mars Smacked by huge asteroid." *Tucson Citizen* (Tucson, Arizona) (June 26, 2008), p. 1.
Flamsteed, Sam. "Where Comets Come From." *Discover Magazine* (Nov 1995), pp. 83-90.
Friedrich, Otto. "The Soviet Empire – A Land Great and Rich in Search of Order." *Time.* (March 12, 1990), p. 46.
Gallant, Roy A., L. Mansinha & D. E. Smylie. "The Sky Has Split Apart" and "The Rotation of the Earth." *Scientific American.* (Dec 1971).
"Germs in Space." *Sky and Telescope.* (April 1991), p. 357.
Gibbons, Ann. "How the Akkadians' Empire Was Hung Out to Dry." *Science*, vol. 261. (Aug 20, 1993), p. 985.
Gibbs, Nancy. "Apocalypse Now." *Time.* (July 1, 2002), p. 41.
Grinspoon, David H. "Venus Unveiled: A Great Volcanic Flood Must Have Resurfaced Earth's Sister World Some 600 million years Ago." *Astronomy.* (May 1957).
"Halley's Gas and Dust Jets." *Sky and Telescope.* (Nov 1988), p. 456.
Hartman, William. "The Changing Face of Chiron." *Astronomy.* (Aug 1990), pp. 47-48.
"Historical Map of the Soviet Union." *National Geographic.* (March 1990).
Horgan, John. "In the Beginning." *Scientific American.* (Feb 1991), p. 124.
Hoyle, Fred & N. C. Wickramasinghe. *Diseases From Space.* Harper & Row, 1979.
"Hubble Detects Long-Sought Comet Population Beyond Neptune." *Space Science Telescope Institute.* Washington, D. C., (June 14, 1995).
"Hubble Spots An Icy World Far Beyond Pluto." *Science Daily Releases.* (Oct 8, 2002).
"Icy mini-planet found in outskirts of solar system." Associated Press, *Tucson Citizen* (Tucson, Arizona), (June 4, 1997), p. 4A.
Vesilind, Priit J. "In Search of Vikings." *National Geographic* Vol. 197, No. 5. (May 2000), pp. 18-19.
Jaroff, Leon. "At Last, the Smoking Gun." *Time*, (July 1, 1991), p. 61.

Kahn, Michael. "Ancient bacteria could point to life on Mars." *Reuters*. (8/27/07).

Kerr, Richard A. "Sea-Floor Dust Shows Drought Felled Akkadian Empire." *Science*, Vol. 279, No. 5349. (Jan 16, 1998).

Kiefer, Walter S. "Forming the martian great divide." *Nature*. June 26, 2008, pp. 1191-1192.

Krager, Jeffrey. "The Gentle Cosmic Rain." *Time*. (June 9, 1997), p. 52.

Lamb, H. F. & others. *Nature*. 373 134 (1995).

Leake, Jonathan. "Mission to Destroy it Before it Get Us." *The Sunday Times*. (Dec 24, 2006).

Lemonick, Michael. "Is the Bible Fact or Fiction?" *Time*. (Dec 18, 1995), p. 64.

Lemonick, Michael. "Score One for the Bible - Fresh Clues support the story of Joshua at the walls of Jericho." *Time*. (March 5, 1990).

MacRobert, Alan. "The Messier Crater Pair." *Sky & Telescope*. (April 1992), pp. 424-425.

Mansinha, L. & D. E. Smylie. "Earthquake and the Earth's Wobble." *Science*. (Sept 13, 1967), p. 1127.

"The Rotation of the Earth." *Scientific American* (Dec 1971), pp. 80-89.

Marinova, Margarita M. "Mega-Impact formation of the Mars hemispheric dichotomy." *Nature*. (California Institute of Technology), Oded Aharonson, et al. June 26, 2008, p. 1216-1218.

McFadden, Lucy. "Death from Above?" *Sky & Telescope*. (Jan 1996), p. 54.

Meese, D. A. et al., "The Accumulation record from the GISPZ core as an indicator of climate change throughout the Holocene," *Science*, 266 (1994), 1680.

Monastersky, Richard, "The Rise of Life on Earth," *National Geographic*, Vol. 193, No 3, (March 1998), pp. 74-75.

"Moon Blast! New Insights into the Birth of our Satellite." *Time*. (Aug 27, 2001), p. 55-56.

Morris, Charles. "Comets Encke and Levy: Something Old and Something New." *Astronomy*. (Sept 1990), p. 67.

Meese, D.A et.al., *Science*, 266, 680, (1994).

Naege, Robert. "Cosmic Rain of Mini-Comets." *Astronomy*. (Sept 1997), pp. 24 and 26.

Nimmo, F. "Implications of an Impact Origin for the Martian hemispheric dichotomy." *Nature.* (University of California, Santa Cruz), S. D. Hart, et al. June 26, 2008, pp. 1220-1223.

"Noah's Flood Hypothesis May Not Hold Water." *Rensselaer Polytechnic Institute Magazine.* (Sept 2002).

Noland, David. "The Asteroid Threat is Out There." *Popular Mechanics.* (Jan 11, 2007).

O'Brien, S. R. & P. A. Mayewski et.al. "Complexity of Holocene Climates as Reconstructed from a Greenland Ice Core." *Science* Vol. 270, Issue 5244. (Dec 22, 1995).

"Ozone Holes." *Associated Press.* (Oct 20, 2000).

"Panel skeptical an Asteroid will Hit the Earth - but just in case..." *Arizona Daily Star* – AP. (July 5, 1991), p. 13A.

"Quake experts find fault with miniseries," *Arizona Daily Star*, AP, (April 11, 2004), p. A6.

Recer, Paul. "Dinosaur Death Reports Claim Asteroid did it." *Tucson Citizen* AP (Tucson, Arizona), (Feb 17, 1997), p. 6A.

Rocker, Simon. "Catch a Falling Star." *Jewish Chronicle.* (March 6, 1998).

Rogers, Adam. "Eking Out Life in the Ice." *Newsweek.* (July 6, 1998), p. 62.

Sanderson, R. "The night it rained fire." *Griffith Observer.* (Nov 1984), p. 2.

"Save the Earth." *Time.* (Feb 1, 1993), p. 56.

Schultz, Peter. & J. Kelly Beatty. "Teardrops on the Pampas." *Sky and Telescope.* (April 1991), p. 391-391.

Spratt, Christopher. "It Came from Outer Space." *Astronomy.* (Feb 1991), p. 68.

Steel, Duncan. "Tunguska at 100." *Nature.* (June 26, 2008), pp. 1157-1159.

Stone, Richard. "Target Earth." *National Geographic.* (Aug 2008), p. 146.

"Studies of meteorite boost life on Mars theory," *L.A. Times*, Washington Post News Series, in *Tucson Citizen*, (March 14, 1997), p. 11A.

Tagliaferri, E. & A. Erlich, et al. "Detection of Meteoroid Impacts by Optical Sensors in Earth Orbit." In *Hazards Due to Comets and Asteroids*, T. Gehrels ed., University of Arizona Press, Tucson. (1995), Section II. Also see *CommSpacTrans* Sec3.4. html-nasa.gov.

Teng, Poh Si. "Antarctic Smackdown." *The Columbus Dispatch.* (Columbus, Ohio), 6/27/06.

"The Bible in Science." *New York Observer.* March 26, 1863. As quoted in *The Scandal of the Evangelical Mind* by Dr. Mark A. Noll, Grand

Rapids, Michigan: William B. Eerdmans Publishing, 1994.
Verschuur, Gerrit. "The End of Civilization?" *Astronomy*. (Sept 1991), p. 54.
Veslind, Priit J. "In Search of Vikings." *National Geographic* Vol. 197, No. 5. (May 2000), p. 18.
Wakefield, Julie. "Cosmic Rain." *Sky and Telescope*. (Aug 1997), p. 29.
Weiss, H., M. A. Courty, W. Wetterstrom, & others. "The Genesis and Collapse of Third Millenium North Mesopotamia Civilization." *Science*, Vol. 261. (Aug 20, 1993), p. 996.
"Were the Dinosaurs Roasted to Death?" *Newsweek*. (March 14, 1988), p. 66.
Whitemire, D. P. & A. A. Jackson. "Are Periodic Mass Extinctions Driven By a Distant Solar Companion?" *Nature*. (April 19, 1984), 308, pp. 713-715.
Wilford, John Noble. "Red Sea Parting may have been Wind Caused, Scientists say." *New York Times* copyright in the *Arizona Daily Star*. (March 15, 1992), p. 11A.
Wood, Bryant. "Did the Israelites Conquer Jericho? A New Look at the Archeological Evidence." *Biblical Archeological Review*. (March/April 1990).
Woodward, Kenneth L. "The Way the World Ends." *Newsweek*. (Nov 1, 1999), p. 68.
Young, Davis A. "Theology and Natural Science," *Reformed Journal*. (May 1988).
Zielinski, G.A., et.al., "Record of Volcanism since 7000 BC from GISPZ Greenland ice core and implication for the volcano-climate system," *Nature* 264 (1994), 948.

Specialized Publications

Abbott, Dallas & Ann Isley. "Extraterrestrial Influences on Mantle Plume Activity." *Earth Science and Planetary Letters*, Vol. 205. (Dec 2002), pp. 53-62.
Abbott, Dr. Dallas, Edward Bryant & others. "Impact Craters as Sources of Mega-tsunami." From the *Report of International Tsunami Expedition to Madagascar* (Aug 28-Sept 12, 2006).
Abbott, Dallas, Edward Bryant & others. "Impact Craters as Sources of Mega-tsunami Generated Chevron Dunes." Abstract of Paper presented

at the *Geological Society of America*, Philadelphia Annual Meeting, Paper No. 119-120, (Oct 22-25, 2006). Appeared also in Geological Society of America. *Abstract with Program*, Vol. 38, No. 7.

"A Cold New World – The Hubble Space Telescope has measured the diameter of a distant world more than half the size of Plato." NASA, Marshall Space Flight Center. *Government Headlines*, (Oct 7, 2002).

Coffman, Jerry L. and Carl A. Von-Hake, editors. *Earthquake History of the United States*, revised edition. US Dept. of Commerce, (1970).

Courty, Marie-Agnes. "Causes and Effects of the 2350 BC Middle East Anomaly Evidenced by Micro-debris Fallout, Surface Combustion and Soil Explosion." In *Natural Catastrophes during Bronze Age Civilization: Archeological, geological, astronomical and cultural perspectives*. Edited by Benny J. Peiser, Trevor Palmer & Mark E. Bailey. Oxford England: Archaeopress Gordon House, (1998).

Franzen, Lars G. "Cosmic Activity as detected from raised bog stratigraphies in Northern Europe and Siberia." Abstract from *Environmental Catastrophes and Recoveries in the Holocene Conference* 8/29-9/02/02. Department of Geography and Earth Sciences, Brunel University Uxbridge, United Kingdom, received (5/13/02).

Jones, Adrian, David Price & others. "Impact Induced melting and the development of large igneous provinces." *Earth and Planetary Science Letters*, Vol. 202, p. 551.

Keifer, W. S. & B.C. Murray. "The Formation of Mercury's smooth plains." *Icarus*, International Journal of Solar System Studies v. 72, pp. 477-491. Cornell University, Ithaca, New York, (1987).

Lewis, John S., G. Watkins, H. Hartman & R. G. Prinn. "Chemical Consequences of Major Impact Events on Earth" Special Paper 190. *Geological Society of America* (1982).

Master, Sharad. "A Possible Holocene Impact Structure In the Al'Amarah Marshes, Near the Tigris-Euphrates Confluence, Southern Iraq." *Meteorites and Planetary Science* Vol. 36, Supplement, p. A 124, (2001).

"Umm al Binni Lake, a Possible Holocene impact structure in the marshes of Southern Iraq: Geological evidence for its age, and implications for Bronze-age Mesopotamia." Presented at *Conference on Environmental Catastrophes and Recoveries in the Holocene*. Brunel University, Uxbridge, United Kingdom, (Aug. 29 – Sept. 2, 2002).

BIBLIOGRAPHY 559

Montgomery, James E. "Ibn Fadlan and the Rusiyyah." *Journal of Arabia & Islamic Studies*, Vol. 3. (2000), pp. 1-15.

"New Geological Data and K-Ar Geochronology of the Magmatic Rocks on the SE Flank of Mount Hermon." *Geological Survey of Israel*; Report No. GSI/41/88, Ministry of Energy and Infrastructure, Jerusalem. (Dec 1988), pp. 4 and 6.

Palmer, Trevor. "Catastrophes: The Diluvial Evidence." Presented at the *Society of Interdisciplinary Studies*. Easthamstead Park, (Sept 19, 1999).

Peiser, Benny J., Trevor Palmer & Mark E. Bailey, editors. "Comparative Stratigraphy of Late Holocene Sediments and Destruction Layers Around the World: Geological, Climatological and Archaeological Evidence and Methodological Problems." In *Natural Catastrophes During Bronze Age Civilizations: Archeological, Geological, Astronomical and Cultural Perspectives*. Oxford England: Archaeopress Gordon House, (1998).

"Natural Catastrophes During Bronze Age Civilizations: Archeological, Geological, Astronomical and Cultural Perspectives." *British Archeological Reports* – 5728. (1998).

Potts, Laramie V. & Ralph R. B. Von Frese. "Impact-induced mass flow effects on lunar shape and the elevation dependence of nearside maria with longtitude." *Physics of the Earth and Planetary Interiors*, Vol. 153. (Nov 2005), pp. 165-174.

Shimron, Aryeh E. & Barbu Lang "Preliminary Geological Map of the SE Hermon Range." *New Geological Data and K-Ar Geochronology of the Magmatic Rocks on the Southeastern Flanks of Mt. Hermon*, Aryeh E. Shimron, Geological Survey of Israel, Report No. GSI/41/88, Jerusalem, (Dec 1988), Fig. 1-2, p. 7.

Shimron, Aryeh E. "Early Cretaceous Magmatism Along the SE Flank of the Hermon Range." *Israel Geologic Society, Annual Meeting, Ramot, Guidebook for Excursians*. (1989), pp. 1-5, 10 and 141.

"The Early Cretaceous 'Nimrod' Volcanic episode on Mount Hermon." *Geological Survey of Israel*; Report No. GSI/41/88, Ministry of Energy and Infrastructure (Jerusalem). (December 1988). Mishirav, "Boreholes for the Location of Scoria in Har Hermon" – 1987.

Strom, R.G. and others. "The global resurfacing of Venus." *Journal of Geophysical Research*, Vol. 99, 10, 899-10,926, (1995).

Yeomans, D. K. & Kiang, T. "The long-term motion of Halley's Comet." *Monthly Notices of the Royal Astronomical Society*, 197: 633-646 (1981).

Documentaries/Television

"Asteroid Impact." Aired on *The Learning Channel* on (5/30/99).
"Comets: Prophets of Doom." Luke Ellis, Director, *Discovery/TLC/History Channel* (2006).
"Cosmic Catastrophes." *Discovery/TLC/History Channel* (2007).
"Doomsday Asteroids." Aired on *PBS* (4/29/1997).
"Fireballs from Space: Comet Impacts Jupiter." *Discovery/TLC/History Channel* (2006)
"The Great Siberian Explosion." Peter Jones III & Charles Flynn, Directors. First aired: 10/14/1980
"Impact Earth." Aired on *The Learning Channel* on 5/30/99.
Koppel, Ted. "Flirting with Disaster - The Dilemma of Looking for Life in Space." *Nightline ABC News*. (May 16, 1997).
"Mega Disasters: Asteroids." Mike Corkle, Director, *The History Channel* (2006)
"Mega-Disasters: Comets."
"Menaces from the Sky."
"Meteor: Fire in the Sky," *The History Channel*, 2005

Internet

"A Taste For Comet Water." *NASA*. http://science.nasa.gov/headlinesy2001. (May 18, 2001).
Abbot, Dallas, W.B. Masse, & others. "Burckle Abyssal Impact Crater." *Black Rock Forest 2005 Research Symposium* http://www.blackrockforest.org/research/symposium 2005.htm. (June 20, 2005).
Abbott, Dallas, Edward Bryant & others. "Report of International Tsunami Expedition to Madagascar." http://www.Ideo.columbia.edu/users/menke/slides/madagascar06/report.pd.
Abbott, Dallas, Edward Bryant & others. "*Report of International Tsunami Expedition to Madagascar.*" http://www.ldeo.columbia.edu/users/menke/slides/madagascar06/ report.pd. (Aug 28-Sept 12, 2006).
Abrams, Michael. "Stone-Age Asteroid May Have Wiped Out Life in

BIBLIOGRAPHY

America." http://www.discovermagazine.com/2008/jan/stone-age. (Jan 14, 2008).

"Ancient impact may explain Mars mystery." *CNN.com*/2008/Tech/space/. (June 25, 2008).

"Are We Drinking Comet Water?" *Astrobiology Magazine*, NASA http://astrobio.net/news. (March 24, 2006).

"The Atomic Bombing of Hiroshima and Nagasaki." Manhattan Engineering District. Avalon Project." http://www.yale.edu/lawweb/avalon/abomb/mpmenu.htm-Chapter 15 Ground shock.

Beradelli, Phil. "Main Belt Comets May Have Been Source Of The Earth's Water." spacedaily.com/reports. (March 23, 2006).

Britt, Robert Roy. "A Comet's Life: Icy Adventures from Birth to Death." http.//www.space.com/scienceastronomy/solarsystem/comet_linear. (May17, 2001).

"Comets, Meteors and Myth: New Evidence for Toppled Civilizations and Biblical Tales." http://www.Space.com. http://www.space.com. (Nov 13, 2001).

"Earth Deemed Older.", *Space.com*, http://www.space.com scienceastronomy/older_earth-020828html. (Aug 28, 2002).

"Frozen 32,000 Years and Now It's Alive." http://www.Live Science com. (2/28/05).

"24 Hours of Chaos: The Day the Moon was Made." *Space.com*. http://www.space.com. (Aug 15, 2001).

"How Impacts Can Trigger Volcanoes." *Space.com*. http://www.space.com/scienceastronomy/asteroids_volcanoes_030204-1.html. (Feb 4, 2003).

"Caloris Basin." http://www.Wikipedia. (Oct 8, 2006).

Chang, Alicia. "Scientists Think Big Object Whacked Mars." http://www.AOL.com/Science News. (June 25, 2008).

"Collapse of Early Bronze Age Civilization: Has the Smoking Gun Been Found?" Organized by Dr. Benny Peiser. See *CC Net Special – The Cambridge Conference*, http://www.abob.libs. uga.edu/bobk/ccc/cc110501.html. (April 2002).

"Columbia University Research Finds Correlation Between Meteorite and Comet Impacts and an Increase in Volcanic Activity Development." *Earth Institute News*. http://www.earthinstitute.columbia.edu/news, Jan 17, 2003.

"Comet Cools Clovis." *NASA*. http://www.Astrobio.net. from a National Science Foundation News release. (Aug 23,2007).

"Earth's Newest Impact Crater." http://www.livinginperu.com/news/4719. (9/15/07).

Encarta 98.

Fernandez, Liubomar & Patrick J. McDonnell. "Meteorite Causes a stir in Peru." *Los Angeles Times*. http://www.latimes.com/news/nationworld/latinamerica/la-fg-meteor21Sep21. (9/21/07).

Floyd, Mark. "Mars meteorite similar to bacteria-etched Earth rocks." Oregon State University, http://www.Phyorg.com. (3/23/06).

Gallant, Roy A. "The Sky Has Split Apart." www.galisteo.com/tunguska/does/ p 4.

"Geology of Venus." *Wikipedia*. http://www.en.wikipedia.org/wiki/venus.

Gusiakov, Viacheslav. "Report of International Tsunami Expedition to Madagascar." *Novosibirsk Tsunami Laboratory*, Russia 630090 http://www.tsun.sscc.ru/proj.htm. (August 28 – September 12, 2006).

Harder, Ben. "Was Moon Born From Planet's Crash into Earth." *National Geographic News*. http://www.nationalgeographic.com/news/2001/08/0820–moonimpact.html. (Aug 20, 2001).

"What Caused Argentina's Mystery Craters?" *National Geographic News*. http://www.nationalgeographic.com/news. (May 9, 2002).

Hartman, William K. "1908 Siberian Explosion." http://www.psi.edu/projects/siberia/.

"Hubble Spots An Icy World Far Beyond Pluto." *Science Daily Releases*. http://www.sciencedaily.com/releases. (Oct 8, 2002).

"Kaboom! Ancient impacts scarred the Moon to its Core, May have created 'Man in the Moon,'" http://www.theallineed.com/astronomy/06022201.htm. (2/10/06). Also in *Ohio State Research*, Ohio State University, Columbus, Ohio, http://www.osuiedu/units/ research (2/8/06). And in *Science Daily*. http://www.sciencedaily.com/releases/2006/02/060210091105.htm. (Feb. 10, 2006).

Keller, Anne reporter. *CNN Today*, http://www.cnn.com (May 29, 1997).

Lubick, Naomi. "Volcanic Accomplice." http://www.Scientific American.com. (March 17, 2001).

"Magma's Makeup Yields New Clues to Catastrophic Eruptions." *University of Rochester Research*. http://www.sciencedaily.com/releases/1998/07/980724080426.htm. (July 24, 1998).

BIBLIOGRAPHY

Marusek, James. "What Would Happen if a Massive Oort Cloud Comet Strikes Earth." *Cambridge Conference*, ccNet, Issue (5/1/2003). And http://www.personals. galaxyinternet.net/tunga/ Venus.htm (6/13/2003).

"Massive Volcanic Eruptions in Siberia Linked to Largest Extinction." *Science*. http://www.rochester.edu/pr/releases/ear/basu2.htm. (7/12/91).

Matthews, Robert. "Meteor Clue to end of Middle East Civilization" and "Scientists - Quakes may have made Earth Wobble," *CNN Online* 12/29/quake.wobble.reut/ index.html (12/29/04).

Matthews, Robert. "Meteor Clue to end of Middle East Civilization." http.// news.telegraph. co.uk. (April 16, 2002).

"Migratory birds know no boundaries." *International Center for the Study of Bird Migration* http://www.birds.org.il/show_item.asp?levelId=457

Milan, Wil. "The 1833 Leonid Meteor showers: A Frightening Flurry." *SPACE.com*. (11/14/01).

Minkel, J. R. "Giant Asteroid Flattened Half of Mars, Studies Suggest Hemisphere-size Crater." *Scientific American*. http://www.sciam.com/articles.cfm?d=giant-asteroid-flattened & prin . . . (June 25, 2008).

"Mystery illness strikes after meteorite hits Peruvian village." (Update) *Agence France-Presse* URL-Physorg.com/news. (9/19/2007).

Nadamuni, Sridhar. "For 'Man in the Moon,' a bump in the head." *The Boston Globe*. http://www/boston.com/news/ globehealth_science/articles/2006/02. (Feb 13, 2006).

"New Class Of Comets May Be The Source Of The Earth's Water." http://www.ifa.hawaii.edu/. (March 23, 2006).

"New Organism raises Mars questions." CNN.com.http://www/cnn.com/2005/TECH/ science /02/23/bacterian.defrost.reat/index.html. (Feb 24, 2005).

"New Clovis – Age Comet Impact Theory." *Newswise*. University of Oregon, Asteroid and Comet Mission News, Science and Technology. http://www.newswise.com/p/articles/ view/530208. (May 21, 2007).

"New Clovis – Age Comet Impact Theory." Iron and Ice, http://www.spacedaily.com/ reports/ New_Clovis_Age. (May 23,2007).

"News – Newly Discovered Star May Be Third Closet." NASA, http://www.jpl.nasa.gov/news/ news-print.cfm?release=2003-072 (May 20, 2003).

Niroma, Timo. "Sodom and Gomorrah." Helsinki Finland. http://www.

eunet.fi./pp/tilmari/ tilmari2.htm.

O'Hanlon, Larry. "Siberian Traps." http://www.dsc.discovery.com/convergence/ supervolcano/others/ others_07.html.

Phillips, Tony editor. "The Tunguska Event – 100 years Later." *NASA Science News*. http://www.science.nasa.gov/rss.xml! (June 30, 2008).

"Possible Extinction Crater Found Under Antarctica." Staff Writers. http://www.spacedaily. com/reports (6/03/06).

Pringle, Heather. "Did a Comet Wipe out Prehistoric Americans?" *New Scientist*. http://www.com.newsf.service. (May 22, 2007).

Ravilious, Kate. "Earth's Volcanism Linked to Meteorite Impacts?" *New Scientist* Issue 2373. http://www.Newscientist.com. (Dec 13, 2002).

"Red dwarf star releases giant burst of light." http://www.chinaview.cn. and http://www.news.xinhuanet.com/english/2000-05/20content_821549. htm (May 20, 2008).

"Risala: Ibn Fadlan's Account of the Rus," *Viking Answer Lady*, http://www.vikinganswerlady.com. Based on the translation of a composite of surviving manuscript versions, passage 80 and 82 part 2.

Salleh, Anna. "Mega-tsunamis More Common Than We Think." *Australian Broadcasting Corporation Science Online*. http://www.abc.net.au/news/newsitems/200611/51790224. htm. (Nov 16, 2006).

"Scientists, Government Decry NBC Miniseries '10.5,'" Reuters New Media on http://www.au.new.yahoo.com. (April 30, 2004).

Solomon, Sean C. "The Resurfacing controversy for Venus: An overview and mechanistic perspective." MIT's *Tectonic History of the Terrestrial Planets*, 03, 1993, http://www.adsabs. Harvard.edu/abs/1993thtp. reptu...s. INTERNET

"Star Search Finds Neighborly Red Dwarf." http://www.Space.com (May 20, 2007).

Stenger, Richard. "Scientists discovers possible microbe from space." *CNN.com*. (Nov 24, 2000).

Szpytman, Jack J. "The Origin and Formation of the Great Lakes Region- the Global Deluge and Ice Age." Grosse Point Woods, Michigan, 1980, copy #98. http://www.universal-flood.net/

Teegarden, B. J., S. H. Pravdo & others. "Discovery of a New Nearby Star." *Astrophysics*. http://www.arXiv.org>astro-ph. (Feb 11, 2003).

"Teegarden's Star." *Sol Station*. http://www.solstation.com/stars/50-02530.htm.

"Teegarden's Star." *Wikapedia*.
Than, Ker. "Comet Swarm Delivered the Earth's Oceans?" *National Geographic News,* http://www.nationalgeographic.com/news/. (Aug 5, 2009).
"The Impact Through the Moon." *Softpedia News-Science*. http://www.news. softpedia. com/news/.
"The Tunguska Explosion." http://www.madladdesigns,co.uk/unexplained/enigmas/tunguska.htm.
"Tsunami – Other historic tsunami." *Wikipedia*. http://www.en.wikipedia.org.
Wade, Nicholas. "DNA Boosts Herodotus' Account of Etruscans as Migrants to Italy." http://www.New York Times.com. (April 3, 2007).
Walton, Marsha. "Trembler big Enough to Vibrate the Whole Planet." *CNN ONLINE*. http://www.story.muddy.wave.jpg. (5/19/05).
Young, Emma. "Ozone Alert." *NewScientist.com*. (Oct 7, 2000).

Personal Communication

Brygider, Dr. Brandon R. Retired, Hofstra University, Hempstead, New York, 2004-2010: Miscellaneous.
Lewis, Dr. John S. Retired Professor of Planetary Sciences, University of Arizona, Dec. 2001-May 2002.
Kelly, Dr. Clifford. Director of Academics, Dr. James Dobson's *"Focus on the Family" Institute*, Colorado Springs, Colorado, May 1, 2001.
Palmer, Dr. Trevor. Dean of the Faculty of Science and Mathematics, Nottingham Trent University, United Kingdom, Jan 9, 2004.
Shimron, Aryeh E. Geologist, previously of Israel's Geological Survey, Jerusalem Israel, Nov 29, 1989.
Von Frese, Dr. Ralph. Professor Dept. of Geological Sciences, Ohio State University, Sept 14, 2006.

Index

Symbols

100 pound 11, 261, 277, 283, 326
100 pounds 49, 50, 175, 361
144,000 348, 352, 378, 380
185,000 Assyrians 103, 153, 182-184, 189, 196, 465
200 million 5, 6, 11, 43, 44, 48, 50, 247-249, 259, 283
1404 BC 155-158
1600 Hebrew furlongs 351, 366
1868 Chilean tsunami 136
2807 BC 13, 125, 134

A

Abaddon or Apollyon 234, 245
Abbott, Dallas 14, 15, 126-133, 136, 137, 376
Abraham 66, 70-72, 81-84, 145, 147, 148, 152, 195
Accad 70, 289. *See also* Akkad
acid rain 53, 54, 155, 159, 219, 257
active comet 45, 46, 92, 169, 253-257, 431, 436-441, 502
active comets 29, 44, 45, 166, 167, 254-256, 436- 441, 452, 492
a day of trouble 183, 189, 190
Aeneid 27, 143
airburst 41, 42, 223, 415, 416, 463. *See also* comet airburst
Aitken Basin 274
Akkad 68, 70, 239-241, 288-290, 293, 296, 302-308, 311, 314-318, 322, 323, 362
Akkadian(s) 66, 68-73, 90, 112, 117, 133, 135, 141, 146, 147, 164, 166, 167, 171, 175, 195, 207, 239-242, 258, 259, 289, 290, 292-315, 321-324, 337-339, 342, 362, 368, 395, 404, 417, 419, 461, 462, 482, 493, 498
Akkadian Empire 68, 239-242, 258, 290, 295, 297, 300-315, 321-323, 337, 338, 362, 368, 498
Alvarez, Luis and Walter 38
Amarah Crater 17, 78, 126, 138, 171, 241, 289, 290, 308, 309, 314, 315, 320, 323, 368
Amarah Impact Crater 140-143, 239-242, 258, 289, 323, 338
Amorites 5, 70, 172, 190, 191, 196, 277, 304, 348, 388
Ancient Near Eastern Texts - Relating to the Old Testament 299, 306, 494
Ancient Records of Assyria and Babylonia 307
angel(s) 43, 47-54, 88, 158, 170, 183, 184, 206, 207, 217, 234, 245, 248, 451,

491, 492
antichrist 43, 100, 277, 311-316, 337, 340-344, 362, 364, 368, 373, 375, 379-382, 385, 388, 389, 451-456, 484, 485. *See also* "little horn"
antipodal 267, 268, 271, 272
Anzu 78, 462
Apophis (asteroid) 450
Ararat 106, 110, 120, 133-137, 142, 143, 216, 419
Ark 105, 106, 121, 122, 135, 137
Armageddon 2, 6, 11, 33, 61, 194, 258, 277, 325-332, 340, 346, 347, 353, 355, 356, 359, 360, 366, 369, 375-379, 381, 391, 449, 469, 484, 485. *See also* Har-mageddon
army of heaven 85, 95, 117
Artemis 78
Ashkenaz (Ashchenaz) 471, 472. *See also* Ishkuza or Ashguza; *See also* Scythians
Ashtaroth 73, 452. *See also* Ishtar
Ashurbanipal 306
Asia Minor 333, 338, 339, 469, 482
Asshurnasirpal II 338
Assyria 183-185, 306, 307, 483
Assyrian Court 6, 295, 338, 483
Assyrian(s) 5, 6, 66-72, 99, 103, 117, 141, 146, 153, 164-168, 182-184, 189-191, 196, 238, 259, 293, 297, 304, 306, 323, 338-342, 348, 462, 465, 470-485, 493
asteroid families 139, 142
Atrahasis 122. *See also* Ut-Napishtim
axis of rotation 117, 177-180

B

"babel" 68, 298
"babylon" (gate, door or house of god) 68, 298, 323
Babylon 11, 71, 73, 192, 260, 287, 293, 296-302, 305, 306, 312, 315, 317, 320, 323, 338, 362, 382, 494, 498
Babylonian 4, 11, 16, 28, 66-81, 90, 94, 95, 104, 117, 118, 122, 123, 140, 141, 166, 168, 169, 293, 295-306, 312, 342, 404, 451, 459-463, 466, 481, 493
Babylonians 66-74, 79, 80, 99, 122, 141, 146, 161, 164, 167, 259, 294, 297, 298, 304, 323, 417, 452, 492, 493, 494
Babylonian Talmud 16, 104, 118
bacteria 50, 52, 233-238, 242-246, 282, 358, 411, 442-448, 499-508
Bailey, Mark E. 62, 77, 425, 438, 461
Baillie, Mike 18, 241
baldness 244, 245
band of destroying angels 207, 491

INDEX

Banias Valley 373, 374, 385
"Baptism of Fire" 59, 233, 391
base surge 375, 412, 413
battle of Armageddon 328, 332, 340, 346, 353-356, 366, 369, 379, 381, 391, 484
battle of "Ezekiel 38-39" 330-332, 346, 356, 363, 366, 387
Beegle, Dewey 7
Biblical Flood 15, 16, 106, 107, 128, 130, 134, 136, 138, 140-142
bilingual dictionaries 294
binary star 397
black cloud 162, 170, 463
Black Sea Flood? 140, 141
"blast" 103, 182-191, 227, 238, 370
blast that killed 185,000 103, 182, 191, 238
Block, Daniel I. 337, 478
bolide 123
"bottles of heaven" 205, 206, 252, 405, 429, 432-435, 494. *See also* "water bowls of heaven," and "water jars of the heavens"
bottomless pit 5, 11, 50, 233-235, 239, 245, 246, 273
"bowl" or "vial " 205
bowls of heaven 210, 429, 431, 432, 434
Boyer, Paul 326, 337, 477, 479
Bozrah 351, 352, 366
breastplates of iron 52, 234, 502-504
brimstone 5, 6, 11, 44-50, 81, 83, 91, 97, 143-148, 152, 174, 185, 186, 191, 199, 227, 237, 247, 249, 253-259, 283, 356, 374, 387, 389, 413, 416, 436, 437, 440
brown dwarf 399, 400
Bryant, Edward 13, 14, 128, 132
Bull of Heaven 28, 78, 123, 207, 240, 460, 461. *See also* Ishkur
Burckle Crater 13, 14, 17, 126-134, 138, 151, 419, 420
Burckle Impact 15, 16, 134, 136, 142, 143
Burckle Impact Crater 142
burning coals 90, 186, 237, 249
burning lamp 81, 82, 148, 195

C

Caesarea Phillippi 373
"caldera" 371
Caloris Basin 271, 274
Canaanites 71, 72, 181, 190, 191, 196, 304
carbon monoxide 49, 187, 219, 254, 441, 465
"central condensation" or "false nucleus." 254

570 INDEX

chemical fingerprint 112, 433
chevron(s) 13-15, 127-132, 136-139, 142, 151, 216
Chicxulub Crater 125, 213, 216, 270, 422
Chiron 59, 60, 264, 279
Chodas, Paul 415, 416
Cimmerians 338, 339, 473, 482, 483. *See also* Gomer
"Clovis" 209
Clube, Victor 16, 24, 62, 77, 79, 102, 107, 153-158, 165-170, 176, 186, 187, 194, 201, 202, 219, 223, 231, 238, 414, 421, 425, 438, 443, 461, 507, 508
coals of fire 50, 91, 164, 185, 189, 209
"collapsed caldera" 371
coma 46, 166, 253-256, 404, 438, 440, 441, 502
comet airburst 416. *See also* airburst
cometary bombardment 62, 145, 257, 277, 355, 410, 420
cometary gods 28, 72, 76-78, 92, 94, 99, 147, 195, 297, 323, 451, 461
cometary hailstones 49
cometary ice 49, 77, 387, 388, 431
cometary impact 25, 29, 32, 44, 97, 107, 141, 145, 146, 161, 228, 229, 276, 292, 308, 329, 353, 386, 492, 507
cometary impacts 29, 33, 37, 39, 41, 53, 58, 63, 97-99, 109, 229, 278, 325, 355, 409, 415, 456
cometary messenger 170, 184, 185, 189, 197, 206, 210-214, 218, 221, 222, 225, 235, 246
Comet Borsen-Metcalf 466
Comet Encke 57, 77, 161, 166-169, 255, 256, 440
Comet Halley 44, 58, 155-157, 166-169, 219, 255, 433, 438, 439, 453
comet hitting the atmosphere 41–42
comet hitting the land 36–41
comet hitting the ocean 40–41
comet impact 16, 18, 25, 26, 31, 36, 39, 78, 79, 96, 106, 121, 123, 130, 152, 161, 176, 208, 220, 232, 236, 241, 242, 252, 262, 277, 316, 338, 356, 358, 413, 417, 419, 443, 445, 504
Comet Kohoutek 219
Comet Linear 31
Comet Machholz 30, 31, 439
Comet Machholz 2 30, 31, 250
Comet Morehouse 219
comets 28–30
Comets: A Chronological History 62
Comet Shoemaker-Levy 423
Comet Shoemaker-Levy 9 9, 30, 57, 59, 198, 218, 221, 250
companion star 396, 399, 403, 406
Comparison of the Bible's story of the Tower of Babel and the Akkadian Empire

322
cosmic bombardment 143, 145, 150, 172, 178, 198, 229, 257, 314, 398, 420, 461, 499
Cosmic Catastrophes 61, 399, 450
cosmic chemical signature 138, 171
cosmic dust 53, 57, 149, 219
cosmic impact(s) 14-18, 25, 26, 29, 30, 34, 35, 40, 50-55, 77, 79, 110, 127-131, 134, 138, 146, 149-152, 164, 178, 189, 197, 198, 208, 209, 221, 241, 249, 263-266, 272, 278, 279, 308, 312, 326, 376, 377, 388, 411-413, 419, 420, 441-444, 451, 455, 459, 490, 499
cosmic winter 39
Courty, Marie-Agnes 15, 130
"creatures of hell" 339
Crick, Francis 236, 444
crustal melting 276, 284, 413
crustal removal 275, 276, 284, 413
crust blasting impact 273, 275, 284
"cryobiosis" 446
Cummings, John 474, 480
cuneiform 67, 122, 241, 291, 292, 293-296, 304-306, 313, 478, 481, 482, 483
"cutting a covenant" 82, 148
cyanide or cyanogen 52, 187, 211, 219, 257, 443

D

"dam" 160, 165, 349
dark dwarf star 397, 406
day of battle and war 196
day of God's wrath 356, 368
day of the Lord 60, 65, 87, 97, 204, 276, 279, 282, 284, 325, 327, 329, 348, 349, 363-369, 373-376, 379-381, 384-387, 390
day of trouble 183, 189-191, 196
day of vengeance 366, 367, 381
days of battle and war 85, 104, 116, 178, 196, 205, 418, 434
Dead Sea 139, 148-150, 350, 351, 366
Deborah and Barak 103, 181, 182, 190, 191, 196
Deccan Plains (India) 377
Deep Impact 449, 450
defiant act 240, 288, 307, 310, 311
Deluge 107
demonic 22, 26, 29, 43-46, 48, 51, 52, 62, 63, 492
demon possessed 5, 23, 29, 43
Diana 78
dirty snowballs 49, 85, 424, 432

572 INDEX

Disciplinia Etrusca 343
Diseases From Space 236, 444
divine wrath 240, 288, 307, 310, 311, 315, 323
Documentaries-Cosmic Impact 450
door of god 299. *See also* "gate of god" and "house of god"
doublet 86, 119, 180, 184, 207, 429, 431
Dumuzi 74, 493. *See also* Tammuz
dust deposits 13, 15, 138, 171, 239-242, 289, 308, 314, 315
dwarf star 395-397, 400-406

E

earthquake Richter scale 35–36
"Ecclesiastes 1:9 and 3:15" 4, 96, 200, 221, 222, 225, 227, 238, 241, 258, 280, 348, 353, 373, 387, 497
"eh-rets" 111, 180
Einstein, Albert 9
end of heaven 65, 87, 88, 147, 180, 203, 384, 396, 402-405, 424, 462
Epic of Gilgamesh 28, 122, 123, 135, 174, 207, 460, 463
cponyms 339, 482
Erech (Uruk) 28, 67, 123, 302-306, 315, 317, 323, 460, 493
eruptive red dwarf 400, 403-406
Etruria (Tuscany) 343, 344
Etruscans 342-346
Euphrates 32, 47, 48, 70, 71, 120, 132, 140, 216, 241, 247-252, 258, 283, 328, 355, 359, 419, 464, 492
European Community 4, 313, 453
European Economic Community 341
evil empire 6, 333, 335, 475
evil storm 79, 459, 460. *See also* "great storm of heaven"
expanse 426-429, 495. *See also* firmament
"extraordinary proof" 15, 134
Ezekiel (prophet) 364
Ezekiel 38-39 6, 7, 96, 330-342, 346, 356, 359-367, 378, 387, 469, 476-481, 484, 485

F

Fadlan, Ibn 334, 470
fault generated earthquake 262, 487, 488
Feiler, Bruce 103
Fenambosy Chevron 131
Fifth Seal 354
Fifth Trumpet 5, 32, 39, 50, 51, 233-236, 238-246, 252, 259, 282, 307, 308, 358, 359, 447, 499-508

# INDEX

Fifth Vial 32, 234, 235, 242, 243, 251, 358, 447, 500, 505, 506
Fire and hail 88, 186, 210. *See also* hail and fire
"fireballs" 144, 441
fire falling from heaven 81, 144, 146
"*Fire in the Sky*" 145
fire mingled with the hail 93
firmament 426-429, 495. *See also* expanse
First Seal 354, 358, 359
First Trumpet 32, 41, 197, 198, 206-211, 215, 223, 232, 248, 251, 359, 505, 506
First Vial 32, 54, 197, 198, 210, 211, 215, 251, 360, 505, 506
flare star 400-405
flashing thunderbird 462
Foes From the Northern Frontier 336, 476
fountains 53, 106-109, 112, 130, 151, 217, 221. *See also* sources
fountains of the deep 106, 109, 151
four horsemen of the apocalypse 354
Fourth Seal 354, 358, 359
Fourth Trumpet 32, 222, 225-227, 232, 251, 416
Fourth Vial 32, 42, 225-228, 251, 416

G

gas jets 44, 255, 256, 438, 453. *See also* jets of gas and dust
"gate of god" 68, 299, 323. *See also* "house of god" and "door of god"
"Genesis 7:11" 106, 108, 112-120, 123, 124, 135, 151, 178, 191, 196, 278, 417, 420, 430, 433, 495
Genesis 7:11- Retranslation 118
Genesis 10:9-10 and 11:1-9 Translations 317
Genesis 11: 1 Translations 301
geological "END GAME" 391
giant comet 16, 107, 264, 279
giant impact 35, 176, 266, 274, 280, 281, 284
"giant impact mountain melting" event 266
giant impacts 266
giant progenitor comet 280
Gibbon, Edward 480
Giotto 255, 439, 453
Gleiser, Marcelo 23, 37, 60, 89
Gog 333-336, 339, 340-346, 375, 467-469, 475, 476, 480-485. *See also* Gugu and Gyges
Gomer 333, 338-341, 473, 477, 478, 482-485. *See also* Cimmerians
Gondwanaland 271
great battle in the mountains 330
great chunks of ice 77, 158, 491. *See also* heavy hail

great day of God Almighty 258, 279, 329, 331, 355, 384
great day of his wrath 357, 364
"Great Dying" 270
great earthquake 11, 21, 252, 260, 261, 276, 277, 329, 356, 357, 360-362, 385
Great Flood 151, 152, 417, 418
great hail 32, 50, 174, 191, 252, 260, 261, 277, 329, 356, 359, 360-362, 387. See also great hailstones; heavy hail; large hailstones; and very heavy hail
great hailstones 50, 97, 146, 147, 174, 175, 237, 277, 356, 389. See also great hail; very heavy hail; heavy hail; and large hailstones
Great Lakes 30, 209, 220, 221, 250
great star from heaven 147
great stones 93, 96, 172, 174, 191, 381, 388
"great storm of heaven" 79, 188, 459, 465. See also evil storm
"great terrors" 92
great tribulation 2, 196, 355, 378, 395. See also tribulation period
Great Whore 298, 312
Gugu 339, 476, 481-483. See also Gog and Gyges
Gulf of Suez 163, 164
Gyges 339, 342, 476, 481-483. See also Gog and Gugu

H

"hail" 174
hail 32, 49, 50, 77, 85, 88, 91, 93, 96, 104, 116, 119, 146, 149, 154, 158, 164, 167, 173-175, 178, 186, 189-191, 196, 203, 206-212, 228, 252, 260, 261, 277, 329, 356, 359-362, 387, 405, 418, 423, 424, 432, 434, 461, 491, 492, 495
hail and fire 96, 146, 173, 206-212, 228, 461, 492. See also Fire and hail
hailstones 11, 28, 29, 49, 50, 79, 87, 93, 96, 97, 146, 147, 172-175, 182, 189, 191, 207, 237, 261, 277, 283, 326, 356, 388, 389, 413, 459-463
hailstones and flames 28, 79, 146, 174, 207, 460, 461
"hairy star" 44
"hairy stars" 28
Har-mageddon 328-330, 355, 387. See also Armageddon
Hazards Due to Comets and Asteroids 223
"heaven" 86, 402, 403, 427-429, 495
heaven of heavens 86, 180, 402, 403, 423, 426, 427
heavy hail 50, 77, 93, 154, 158, 167, 173, 207. See also great hailstones; very heavy hail; great hail; large hailstones; and great chunks of ice
Hellas Basin 274
Herodotus 238, 343-345, 471, 475, 476
Hezekiah 5, 103, 183, 189-191, 196, 238, 283
Hierapolis 474

INDEX 575

hierodule of heaven 73. *See also* Inanna
hieros-gamos 69, 296, 310
Hills Cloud 252
Hills, Jack G. 425
hills (mountains) melted like wax 278
Hislop, Rev. Alexander 69, 493
"historical archetypes" 484
Hodge, Charles 8
Holocene Impact Working Group 14, 15, 126, 128, 130, 133
"holy marriage" 69, 296, 310
Holy Roman Empire 313, 337, 341, 342, 380, 453
"hornets" 383
Horus 168
host of heaven 72, 75, 76, 84, 85, 90, 94, 95, 117, 180, 186, 297, 366, 426, 433, 451-454
hot thunderbolts 158, 491, 492
"house of god" 299, 305, 323. *See also* "gate of god" *and* "door of god"
Hoyle, Fred 52, 236, 237, 444-446
Hulah Basin 349, 351, 379
Hulah Lake 350. *See also* waters of Merom
Hymnal Prayer of Enheduanna: The Adoration of Inanna in Ur 28, 73, 460

I

ice core data 139
Imdugud 207, 460, 462. *See also* Ninurta
impact crater 13, 14, 17, 18, 51, 107, 126, 127, 145, 150, 176, 233-235, 239, 243-246, 271, 413, 419, 421, 441, 464, 466, 507
impact driven mega tsunami 135
impact driven tsunami 16, 115, 124
impact earthquake 114
impact ejecta 131
impact generated earthquake 163, 262, 276, 361, 487
impact generated tsunami 40
impact glass 14, 127, 129
impact spherules 14, 127, 129
Impact: The Threat of Comets and Asteroids 25, 202, 215
Inanna 28, 68-74, 78, 81-84, 168, 207, 296-298, 306-312, 337, 395, 452, 460, 493. *See also* hierodule of heaven; Ishtar; mother goddess; Queen of Heaven; *and* vulva of heaven
Indian Ocean 13, 14, 107, 115, 120, 125-132, 142, 152, 216, 419
Indonesian Earthquake 40, 117, 121

Indonesian tsunami 9, 14, 16, 109, 132, 135, 418
infrared dwarf 399
infrared emitter 403
In the Beginning 101, 294
Ipuwer Chronicle 160, 168
iridium 113, 149, 171, 208, 209
Isaiah (prophet) 364
"Isaiah 24:18-20" 178, 179, 196, 263, 277, 278, 413
Ishkur (Adad in Akkadian) 28, 78, 90, 207, 240, 460, 461, 466. *See also* Bull of Heaven
Ishkuza or Ashguza 472. *See also* Ashkenaz (Ashchenaz); *and* Scythians
Ishtar 68, 72, 73, 84, 94, 95, 297, 306, 337, 395, 452, 493. *See also* Inanna

J

jacinth 45, 46, 247, 249, 253, 254, 283, 436, 440, 441
Jacobsen, Thorkild 297, 462, 466, 494
Jenkins, Jerry 2
Jericho 156-158, 165
jets of gas and dust 44, 255, 439. *See also* gas jets
Job 38:22-23 85, 104, 116, 118, 178-180, 189, 190, 196, 203, 252, 366, 395, 418, 423-426, 430, 431, 434, 495
Jordan River 165, 349-351, 366, 372
Jordan Valley 349
Josephus 148, 238, 470-473
Joshua's Great Victory 5, 103, 153, 172, 196, 207, 278, 353

K

"Ka-dingir-ra" 299. *See also* "gate of god" *and* Babylon
karst 350
Kelly, Clifford 20
Khima 104, 118
Kiev 334-336, 469, 470, 477
King Belshazzar 295
King Darius 94
King David 238, 283, 340, 484
King of Heaven 94, 297
King Shalmaneser 295
Kish 123, 140
Kramer, Samuel Noah 69, 70, 71, 175, 188, 288, 293, 303, 304, 307, 395, 494
Kuiper Belt 203, 252, 279, 425, 495
Kur 73, 168

INDEX

L

LaHaye, Tim 2
Lake Baikal 219, 220
Lake Van 137, 138, 142
Lamentation Over the Destruction of Sumer and Ur 28, 174, 175, 207, 460, 461, 463
Lamentation Over the Destruction of Ur 175, 187, 459, 463, 465
Lamont-Doherty Observatory 14, 126
land of Shinar 290-293, 296, 298, 302-306, 312, 317, 318, 493
large hailstones 49, 172, 173. *See also* great hail; great hailstones; heavy hail; *and* very heavy hail
"Late Heavy Bombardment" 113
Left Behind 2, 26
Leonid meteor shower 446
Levitt, Zola 479
Lewis, John S. 24, 26, 125, 137, 143, 146, 172, 177, 193, 202, 229, 236, 253, 257, 413, 414, 443, 507
"lie in wait" 115-119, 135, 151, 178, 179, 196, 216, 263, 278, 417, 430, 434
"lightning" 174
Lindsey, Hal 2, 33, 475, 478-480, 489, 490
"little horn" 451, 452. *See also* antichrist
locusts 51, 154, 161, 233-235, 243-246, 259, 358, 383, 500-508
"long day" 177, 178
"loosed" 16, 116-119, 131, 151, 178, 179, 196, 247, 252, 263, 276-278, 417, 420, 430, 434
Lord of Hosts 36, 65, 84-87, 90, 91, 103, 117, 179, 229, 263, 264, 276, 294, 298, 352, 384, 385, 390, 403, 424, 432, 466
Los Alamos National Laboratory 15, 133
Lowth, Bishop 477, 480
Luckenbill, Daniel David 307
Ludu 339, 476, 482, 483. *See also* Lydia
"lurk" 115, 116
Lydia 339, 342-346, 474-476, 481-484. *See also* Ludu

M

Madagascar 14, 129-132, 136-138
Madagascar chevrons 129, 131, 138
Magog 327, 333-346, 374, 389, 467, 469-485. *See also* Magugu
Magugu 339, 476, 482. *See also* Magog
Maimonides 10
"Main Belt Comets" 113
Manicouagan Crater 218

"Man on the Moon" 267, 270
Marduk 74, 94, 168, 297
Marsh Arabs 18
Maschiach ben David 383
"mascon" 269
Masse, Bruce 15, 133, 134
Massoretes 302
Master, Sharad 17, 123, 140, 239, 290
Mega Disasters: Comet Catastrophe 151
mega impact(s) 35, 266, 273-275, 283, 284, 384, 390
mega tsunami(s) 13, 109, 110, 114, 127, 128, 132, 135-137, 417, 419
Megiddo 328-330, 346, 348
"melt" 31, 34, 60, 87, 90, 252, 265, 268-272, 281-285, 350, 384, 390, 391, 413, 433, 437
melted glass 128, 131
melt mountains 34
Meshech 295, 333, 338-340, 467, 473, 477, 478, 482-485
messenger 48, 170, 183-185, 189, 190, 197, 206, 207, 210-214, 217, 218, 221, 222, 225, 233-238, 245-248, 251, 252, 260, 491, 496
messengers 28, 47, 48, 85, 88, 158, 184, 186, 189, 206, 207, 241, 246-248, 250-252, 411, 491, 492
Meteor Crater 17, 37
micro fossils 14, 131
"mighty ones" 85, 87, 88, 179, 180, 203, 368, 384, 424. *See also* "sanctified ones"
migratory birds 366
ministers 86, 88, 150, 184, 186, 207
ministers of flaming fire 150
ministers of wind and flaming fire 86
"molten rock upward" 215
Morrison, David 40, 61, 211, 399, 430
Moses 82, 93, 96, 162, 164, 379
mother goddess 72, 73, 343, 452. *See also* Inanna
Mother of Harlots 298, 312
mountain assembly of troops 329, 330
mountain melting mega impact 266, 273, 284
mountain rendezvous of troops 258, 329, 346, 378
mountains melted 278
mountains of Ararat 106, 110, 120, 133-137, 142, 143, 216, 419
mountain to disappear 32, 34
Mount Hermon 259, 346-352, 366, 369, 371-373, 378-380, 385
Mount Pelee 370
Mount St. Helens 370, 372, 374

INDEX 579

Mount Zion (Sion) 347, 348, 352, 366, 378, 380
"mouth be one" 301. *See also* "one lip"
mouths 5, 11, 29, 37, 44-48, 95, 132, 147, 195, 199, 247, 249, 254-259, 283, 411, 436-441, 461. *See also* vents
Muller, Richard 396, 398
multiple cosmic impacts 420
multiple impacts 9, 30, 31, 114, 198, 201, 231, 250, 410, 420-423
Mystery Babylon 298, 312, 362

N

Nachmani, Rabbi bar 104, 118, 119
Nanna (Sin) 297
Napier, Bill 16, 24, 62, 77, 79, 102, 107, 154-158, 165-170, 176, 186, 187, 201, 202, 219, 223, 231, 238, 414, 421, 425, 438, 443, 461, 507, 508
Nebuchadnezzar 94, 95
Nemesis 397, 398
Neo-Babylonian Empire (612-536 B.C.) 297
Newe Ativ Caldera 371-373, 385
Newe Ativ Volcanic Complex 372, 385
New Madrid Earthquake 161, 165, 262
Newton, Sir Isaac 89, 440
nickel 14, 129-131, 149, 171, 219
Nimrod 68, 69, 305-307, 312, 317, 322
Ninurta 78, 90, 122, 168, 207, 462-466. *See also* Imdugud
nitric acid or nitrous acid 53, 211, 219, 443
Noah's Flood 15, 71, 101-104, 113, 122-125, 140, 152, 191, 196, 216, 290, 313, 417-419, 433
Noll, Mark A. 7
Novgorod 334, 335, 470
Novosibirsk Tsunami Laboratory 128, 132
nuclear bombs 33–36
nuclear winter 18, 241
nuée ardente 369-375, 385

O

ocean impact 107, 125-128, 133, 136, 151, 213-217, 228, 419
ocean impacts 16, 105, 106, 112, 114, 125, 417, 419
Olivet Prophecy 98
"one commander" 300-302, 315, 317, 322
"one gathering" 301, 302, 321
"one government" 300, 301, 321
"one lip" 300-302, 317, 319, 321. *See also* "mouth be one"
Oort Cloud 85-87, 104, 116, 119, 135, 180, 189, 190, 196, 203, 252, 273, 279,

367, 384, 397, 398, 402-405, 411, 418, 423-431, 434, 495
orbital backtracking 78, 124, 139, 142, 155, 158, 166, 167
ordinances of heaven and Earth 86, 103, 327, 498
oxygen isotope ratio 140

P

Palmer, Trevor 105, 119
Pasteur, Louis 243, 448
"pattern of world history" 308
Peiser, Benny 16, 104, 106, 149
pestilence 29, 52, 237-239, 242-244, 358, 444, 447, 448
Phaethon 168, 465
Philo 162
pillar of a cloud 165, 166, 169, 170
pillar of fire 166, 169, 191
Pitman, Walter 140
plagues of the Exodus 43, 77, 154-160, 165, 171, 194, 229, 379, 410
"planetary scale impacts" 35, 274, 284, 413
Pliny 168, 474, 475
Potok, Chaim 101, 294
Potts, Laramie 266-272
Primary Chronicle 334, 335, 470
Pritchard, James 300, 306, 494
Proxima Centauri 396, 401
Psammetichus I 483

Q

Queen of Heaven 28, 68-78, 82-84, 94, 117, 146, 207, 296- 298, 308, 312, 337, 362, 395, 451, 460, 493. *See also* hierodule of heaven; Inanna; Ishtar; *and* vulva of heaven

R

rain of fire and brimstone and of iron 145
Rain of Iron and Ice 24, 125, 193, 202, 236, 253, 257, 413
"raptured" 378, 379
Reagan, Ronald 3, 335, 468, 475
red dwarf 399-406
Red Sea 5, 161-165, 170, 191, 368
red tide 159
Reed Sea 5
reel to and fro like a drunkard 179, 264, 276, 277
retene 208

"Revelation 20:8" 340, 485
River Zered 351, 366
Rochechouart Crater 218
Rogue Asteroids and Doomsday Comets 22, 38, 107, 262
Roman Empire 311, 313, 337, 341-344, 380, 453, 480
Rus 334, 335, 336, 469, 470, 476, 477, 480
Russians 6, 333, 334, 335, 346, 380, 453, 469, 470, 478, 480
Ryan, William 140

S

"sacrifice" (upon the mountains of Israel) 330, 332, 351, 365-367, 379, 380
"safety" 455
"sanctified ones" 87, 179, 180, 366, 384, 424. See also "mighty ones"
Sargon (true king) 296, 297, 301, 302, 305-307, 310, 312, 316, 317, 322, 337, 338, 362
Sargon Chronicle 305
Sargon II 338
"save" or "saved" 378, 383
Scandinavian 334-336, 469, 470, 476, 480
Scofield Reference Bible 477
Scythians 333-335, 469-480. *See also* Ashkenaz (Ashchenaz); *and* Ishkuza or Ashguza
sea of molten rock 281, 391
Second Coming 332, 340, 367, 484
Second Seal 354, 358
Second Trumpet 32, 212-216, 226, 232, 251, 419
Second Vial 32, 54, 214-216, 251
Semiramis 68
Sennacherib 183-189, 227
Septuagint 65, 124
Seven Seals 325, 327, 353-355, 358-362, 391
Seventh Seal 359
Seventh Trumpet 32, 232, 252, 258-264, 276-283, 329, 331, 355, 356, 359-363, 366, 367, 377, 384, 390
Seventh Vial 32, 50, 252, 258-265, 273, 276-279, 282, 283, 329, 356, 358-363, 384
Seven Trumpets 25, 29-33, 44, 51, 63, 95, 96, 193, 197-205, 251, 252, 311, 325-327, 353-355, 358-363, 391, 398, 410, 420, 455, 456, 496
Seven Vials 25, 29, 32, 33, 44, 51, 63, 95, 96, 193, 197-202, 252, 312, 327, 353, 354, 358-363, 410, 420, 456, 496
shake, shaken or shaking 4, 18, 21, 22, 34-39, 43, 58, 60, 63, 96-98, 116, 117, 146, 147, 164, 179, 196, 231, 233, 261-264, 276-278, 313, 326, 357, 363, 366, 368, 385, 388, 391, 393, 413, 459, 461, 487-490

shake the heavens and the earth 263, 264, 276, 385
Sharkalisharri 305
Shimron, Aryeh E. 347, 371, 372
Shinar 290, 291, 293, 296, 298, 302-306, 312, 317, 318, 322, 493
shocked quartz 125, 137, 138, 149, 171
"shooting stars" 28
Shuruppak 140
Siberian Traps 270, 377
sign of the Son of Man in heaven 368, 393, 394, 398
signs and wonders 91-94, 97, 158, 382, 410, 452
signs of heaven 77, 92
Sisera 94, 181, 210
Sixth Seal 37, 59, 355-367, 384
Sixth Trumpet 5, 6, 32, 44-52, 241, 246-259, 280, 283, 328, 355, 359, 436, 491, 492
Sixth Vial 32, 48, 241, 248, 249-252, 258, 283, 328, 329, 355, 358, 359, 361
sling stones 204, 207, 385, 460, 462, 466
Smith, Chuck 479
smoke of a furnace 145, 152, 258
smoking gun 17, 107, 142, 143, 242, 314, 338, 419, 422, 436
Sodom and Gomorrah 18, 81-84, 101, 102, 143-152, 191, 196, 227, 258, 290, 313, 416
"sources" 112-119, 135, 142, 151, 178, 196, 216, 221, 399, 417, 430, 433, 434, 494. *See also* fountains
sources of the great deep 112, 114, 116, 135, 178, 196, 430, 433
spherules 14, 57, 127, 128, 129, 149, 171, 209
spherules of melted or fused glass 128, 149
"star" 26–28, 394
Star of Bethlehem 394, 395, 399
Star out of Jacob 393-395, 398, 406
starry host 87, 246, 252, 451
starry hosts 65, 253
stars falling from heaven 11, 356, 384
Steel, Duncan 22, 38, 39, 107, 167, 262, 263
storehouses of the hail 85, 104, 119, 189, 190, 196, 203, 252, 405, 423
storehouses of the snow 85, 104, 119, 189, 190, 196, 203, 252, 423, 495
sulfur 27, 28, 46, 49, 50, 53, 143-145, 149, 155, 174, 185, 211, 219, 227, 247, 249, 253, 254, 257, 387-389, 416, 437, 441, 442, 447, 503
sulfuric acid 53, 219, 443
Sumer 28, 67-74, 140, 174, 175, 207, 293, 302-306, 317, 318, 322, 362, 460-463, 493
Sumerian 28, 66-73, 76-83, 90, 112, 117, 122-124, 135, 140, 147, 166, 168, 175, 187, 195, 207, 239-241, 288-310, 315, 322, 323, 337, 340, 451, 459-

INDEX

462, 465, 466, 484, 485, 493
Sumerian/Babylonian Cometary Gods 298, 459
Sumerians 66-73, 79-81, 99, 122, 123, 133, 141, 146, 161, 164, 167, 240, 259, 293-295, 296, 299, 303-306, 395, 404, 417, 419, 452, 460-463, 493, 494

T

Tammuz 68, 74, 75, 493. *See also* Dumuzi
Tarquinius Superbus 343
Taurid meteor stream 77, 167
tektites 125, 137-139, 171
tempest 88, 94, 184-186, 189, 210, 407. *See also* whirlwind
temple tower 71, 192, 290, 291, 296-300, 305, 306, 309, 315, 338, 382
temple towers 67, 68, 293, 309
"terrified by the signs of the heavens" 92
The Antiquity of the Jews 238, 470, 471, 473
the "blast" 103, 182, 185-189, 238
The Book of Ezekiel 337, 478
The Cosmic Serpent 25, 79, 153, 156, 167, 187, 201, 238, 508
The Cosmic Winter 25, 102, 107, 155, 156, 167, 176, 186, 202
The Curse of Agade 175, 288, 307
The Curse of Akkad 239-241, 289, 307, 308, 311, 314, 315, 323
The Destiny of Nations 474, 480
the Exodus 43, 51, 77, 87, 92-96, 103, 146, 153-160, 164-168, 171, 173, 184, 190, 191, 194-196, 207, 211, 222, 225, 229, 238, 244, 283, 326, 327, 379, 380, 382, 461, 491
the Flood 16-18, 104-112, 115, 117-124, 130, 134-138, 141, 142, 150, 151, 178, 196, 278, 290, 417-420, 463
"the harlot of heaven" 297
The History 145, 343, 471, 476
The Late Great Planet Earth 2, 33, 475, 478-480
The New Bible Dictionary 481, 494
The Origin of Comets 62, 77, 425, 438, 461
the pattern 51, 189, 190, 192, 229
The Prophet and the Astronomer 23, 60, 89
The Scandal of the Evangelical Mind 7
"The Sumerian King List" 306
the "three woes" 232
the Tower 71, 287, 290-304, 307-309, 311-317, 322, 498
The Treasures of Darkness - A History of Mesopotamian Religion 297, 462, 466, 494
The Two Babylons 69, 493
Third Seal 354, 355, 358
Third Trumpet 32, 53, 217-222, 232, 251, 443

Third Vial 32, 221, 222, 251
Tiamat 73, 168
Tigris 70, 120, 132, 216, 419
time of Jacob's trouble 196, 355, 382, 395
times of trouble 85, 104, 116, 178, 180, 189, 190, 196, 203, 252, 418, 423, 424, 434, 495
Togarmah 295, 333, 338, 339, 477, 478, 482, 484
Tollman, Edith and Alexander 107
Tower of Babel 71, 290-296, 301, 303, 307-315, 317, 322, 323
Tower of Babylon 287, 323, 498
Translations of the Story of the Destruction of the Tower of Babel (Genesis 10:9-10 and 11:1-9) 317
tribulation period 196, 327, 330-332, 340, 355, 358, 360, 367, 378, 379, 452, 484. *See also* great tribulation
Troyes, Bishop 334, 470
tsunami 9, 13-16, 40, 107, 109-115, 119-137, 141, 142, 151, 161, 163, 177, 191, 213, 410, 416-420
tsunami deposits 13, 14, 125, 127, 130, 131, 136, 137, 142, 151, 420
 125, 140
tsunamite(s) 40, 125, 137, 140, 419
Tubal 295, 333, 338-340, 467, 477, 478, 482-485
Tunguska 41, 42, 54-58, 146, 161, 166, 171, 178, 185, 208, 209, 220, 223-227, 415, 463
Typhon 168

U

Unger, Merril F. 480
Ur 28, 66-71, 73, 83, 123, 140, 147, 174, 175, 188, 195, 207, 297, 303, 459-461, 463, 465, 493
Ut-Napishtim 122. *See also* Atrahasis

V

Valley of Gibeon 5, 93, 103, 172
Valley of Jehoshaphat 352
vents 44, 45, 47, 48, 147, 199, 249, 254-257, 283, 411, 436-441, 492. *See also* mouths
Verschuur, Gerrit L. 25, 61, 102, 107, 120, 121, 202, 215, 287, 288
very heavy hail 93, 154, 158, 167, 173, 207. *See also* great hailstones; heavy hail; great hail; large hailstones; and great chunks of ice
very high nickel content 131
"vial" 205, 210
Viking(s) 334-337, 469, 477, 480
Virgil 27

INDEX

Vladimir, Prince 334, 470
Von Frese, Ralph 266-272
vulva of heaven 73. *See also* Inanna

W

Walking the Bible 103
"wandering stars" 28
"water bowls of heaven" 210, 431, 432. *See also* "bottles of heaven"
"water jars of the heavens" 205, 429, 432, 434, 435, 495. *See also* "bottles of heaven"
waters of Merom 350. *See also* Hulah Lake
weapons of indignation 180
When Time Shall Be No More: Prophecy Belief in Modern American Culture 326
Whipple, Fred 167, 431, 438
whirlwind(s) 94, 161, 164, 186, 204, 210, 390, 441, 497. *See also* tempest
Whore of Babylon 73, 298
Wickramasinghe, N. Chandra 236, 444-446
Wild Bull of Heaven 78, 123
Wilkes Land Crater 270, 271
Wilkes Land Impact 270, 271, 280
windows of heaven 106, 108, 115, 116, 151, 178
winepress 349-352, 366, 367, 379, 380, 391
"winged" goddess 306
Wiseman, D. J. 305, 307
Wolkstein, Diane 71, 395
Wooley, Sir Leonard 140
wormwood 53, 217, 218

Y

Yamauchi, Edwin 336, 476
Yeomans, Donald K. 62, 155, 169, 395, 439
Young, David A. 110

Z

Zagros Mountains 120
ziggurat 67, 71, 293, 319, 323. *See also* temple tower *and* "Ka-dingir-ra"
Ziusudra 122

Job 38:30 "The waters are hid (the ice inside a comet's rocky crust) as with a stone (a comet's rocky crust), and the face of the deep (the mass of water in a comet) is frozen."

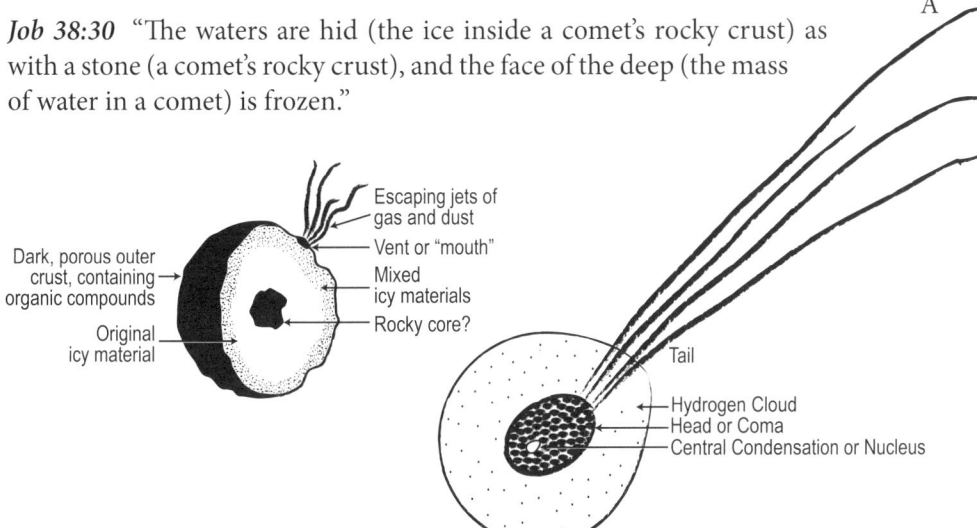

Revelation 9:17-19 ". . . having breastplates (an active comet's coma) of fire (the fiery white central condensation in the coma), and of jacinth (comas are blue), and brimstone (comas contain brimstone), and the heads . . . as the heads of lions (fuzzy appearance of an active comet's head or coma) and out of their mouths (vents in a comet's crust) issued fire and smoke and brimstone (the gas and dust that escapes from an active comet's vents or mouths) . . . For their power is in their mouth (gas and dust jets that escape through a comet's mouth or vent) and in their tails (extension of the gas and dust jets) for their tails (an active comet's long curving tail) were like unto serpents and had heads, and with them they do hurt".

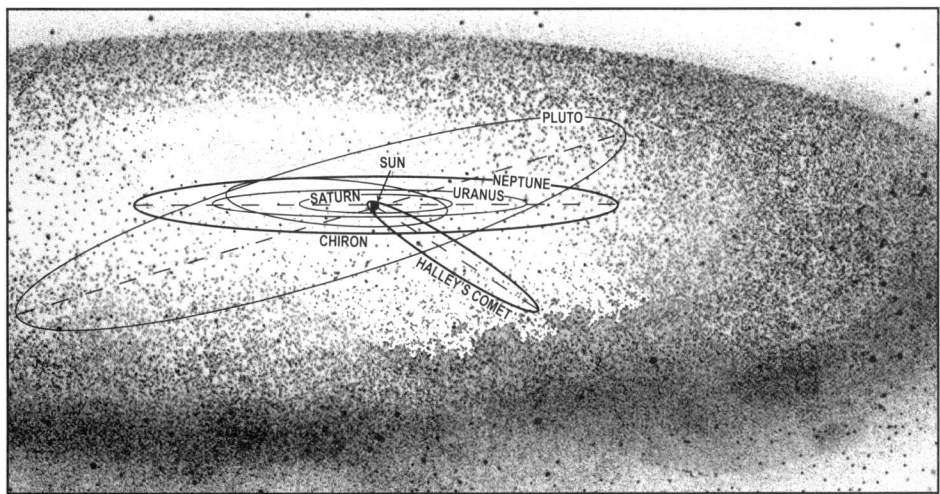

THE KUIPER BELT of comets which develops into the **OORT CLOUD**, the spherical reservoir of comets at the end of the solar system that encloses the entire solar system. *Job38:22-23* (*NIV*) "Have you entered the storehouses of the snow (comets) or seen the storehouses of the hail (comets) which I reserve for times of trouble, for days of war and battle?"

B

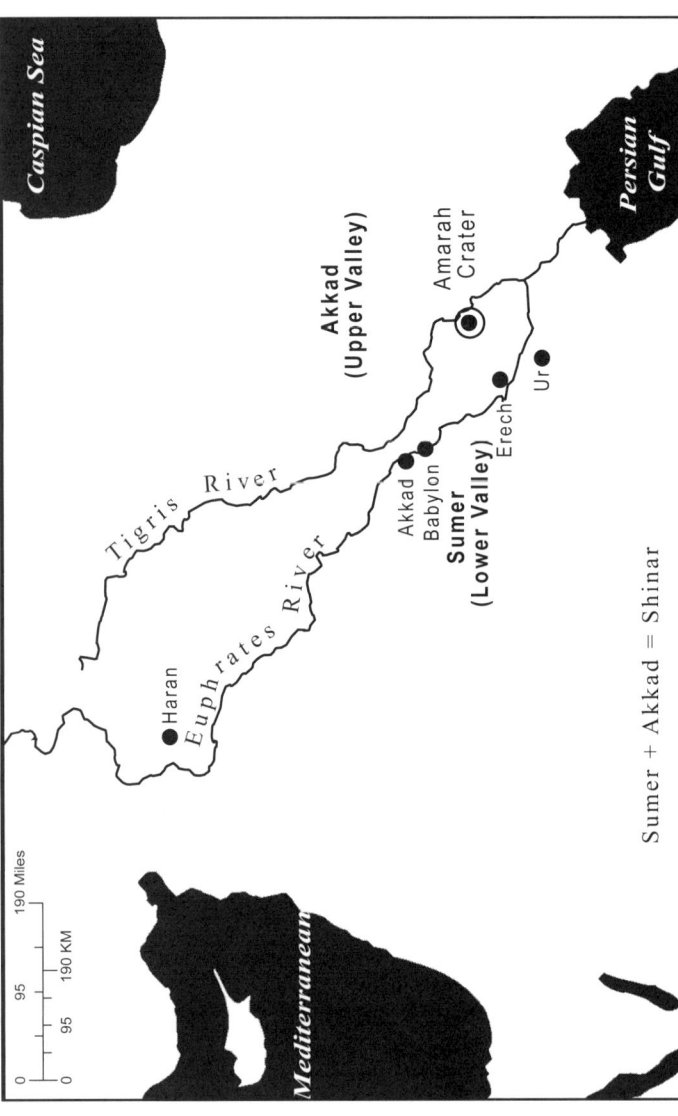

IMPORTANT SUMERIAN AND AKKADIAN SITES AND THE BIBLE *Genesis 10:10* refers to "Babel" (*Babylon-NIV*), "Erech," "Akkad," and "the land of Shinar." // *Genesis 11:28-31* refer to "Ur." // *Genesis 11:31, 32* and *12:4-5* refer to "Haran." This map also shows the location of the **Amarah Crater**. This two mile in diameter crater provides physical evidence that a cosmic impact took place in the Holy Land during Biblical times. The crater was discovered after Saddam Hussein had the lakes drained in Southern Iraq in retaliation against the Marsh Arabs in the area. The Sumerian text *The Curse of Akkad* and the Bible's story of the destruction of the *"Tower of Babel"* both record the occurrence of this catastrophic event which ties to the sudden abandonment of Akkadian sites, and collapse of the Akkadian Empire.

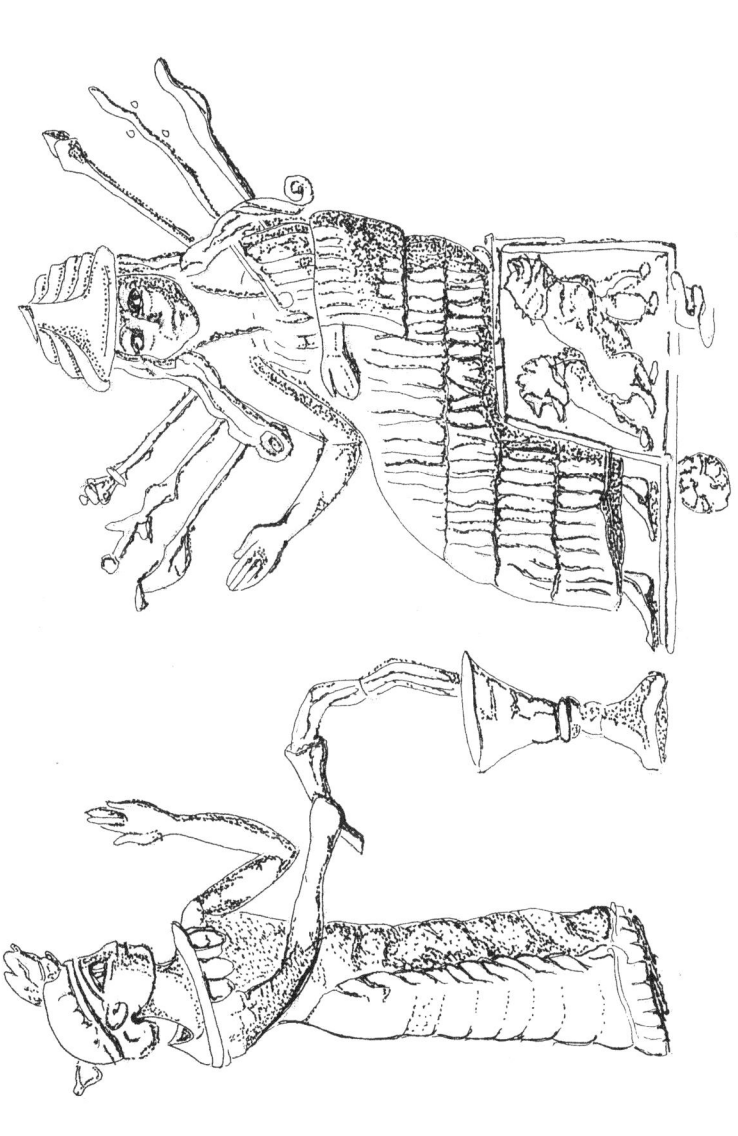

INANNA – THE QUEEN OF HEAVEN (Akkadian period seal, Nephrite 4.3 cm, c. 2330-2150 BC) Here the Sumerian "Queen of Heaven" Inanna wearing a multi-horned headdress is sitting on a throne decorated with crossed lions. One female worshiper pours a drink into an offering stand. *Jeremiah 44:25* " ... we have vowed to burn incense to the queen of heaven, and to pour out drink offerings unto her." The Akkadians, Babylonians and Assyrians called this all powerful goddess Ishtar; the Canaanites called her Ashtoreth. Inanna/Ishtar also relates to the *"whore of Babylon"* referred to in *Revelation 17*.

C

BURCKLE CRATER Burckle Crater is a recently discovered 18 mile wide impact crater in the Indian Ocean. The giant impact that produced this crater occurred around the time of the *Biblical Flood*. This gigantic impact would have also caused a massive earthquake and a series of large aftershocks that lasted for months, which in turn generated a series of towering tsunami waves that went out in all directions for months. Mile high tsunami waves would have raced north toward the site of *Noah's Flood* in the Tigris-Euphrates Valley in Iraq, before slamming into the *"mountains of Ararat"* (Urartu/Armenia) – *Genesis 8:4*. Very thick tsunami deposits relating to this impact have been found inland in Africa, Australia and India.

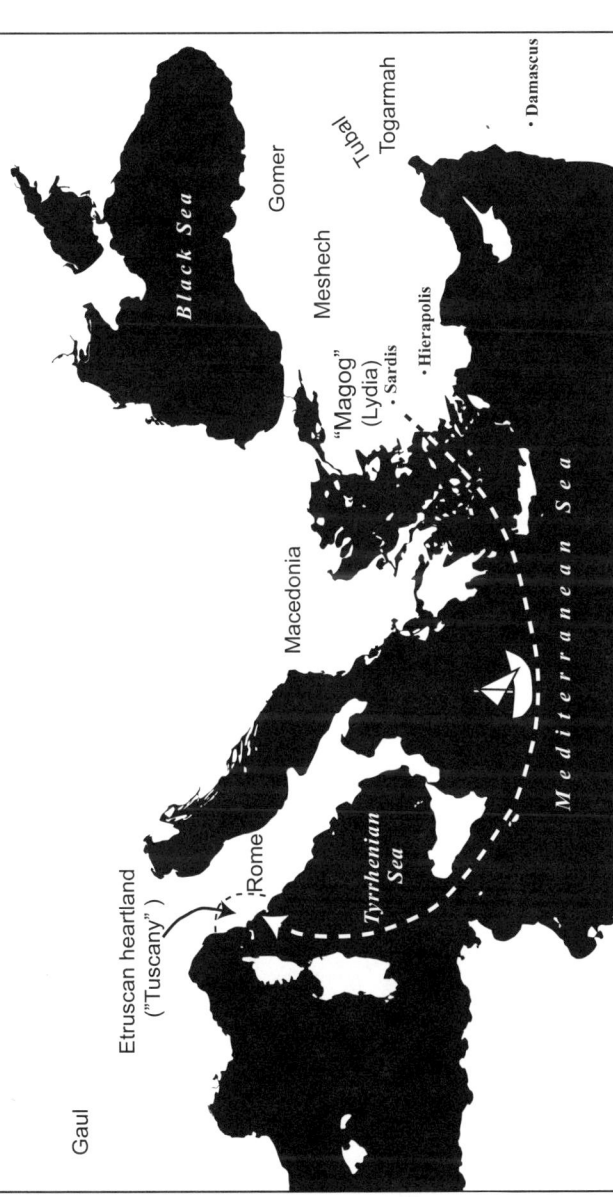

ANCIENT NATIONS OF EZEKIEL 38/39 Ezekiel 38/39 makes reference to a group of ancient nations that are all found in Asia Minor: Magog (Lydia), Meshech, Tubal, Togarmah and Gomer. The correct identification of these nations comes from the "Royal Court Records of the Assyrians"; the primary source on this subject. These same Assyrian Court records are referred to in the Bible (*Ezra 4:15, 19*). **LYDIAN MIGRATION TO ITALY** An early Lydian dynasty experienced a prolonged famine which caused half of the Lydian population to migrate to the west coast of Italy, just north of Rome. The Romans called these Lydian migrants "Etruscans." The Etruscans settled in an area called Tuscany and they helped found the city of Rome and jump start the Roman civilization. Hundreds of years after this migration, the Lydian population in Asia Minor came to be ruled by a militant ruler whom the Greeks called "Gyges of Lydia." To the Assyrians this same leader was known as "Gugu King of Ludu," and "Gugu of Magugu." Ezekiel referred to this same leader as "Gog of Magog" where "Magog" means "the land of Gog."

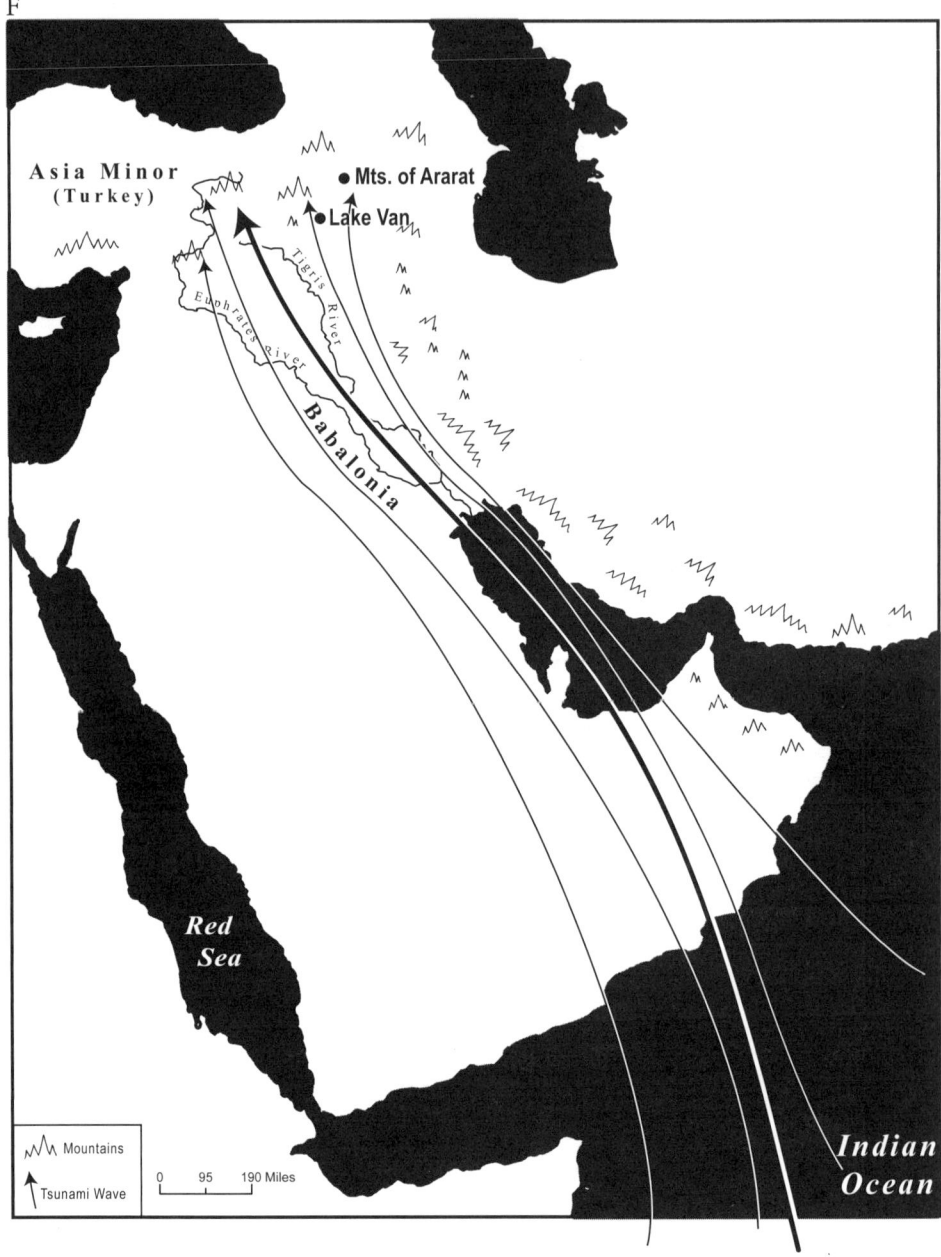

THE FLOOD Mega-tsunami waves generated by the gigantic **Burckle Impact** in the Indian Ocean (see previous diagram) bring a succession of towering walls of water over a period of months that would have raced up the Tigris-Euphrates Valley until they washed over (*Genesis 7:19*) and were stopped by the "*mountains of Ararat*" (Uartu/Armenia) – *Genesis 8:4*. Core samples taken from Lake Van at the base of these mountains provide some tentative evidence of this catastrophic event.

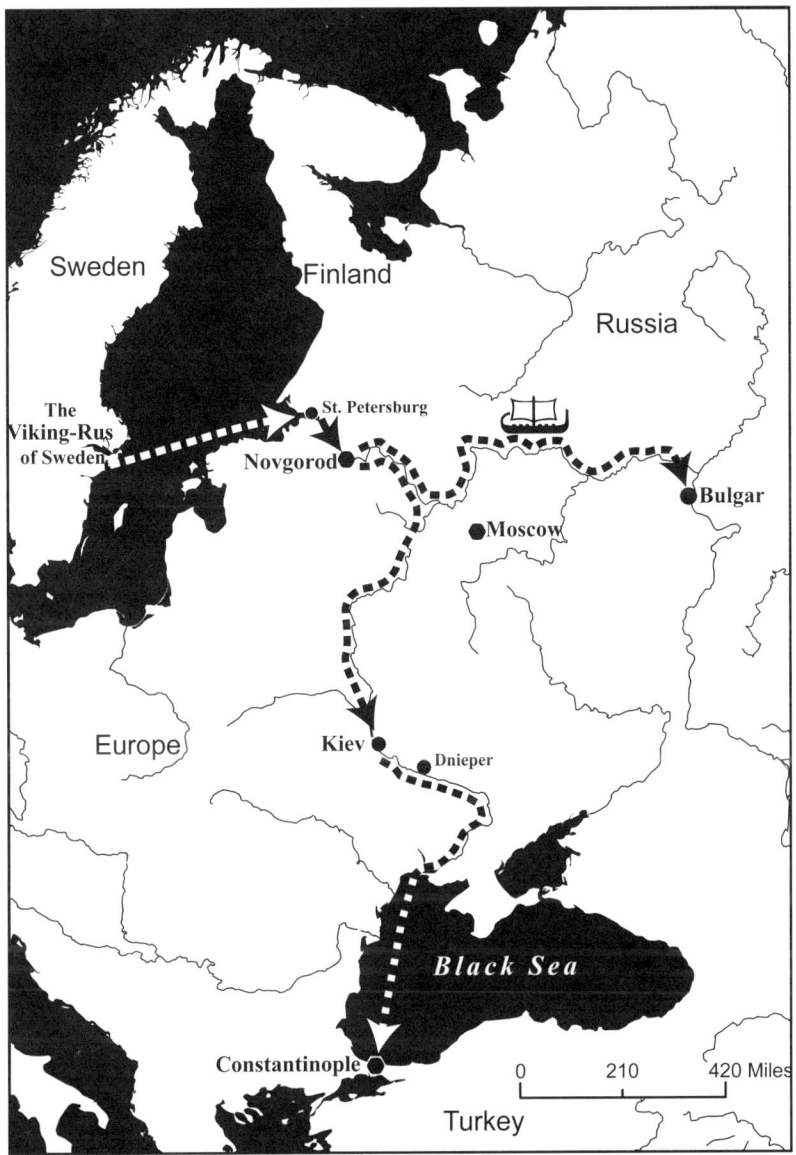

THE VIKING ORIGINS OF RUSSIA Groups of Viking traders called the "Rus" traveling in shallow draft ships moved along the great rivers of Russia to establish trading posts which became the region's major cities (for example: Novgorod, Bulgar, Kiev and Moscow). According to the 12th century document known as the *Primary Chronicle* in 852 AD the land around Kiev was named "Rus" and the inhabitants called "Russes." The renowned Arab Chronicler *Ibn Fadlan* (921 AD) tells how he met a group of tall, blond haired Swedish traders called the "Rus" along the Volga River near the Bulgar capital. The Rus came to rule over the local Slavic people and found the Russian state. The belief that Russian origins involve Magog and/or the Scythians (see previous diagram) is contrary to historical fact. Russia is **NOT** the so-called "*Evil Empire*" of *Ezekiel 38/39*.

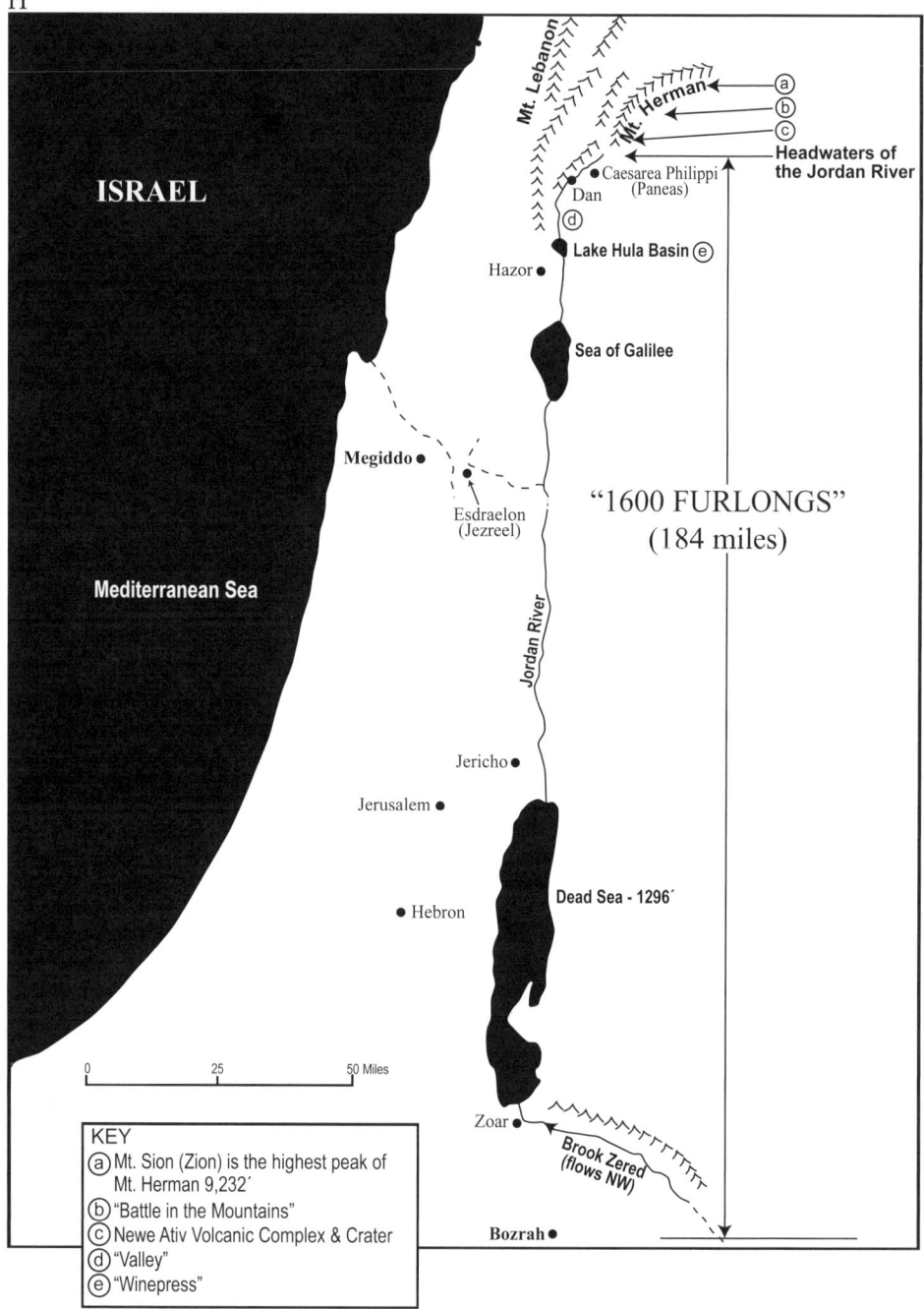

BLOOD TO FLOW A DISTANCE OF 1600 FURLONGS "And the winepress was trodden without the city, and blood came out of the winepress, even unto the horses' bridles by the space of a thousand and six hundred furlongs." *Revelation 14:20.* (Hebrew furlong = 606 ft. = stadion = 1/8 Roman mile *606 x 1600/5280 = 183.7)

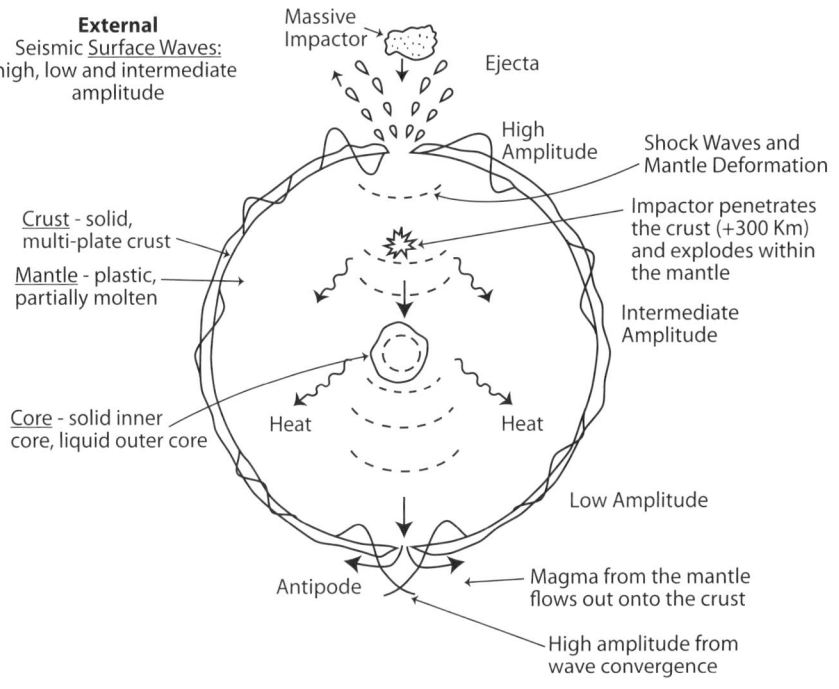

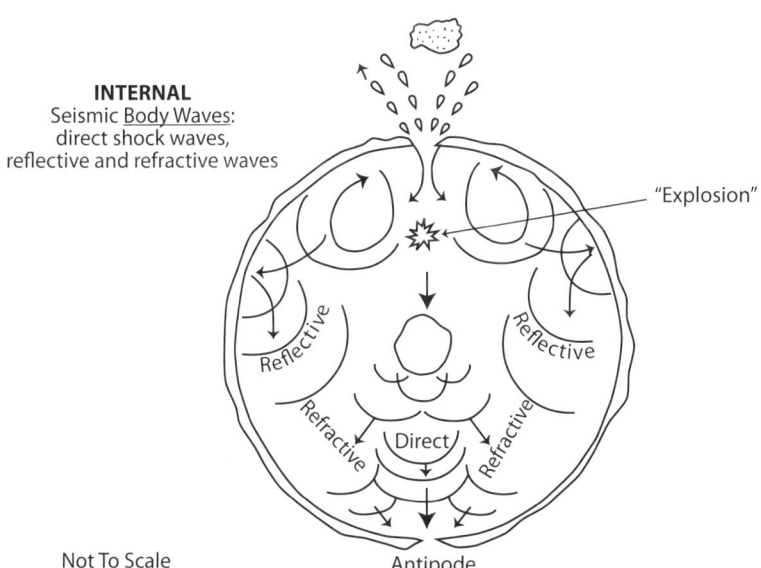

MASSIVE COSMIC IMPACT *Revelation 16:18-20* "... there was a great earthquake, such as was not since men were upon the earth, so mighty an earthquake, and so great... and the cities of the nations fell (high amplitude global shock waves from the earthquake and aftershocks caused by the impact) ... And every island fled away, and the mountains were not found (magma flow and crustal melting as a result of the intense heat produced by the impactor exploding within the earth's mantle)."

About the Author

Jeffrey Goodman received a professional degree in Geological Engineering from Colorado School of Mines, a MA in anthropology from the University of Arizona, a MBA from Columbia University Graduate School of Business, and a Ph.D. in anthropology from California Coast University. He was accredited by the former Society of Professional Archeologists from 1978 to 1987.

His books, *American Genesis* and *The Genesis Mystery*, included accounts of his discovery of an "early man" site in the mountains outside of Flagstaff, Arizona. Four seasons were devoted to excavating this site, the last of which included archeologists from the University of Alberta. In part, *The Comets of God* tells of the "linguistic and scientific discoveries" Goodman made within the pages of the Bible.

Please visit his Web site at

www.thecometsofgod.com